SSP
५।४।२००३

BIOENERGY FROM SUSTAINABLE FORESTRY: GUIDING PRINCIPLES AND PRACTICE

FORESTRY SCIENCES

Volume 71

The titles published in this series are listed at the end of this volume.

Bioenergy from Sustainable Forestry

Guiding Principles and Practice

edited by

J. Richardson
IEA Bioenergy Task 31,
Ottawa, Ontario, Canada

R. Björheden
Växjö University,
Växjö, Sweden

P. Hakkila
VTT Energy,
Jyväskylä, Finland

A.T. Lowe
NZ Forest Research Institute Ltd.,
Rotorua, New Zealand

and

C.T. Smith
Texas A&M University,
College Station,
Texas, U.S.A.

KLUWER ACADEMIC PUBLISHERS
DORDRECHT / BOSTON / LONDON

A C.I.P. Catalogue record for this book is available from the Library of Congress.

ISBN 1-4020-0676-4

Published by Kluwer Academic Publishers,
P.O. Box 17, 3300 AA Dordrecht, The Netherlands.

Sold and distributed in North, Central and South America
by Kluwer Academic Publishers,
101 Philip Drive, Norwell, MA 02061, U.S.A.

In all other countries, sold and distributed
by Kluwer Academic Publishers,
P.O. Box 322, 3300 AH Dordrecht, The Netherlands.

IEA Bioenergy T31: 2002:02

Cover image: Martin Holmer.

Printed on acid-free paper

TABLE OF CONTENTS

PREFACE

The forests of the world represent a vast potential source of raw material containing energy. This energy source is completely renewable and in principle is almost neutral with regard to balances of greenhouse gases which are implicated in global climate change. In the boreal and temperate forest regions at least, the potential remains largely untapped, but developments are taking place which are likely to increase the use of biomass for energy from plantations and naturally-regenerated forests. Bioenergy, as one of the renewable energy forms, is receiving greater emphasis in many countries as a means of improving carbon balances and mitigating global climate change. In other parts of the world, the procurement and use of woodfuel are very important in the lives of ordinary people. Unless management of the source of bioenergy in plantations and naturally regenerated forests is carried out on a truly sustainable basis, these developments could bring environmental and socio-economic problems.

The editors of this book undertook the task of producing it in the hope that we could thereby help to ensure the sustainability of bioenergy production from the forest. We did not intend to give detailed answers for every situation. Rather the book provides guidelines to the general principles that must be considered. It goes further by showing how a truly sustainable bioenergy system integrates and even transcends the totality of its ecological, silvicultural, technical, economic, social and institutional components. We have tried to suggest the kinds of practices that can follow from general principles to promote sustainability.

The scope of the book extends to naturally-regenerated forests and plantations from which bioenergy is only one of several products. Others may include timber, wood fiber for pulp or chemicals, water or a static role in environmental protection, recreation or shelter. There is less emphasis on plantations of trees or other crops such as switchgrass or short-rotation willow coppice grown expressly for energy production. The eco-geographic focus is on the boreal and temperate regions, particularly in countries of the Organization for Economic Co-operation and Development (OECD), recognizing also the vital importance of woodfuel in other parts of the world, including tropical and sub-tropical regions.

The book is organized around criteria for assessing the sustainability of forest management and harvesting systems for bioenergy and other forest products. These criteria are the subject of current global debate and include:

- Productivity: ecosystem productivity, and forest products including biofuels.

- Technology and economics: technology of harvesting, system logistics and efficiency, biomass fuel costs, macro- and micro-economics, non-commodity values.

- Environment: soil and water conservation, carbon balances, biodiversity, habitat.

- Social issues: employment, culture, recreation, aesthetics.

- Legal and institutional factors: taxation, energy and forest policy, environmental regulations.

It is necessary to develop a conceptual framework for evaluating the relative advantages and disadvantages of using forest biomass rather than fossil fuels for energy. This framework incorporates perspectives of increased intensity of forest management, concerns about the sustainability of management practices, and contributions to the control of global carbon cycles. Forest biomass which is not suitable for industrial raw material may be considered as a reserve of renewable energy. This resource is described in terms of its sources, its quantity and its fundamental fuel properties including behavior during the combustion process and formation of emissions.

Natural and planted forests from which fuel is produced must be managed sustainably. The concept of sustainable forest management includes biological, silvicultural and technical aspects. Specific silvicultural interventions and management strategies can result in production of forest fuel from different forest types and conditions. There may be benefits from integration of energy production with other objectives of forest management. Practical and efficient techniques, strategies and equipment must be available for harvesting, removing fuelwood from the forest and transporting it to an energy conversion facility. Quality control of harvested forest fuel during drying and storage is important.

The high cost of procurement is the most serious barrier to large-scale use of forest energy. In order to improve competitiveness of fuel derived from small-sized trees and logging slash by modification and development of harvesting and transportation systems, cost structures and their effect on economic sustainability should be known. The use of forest energy is often more attractive from the viewpoint of the local or national economy than from that of a business enterprise.

Use of forest biomass for energy production can affect the environmental sustainability of forest management systems. Forest bioenergy production should not compromise soil and water conservation, long-term site productivity, or biodiversity and habitat quality. Soil conservation considerations include the sustainability of soil physical, biological and chemical properties, organic matter and nutrient availability. Operations can be designed to minimize negative impacts of harvesting on important environmental values.

In the present age, people are an increasingly important component of any aspect of forest management and use. Discussion of sustainability of forest ecosystems includes public perceptions and values relating to forests and their use for forest energy production. These perceptions and values must be identified and addressed, and means developed for involving the public in decision-making and strategies for convincing demonstrations of sustainability. Forest energy production has a definable relationship with rural employment, and can have a beneficial effect on community development and life in different cultures, particularly those of remote areas and aboriginal peoples.

Finally, national and regional policies and regulations, including taxation policies, also have a large effect on forest energy production and use. Such legal and institutional factors influence the economic and environmental sustainability of forest energy production.

'Bioenergy from Sustainable Forestry' is a product of co-operation under the auspices of the International Energy Agency (IEA). IEA Bioenergy is an international collaborative agreement set up in 1978 to improve international co-operation and information exchange between national bioenergy research development and demonstration programs. IEA Bioenergy aims to realize the use of environmentally-sound and cost-competitive bioenergy on a sustainable basis, in order to provide a substantial contribution to meeting future energy demands. Task 18 ('Conventional Forestry Systems for Bioenergy'), one of more than a dozen multi-year projects under IEA Bioenergy, brought together and coordinated the work of the international experts who contributed to this book. The breadth and depth of their background and expertise ensures complete and thorough treatment of the many facets discussed. The approach taken is integrative and holistic, and emphasizes the inter-relatedness of the different criteria of sustainability. For that reason, collective credit is given to authors who contributed to the preparation of each chapter and specific individual contributions are not identified.

The preparation of this volume required the combined efforts of many individuals. The editors are deeply and sincerely appreciative of each of their contributions. The 27 authors who drafted sections individually or collectively and then worked to integrate them into unified chapters are named with each chapter. Contributors to individual chapters provided peer reviews of each others' contributions. In addition, further peer reviews of sections, chapters or groups of chapters were conscientiously and critically provided by Mike Apps, Dave Brand, Jean-François Gingras, Shaun Killerby, Andy Lamb, Tony Lemprière, Joe Robert, Bryce Stokes and Jocelyn Tomlinson. The editors are particularly indebted to Ruth Gadgil of the New Zealand Forest Research Institute Ltd. for ideas and technical support in editing and final text preparation. Rose O'Brien and Lynn Collier, of Rotorua, New Zealand also provided invaluable technical support in final text preparation. Teresa McConchie of Natural Talent, Rotorua, New Zealand provided immense assistance with a number of the figures. The professional support and patience of Ursula Hertling, Mary Kelly and colleagues of Kluwer Academic Publishers is gratefully acknowledged.

The idea for this book first appeared in May 1998. From the start, it received the firm endorsement of the Task 18 national team leaders in 12 participating countries and of the many experts from those countries who collaborated in the Task and whose work in many cases is referenced herein. Although Task 18 ended before the book was completed, the strong support continued in the succeeding Task 31 'Conventional Forestry Systems for Sustainable Productin of Bioenergy'.

Finally, the production of this book would not have been possible without the commitment of the members of the IEA Bioenergy Executive Committee and the financial support of the participating countries - Australia, Belgium, Canada, Denmark, the European Commission, Finland, the Netherlands, New Zealand, Norway, Sweden, the United Kingdom and the United States of America.

We hope that the consuming passion of the editors of this volume for bioenergy, forest management and the environment will educate, inform and inspire its readers to adopt the guiding principles and practices of bioenergy from sustainable forestry.

J. Richardson
R. Björheden
P. Hakkila
A.T. Lowe
C.T. Smith

LIST OF CONTRIBUTORS

G. Andersson
SkogForsk
Uppsala
SWEDEN

P. Angelstam
Swedish University of Agricultural Sciences
Grimsö
SWEDEN

A. Asikainen
University of Joensuu
Joensuu
FINLAND

R. Björheden
Växjö University
Växjö
SWEDEN

N. W. J. Borsboom
Forestry Consultant
Texel DenBurg
THE NETHERLANDS

M. Breuss
University of Agricultural Sciences
Vienna
AUSTRIA

J. A. Burger
Virginia Polytechnic Inst. & State University
Blacksburg VA
USA

H. M. Eriksson
Skogsstyrelsen
Jönköping
SWEDEN

P. Hakkila
VTT Energy
Jyväskylä
FINLAND

P. W. Hall
NZ Forest Research Institute Ltd
Rotorua
NEW ZEALAND

J. P. Hall
Natural Resources Canada
Ottawa ON
CANADA

B. Hektor
SLU
Uppsala
SWEDEN

S. Helynen
VTT Energy
Jyväskylä
FINLAND

J. B. Hudson
Forestry Contracting Association
Aberdeenshire Scotland
UK

R. Jirjis
Swedish University of Agricultural Sciences
Uppsala
SWEDEN

B. McCallum
Ensight Consulting
Hunter River PEI
CANADA

D. J. Mead
Forestry Advisor
Takaka
NEW ZEALAND

G. Mikusinski
Swedish University of Agricultural Sciences
Grimsö
SWEDEN

D. G. Neary
USDA Forest Service
Flagstaff AZ
USA

I. Nousiainen
VTT Energy
Jyväskylä
FINLAND

J. Nurmi
Finnish Forest Research Institute
Kannus
FINLAND

M. Parikka
Swedish University of Agricultural Sciences
Uppsala
SWEDEN

R. J. Raison
CSIRO Forestry and Forest Products
Kingston ACT
AUSTRALIA

E.M. Remedio
University of San Carlos
Cebu City
PHILIPPINES

J. Richardson
IEA Bioenergy Task 31
Ottawa ON
CANADA

A. Roos
Swedish University of Agricultural Sciences
Uppsala
SWEDEN

C.T. Smith
Texas A&M University
College Station TX
USA

G. F. Weetman
University of BC
Vancouver BC
CANADA

CHAPTER 1

RATIONALE FOR FOREST ENERGY PRODUCTION

H. M. Eriksson, J. P. Hall and S. Helynen

Modern industrialized societies function within a series of trade and political agreements with a variety of countries, trading blocs, and economic and political alliances. Bioenergy use with its associated activities, production, harvesting, and utilization, is integrated with political, economic and environmental agreements. Recently, many countries have developed an additional range of environmental agreements, conventions and protocols designed to address issues of sustainability of resources and communities. Since the publication of *Our Common Future* in 1987 (WCED 1987) these issues have become increasingly prominent in society and now rank with trade and defense issues in importance.

The management of planted and natural forests is continuously evolving because of developing knowledge and changes in public attitudes and policy. In most cases, there have been growing demands on forests, not only as a source of wood and fiber for society, but also in terms of environmental services, ecological functions and biodiversity conservation coupled with a wide range of consumptive and non-consumptive benefits. Sustainable development involves the balancing of these benefits today and in the future (Maini 1990).

Recent global events affecting the energy sector have promoted the increased use of biomass for energy production, particularly in industrialized countries. One of the most prominent and rapidly developing of these events has been the increasing recognition of the local and global environmental advantages of bioenergy as applied to the issue of global warming. The trend towards cleaner, greener, smaller and more decentralized energy production facilities is having a significant positive impact on the demand for biomass energy. Bioenergy production raises issues related to retention of forest cover, slowing of deforestation, regeneration of natural forests, engagement in intensive forest management, and improvements in management of agricultural and rangeland soils. Biomass energy plantations can be developed in rural areas where forest resources are often available for management. Furthermore, bioenergy production can be based on forest

Richardson, J., Björheden, R., Hakkila, P., Lowe, A.T. and Smith, C.T. (eds.). 2002. Bioenergy from Sustainable Forestry: Guiding Principles and Practice. Kluwer Academic Publishers, The Netherlands.

biomass that would otherwise be unmerchantable, harvesting being integrated with that of traditional timber products. In some cases forest industry may be able to undertake co-generation based on wood waste, thus becoming energy self-sufficient, and opening opportunities for the export of surplus energy. At a world level, biomass energy use represents nearly one billion tonnes of oil equivalent (toe), a level comparable to that of natural gas, coal and electricity consumption. Biomass is the largest renewable energy source in use today, about twice as large as hydropower (WEC 1993).

Use of forests as a source of 'green energy' in developed countries is becoming increasingly important. For the future, bioenergy offers cost-effective and sustainable opportunities that have potential, under a biomass-intensive scenario, for meeting about half of the world energy demand during the next century. A large part of the required reduction of carbon emissions from fossil fuels could be achieved through replacement with biofuels (IPCC 1996). Bioenergy production based on biomass from conventional forestry systems therefore conforms to international agreements and national/international initiatives that promote sustainability.

1.1   INTERNATIONAL ISSUES AND BIOENERGY

The International Energy Agency (IEA) is an autonomous body established in November 1974 within the framework of the Organization for Economic Cooperation and Development (OECD) with a mission to implement an international energy program. The IEA Bioenergy Agreement, signed in 1978, supports and facilitates a comprehensive program of cooperation in bioenergy activities among the 17 members of the Bioenergy Agreement (Brown 1992). Elements of cooperation, collaboration, communication, and planning for emergencies in energy supply characterize the activities of the participants. The strategy of bioenergy programs is to assist participating countries in expanding the use of economical, environmentally-sound bioenergy technologies. IEA Bioenergy has a mission to facilitate, coordinate and maintain bioenergy research, development and demonstration through international cooperation and information exchange. IEA Bioenergy pursues a vision of realizing the use of environmentally and economically sound bioenergy on a sustainable basis to provide a substantial contribution to meeting future energy demands.

Estimated recoverable reserves of coal, oil and natural gas have remained quite stable during the last decade due to improved extraction technology. However, the demand for energy supply from this stable, finite source is expanding, with associated price increases inevitable. Many international organizations and companies have estimated that the contribution of renewable resources to energy production will grow significantly in the future, particularly as the price of fossil fuel energy increases (Hall and Scrase 1998).

Bioenergy will be the main source of renewable energy for several decades until wind and solar energy play a greater role (Shell 1996). Weak economic competitiveness has limited the use of wood energy in most countries. Firewood is used for domestic cooking and heating, and by-products and wood waste from primary and secondary forest industries are utilized for energy production. Wood energy from other sources is usually competitive for large-scale or extensive use only when promoted by fiscal or legislative actions. Research and development efforts have been focused on improvement of the competitiveness of wood energy, since it is a renewable resource with significant environmental advantages over fossil fuels.

Growing public expectation for goods and services to be provided from forests has led to the concept of sustainable forest management as the goal in the management and conservation of forests. The criteria for sustainable forest management include many issues central to conventional forestry bioenergy production systems, such as biodiversity, protection of soil and water resources, the carbon cycle, and socio-economic aspects of community sustainability (Brand 1997, 1998). Concepts central to those proposed as Criteria and Indicators for sustainable forest management and as protocols associated with certification of management operations and forest products are also central to the definition of sustainable bioenergy production and use. The negotiation of the United Nations Convention on Biological Diversity in 1992, the United Nations Framework Convention on Climate Change in 1992, and its associated Kyoto Protocol in 1997 raise important considerations in the use of biomass for the production of energy.

1.1.1 Convention on Biological Diversity

More than 100 countries signed the Convention on Biological Diversity at the United Nations Conference on Environment and Development at the 1992 Earth Summit in Rio de Janeiro (United Nations 1992b), and by 1996 it had been ratified by over 170 countries. The Convention defines biological diversity as *the variability among living organisms from all sources, including terrestrial, marine, and other aquatic ecosystems and the ecological complexes of which they are part.* This definition includes diversity within and between species, and of the ecosystems containing them. The convention addresses the conservation of biological diversity, its sustainable use, and the equitable sharing of benefits from the use of genetic resources. In ratifying these authoritative and non-legally-binding statements of principles, countries have pledged to conserve biodiversity and, at the same time, to use these biological resources in a sustainable manner. Efforts are being made to improve the understanding of ecosystems, increase resource management capability, and promote understanding of ecosystems while working internationally to achieve these goals.

Biodiversity conservation is a key element of sustainable forest management. Harvesting of wood products from native forests may lead to more intensive utilization, including removal of currently unmerchantable material. Changes in forest structure, composition and landscape connectivity may affect forest biodiversity, so care must be taken to consider management for biodiversity, as well as other products and services, through protection of unique ecosystems and critical habitats.

Biodiversity considerations are receiving increased emphasis in the management of planted forests since the choice of crop species is now often focused on a small number of species or clones. Countries have adopted regulations to ensure that production plantations are composed of mixtures of clones, genotypes or species. Diverse forests are needed to maintain the health of planted stands, for good wildlife habitat, and to maintain aesthetic values. Management activities must also protect endangered species, and prevent, or limit, the fragmentation of forests. Governments are increasingly interacting with public and private organizations to establish forest biodiversity programs within a framework of sustainable forest management. This is reflected in the establishment of forest conservation areas, parks, ecological reserves, and undisturbed natural areas.

1.1.2   United Nations Framework Convention on Climate Change and Kyoto Protocol

The Framework Convention on Climate Change (FCCC), (United Nations 1992a) was signed at the 1992 Earth Summit in Rio de Janeiro by 154 countries. These countries agreed to implement measures relating to the monitoring and reporting of national emissions of greenhouse gases (GHGs), to undertake actions to reduce these emissions, and to build our global knowledge of climate processes, impacts, and responses to change. The goal of the agreement was stabilization of the concentration of GHGs at a level that would avoid dangerous anthropogenic interference with global climate. This level should allow ecosystems to adapt naturally to climate change, ensure that food production is not threatened, and enable development to proceed in a sustainable manner. All signatories committed to promotion and cooperation in the development, application and dissemination of technologies for controlling or preventing anthropogenic emissions in energy, transport, industry, agriculture, forestry, and waste management sectors. They also committed to the protection and enhancement of greenhouse gas sinks and reservoirs. Countries may establish mechanisms for the provision of financial resources to achieve these ends.

To further the objective of the FCCC, in 1997 the parties to the Convention adopted the Kyoto Protocol that established emission reduction or limitation targets for 2008-12 in the industrialized countries (United Nations 1997). Covering six greenhouse gases, these targets will become legally binding when the Protocol enters into force upon ratification by a sufficient number of parties. Negotiations continue to elaborate the details of the Protocol. Carbon dioxide from the combustion of fossil fuels is the major anthropogenic GHG causing global warming. The approach of IEA Bioenergy has been to substitute renewable sources of biomass, including woody biomass derived from conventionally managed forests, for fossil fuels used to generate energy. In a sustainable forestry system, carbon dioxide emitted from wood-based energy production is recycled back to biomass as the forest grows, and net carbon dioxide emissions from energy production are limited to emissions resulting from harvesting, processing, transportation, and other operations of the production process. In principle, bioenergy production results in no net emissions, being a zero-sum process. In practice, net GHG emissions from different harvesting and combustion combinations depend on many variables including the nature of the fuel used, distance from the source of supply, and the efficiency of the infrastructure.

One approach to the reduction of greenhouse gas emissions is the substitution of biomass for fossil fuels in industrial processes, particularly electricity generation and transportation. Net emissions from a modern integrated gasification combined cycle (IGCC) plant producing electricity from biomass have been calculated to be 5% of those from an average coal-fired power plant (Mann and Spath 1999). Net carbon dioxide emissions from the use of liquid biofuels from wood for transportation are 65-75% of those estimated for transportation fueled by fossil oil products (Delucchi 1999). These examples indicate that specific fuel substitution alternatives and local conditions are critical to estimates of the potential of wood energy for reduction of net carbon dioxide emissions.

Woody biomass and forest ecosystems may act as carbon sinks. This role can be maintained and/or enhanced by appropriate forest management practices. The use of forest biomass for energy thus provides an opportunity for sequestering carbon. The dual role of wood as a substitute for fossil fuels and as a carbon storage system, and the net contribution of wood energy to global carbon balances, are important issues being addressed by IEA Bioenergy. Forests are likely to be greatly affected by future climate change which is predicted to occur more rapidly than changes experienced in the past. Countries participating in the Kyoto Protocol have focused on mitigation, adaptation, and reduction of uncertainty to give policy makers a wider range of responses. This has prompted the initiation of studies on climate policies and the development of measures including voluntary actions, unilateral agreements, reform of energy subsidy programs, the increased use of renewable sources of energy, and the implementation of domestic and international tradable permit systems. If an increase in energy production from biomass were accompanied by reduced demand for fossil-fuel-based energy systems, the goal of reducing the rate and impact of climate change might be achievable.

Governments can choose policies that reduce the environmental impacts of energy generation and meet conservation and environmental goals other than those related to climate change. Actions resulting from policies designed to reduce net GHG emissions are likely to be more effective when coupled with other related policies. Proposed alternatives for limiting GHG emissions include: establishment of institutional and structural frameworks for strategies; tradable emission permits; agreements with industries; energy efficiency standards; and stimulation of new technologies. Intensified government and private sector research is required if a switch to renewable sources of energy such as biomass is to be accompanied by responsible consideration of environmental concerns, competition with other land uses, and rural employment and income.

The FCCC and subsequent Kyoto Protocol (United Nations 1992a, 1997) have led to the development of methods for carbon accounting and emission accounting in many aspects of the economy. Utilization of forest bioenergy will require careful carbon accounting in forest management practice, energy production and consumption systems. Increased use of biomass for energy contributes to reduction of GHG emissions by reducing fossil fuel use and by storage of carbon in forest ecosystems. The accounting process treats the forest as a closed system in which harvested biomass is replaced through regeneration and/or replanting to retain the carbon pool over time.

Global carbon balances

In the atmosphere, carbon dioxide (CO_2) reflects longwave radiation emitted by the earth, thereby preventing heat from leaving the atmosphere. Increased concentrations of GHGs warm the atmosphere and the biosphere, which in turn inject more energy into global weather systems, thus increasing the frequency of extreme weather patterns. The resulting cooling of the stratosphere is expected to delay the redevelopment of the ozone layer above the polar zones. Carbon dioxide is the main concern, since changes in its concentration contribute most to the risk of climate change. Atmospheric concentrations of other greenhouse gases, such as methane (CH_4) and nitrous oxide (NO_2), in the atmosphere are also increasing due to anthropogenic influence.

From 1850 to 1998, approximately 270 Gt (gigatonnes or 10^9 tonnes) of carbon have been released into the atmosphere by combustion of fossil fuel, cement production and change in land use from forestry to agriculture and urbanization (IPCC 2000). As a result, atmospheric CO_2 concentration has increased about 28% from 285 to 366 ppm (IPCC 2000). Annual global emissions from fossil fuel combustion and cement production between 1989 and 1998 were about 6.3 Gt, and from land use change another 1.6 Gt. Most of this was absorbed by the atmosphere and the oceans, the remainder being taken up by terrestrial ecosystems. Carbon uptake by terrestrial ecosystems is influenced by changes in fertilizer application rates and organic matter decomposition rates associated with increased nitrogen deposition, CO_2 fertilization, and afforestation where forests are replacing former agricultural lands. It is also influenced by natural disturbances such as insect epidemics and fire, which are expected to be exacerbated by climate change.

The role of forestry in climate change mitigation

Forestry may reduce net carbon dioxide emissions in a number of ways (Madlener and Pingoud 1998). Fossil fuels could be replaced with biofuels that are recovered from existing forests, as well as from intensified production on lands currently under extensive management. Reforestation of previously unregenerated, cutover land can replace previous net loss of carbon with a steady-state stock of carbon, or result in the temporary sequestration of carbon. Afforestation of previously barren land will increase the pool of carbon in forest biomass and soil, and result in temporary net uptake of carbon. Most of the carbon stored in biomass will be lost with harvesting, as only a minor proportion ends up in long-lived wood products, and new wood products often replace old worn-out items. If regeneration is successful and the new land use is permanent, an increase in the average carbon pool over the rotation period will result.

Temporary increases in carbon pools in biomass and soil organic matter may result from silvicultural practices designed to increase site productivity (Johnson 1992). Wood products can be substituted for more energy-demanding products, thereby reducing the need for fossil fuels, and for other products, such as plastics, which give rise to higher GHG emissions during their life-cycle. Wood products can be used as biofuels at the end of their lifetime, thereby substituting for fossil fuels. There is minor potential for increasing the carbon pool in wood products through preservation techniques or the recycling of wood and wood products.

Forest management may also cause net CO_2 emissions if harvested land is not reforested, or where site disturbance causes reduction in the amount of soil organic matter. Silvicultural practices designed to improve tree establishment methods should result in greater accumulation of carbon in soil and biomass. Drainage of wetlands may reduce soil organic matter content as a result of peat mineralization.

The net effect of land-use change on the atmosphere depends on a number of factors. These include concomitant changes in the net fluxes of other greenhouse gases (especially CH_4 and N_2O), or changes in GHG emissions resulting from changes in the fossil fuel energy needed to maintain new land-use practices. Non-GHG-related changes in the flux of radiant energy may add to, or reduce, CO_2 emissions.

Land-use change has substantial potential for increasing or reducing net emissions of GHGs because large amounts of land are used for crops, agroforestry, pastoral farming, and forestry. The associated impacts of land-use change in terms of altered crop, animal, and tree production, water yield, biodiversity, energy use, and socio-economic effects can be substantial. They range from negative to positive and are often a mixture of both. Non-climate costs and benefits associated with land-use change need to be considered because they may overwhelm climate-based considerations on local or national scales. All these factors should be considered during policy formulation.

Fossil fuel substitution versus carbon sequestration - competing alternatives?

Article 3.3 of the Kyoto Protocol states that the net change in the stock of carbon caused by afforestation, reforestation, and deforestation must be accounted for in national commitments (United Nations 1997). Accounting for change in carbon stocks is an important issue since it may positively or negatively affect the potential for increased use of bioenergy. Options for substitution of bioenergy for fossil fuels and the sequestration of carbon need to be examined on local and global scales. For fossil fuel substitution, a specific price of carbon can be based on the extra cost of the best alternative energy system. Most carbon sequestration options are based on land-use change scenarios where the result is increased storage of carbon over both short and long periods.

For example, if the cost per unit of carbon associated with fossil fuel substitution is compared with carbon sequestration over a 50-year period, the sequestration option may be appear to be relatively less expensive (Table 1.1-1). If the period is 300 years the fuel substitution option may be cheaper. A decision based on the 50-year perspective would mean that future generations would have to bear long-term maintenance costs. If the future generation decided to return to a less expensive land use, carbon sequestered previously would be lost. If future land availability is limited and no additional inexpensive carbon sequestering options remain, land may be converted to intensive bioenergy production to substitute for fossil fuels, which would cause more sequestered carbon to be lost. Costs per unit of carbon for a sequestration option are therefore dependent on long-term maintenance costs and land use stability.

Table 1.1-1. Relative cost per kg C not emitted or sequestered for two mitigation options.

Option type	50 years	100 years	300 years
Fossil fuel substitution	Above average	Below average	Below average
C-sequestration in biomass or soil	Low	Average	High

One way of gaining the potential GHG mitigation advantages of afforestation while avoiding the price-increasing effect of a carbon credit on biofuel is to establish a maximum value for the carbon credit received for land. In most managed forests total biomass, including roots, generally contains more than 50 t of carbon per hectare at the end of a rotation. Some reduction of the expected rotation average can be justified by uncertainty about actual future amounts of carbon sequestered in standing timber due to risk of windthrow, insects, pathogens or fire, especially since such risks increase with stand age. The maximum credit allowed might therefore be set below the expected rotation average. A uniform maximum carbon credit could be established for all countries, since the average carbon content of managed forests is fairly similar. If a uniform-value approach were adopted, harvesting and utilization of wood products would not affect the carbon credit providing that the forestry land use was maintained.

Land management can generate reasonable financial returns if financial accounting is not directly coupled to increase in the carbon stock. If the forest value is determined by both the value of carbon stocks and wood products, the total value of timber, pulp and biofuel will only be recovered if product values exceed the value of sequestering carbon. Income from the harvest of wood products can be earned at the end of each rotation, whereas the carbon sequestration value for any land use is limited once the carbon stock reaches a steady state. In the context of managed forests, GHG mitigation potential would be aided by successful forest regeneration practices at the end of each rotation.

The long-term mitigation benefit per unit area of land is higher if biofuel is produced and used to replace fossil fuel, than for carbon sequestration alone. Afforestation may increase total forest area and increase the potential for producing biofuel.

The use of fossil fuels will continue at current or higher levels unless their relative price is increased. Increased energy prices will change the basis for enhancement of energy-saving measures and use of alternative energy production systems including biofuels (Heikkinen 1998, Lönner *et al.* 1998, Savolainen *et al.* 1998, Jelavic *et al.* 1998, Barbier and Schwaiger 1998, Bradley 1998). If energy prices are increased, biofuel production for export may be an attractive option for countries with large areas of degraded or extensively managed land, and low domestic energy consumption. Markets for biofuel production and trading would develop with increased establishment of bioenergy plantations grown for export and for designated market demands. These plantations could provide a steady income to landowners and help to maintain the viability of rural economies. Markets for biofuels, like those for timber and pulp, will have to meet sustainability criteria for forest management, including maintenance of biomass stocks, conservation of site productivity, and biodiversity, possibly through a formal certification system.

Co-firing of wood with other energy sources is a convenient way for an energy producer to integrate the use of wood-based fuels into the energy-generating process (EUREC 1996). Variable quality makes pricing and standardization of biofuels more complicated than with fossil fuels. Wood-based fuels can be upgraded to pellets, briquettes, powder, and bio-oils, which have greater homogeneity and higher volumetric energy content, but this processing will incur an extra cost. Wood is more suitable than perennial grasses or agricultural residues for many energy production processes, because it contains smaller amounts of alkali metals, sulfur, nitrogen, chlorine, and ash. A high volatile content (80%) makes gasification an especially suitable process for wood-based energy production. Although wood-based fuel resources are seldom adequate for large scale energy production, co-firing of wood with fossil fuels can be an efficient way of producing energy in large existing units.

1.1.3   Air pollution conventions and protocols

Atmospheric pollution is a potential adverse effect associated with the use of fossil fuels and bioenergy. Two classes of compounds resulting from biofuel combustion are oxides of sulfur and oxides of nitrogen (SOx and NOx), both of which acidify rain. Acid rain damages public health, forests, lakes, rivers and man-made structures (Johansson *et al.* 1993, Hall *et al.* 1998). Sulfur dioxide emissions, a major cause of atmospheric pollution, are largely due to combustion of coal or oil, which contain 0.5-3.0% sulfur. The sulfur concentration of wood is less than 0.1%. Sulfur can be removed from flue gas, but this process is not required in wood-based energy production systems, thus lowering both investment and operation costs. Alkali metal compounds in wood ash are able to capture a proportion of the sulfur released during combustion, especially when wood is co-fired with coal.

Oxides of nitrogen also contribute to the acid rain problem. Emissions arise from combustion of both biomass and fossil fuels, which contain similar amounts of nitrogen. The nitrogen concentration of forest biofuel varies according to tree species, the presence of foliage and bark, and the fertility of forest soil. Environmental regulations concerning NOx emissions from wood-based energy production usually ensure that appropriate and currently-available combustion and flue gas cleaning technology result in harmless emission levels.

Particle emissions, which contribute to air pollution, are also associated with the combustion of solid fuels. If small-scale wood-based energy production plants replace a large-scale plant that has efficient particle separation, net particle emissions tend to increase. Small combustion units, such as wood-burning stoves or fireplaces, can also emit large amounts of carbon monoxide and hydrocarbons, some of which may be carcinogenic. This can be minimized through complete combustion, using modern equipment and environmentally sound operational practices. The metal content of wood from unpolluted areas is low compared with most fossil fuels, and gives forest fuel an additional advantage. Large handling and transportation volumes are typical of biomass-based fuels, so emissions associated with harvest and transport have to be accounted for when estimating GHG emissions.

Air quality criteria are particularly relevant to the use of forest-based fuels for bioenergy, and may be a major constraint on increased use. Improvements in combustion technology, and the reduction of emission of particulate matter, SOx and NOx must be ensured to maintain and improve air quality.

1.1.4 Criteria and Indicators of sustainable forest management

Substantial effort has been made at regional, national and international levels to define and implement sustainable forest management. This effort has focused on agreement about the environmental, social and economic values desired from forests by the public. The development of *Criteria and Indicators* for sustainable forest management represents a political consensus as to how sustainable forest management can be defined, implemented and measured. At the national or international level, Criteria and Indicators have been developed through the Montreal Process, the pan-European Helsinki Process, the Tarapoto Proposal which arose within the framework of the Amazon Cooperation Treaty, and the International Tropical Timber Organization (ITTO) Criteria and Indicators for sustainable tropical forest management (Anon. 1994, Anon. 1995, Helsinki 1994, United Nations 1995, and Carazo 1997).

Most definitions of sustainable forest management contain several criteria (5 to 8 values) that are quantified by measuring 25 to 75 indicators used to measure progress toward achieving the criteria. Environmental values represented by the criteria include the conservation of biodiversity, protection of soil and water quality, the maintenance of ecosystem health and productivity, and the forest contribution to global ecological cycles. Socio-economic values include the benefits of forests to society, and the responsibility of society and government for sustainable forest management. Forest biomass removed for energy production can be viewed as simply another forest product, and should be evaluated by the Criteria and Indicators of sustainable forest management used for conventional forestry operations.

The development of Criteria and Indicators led to two major initiatives influencing the development of systems for evaluating compliance of forest management practices with goals set by managers of private, public or industrial forests. These initiatives have resulted in recognition of the concept of *Adaptive Forest Management* and development of procedures for the *certification* of the sustainability of forest management systems and forest products.

Adaptive Forest Management is a dynamic planning process. It recognizes that the future cannot be predicted perfectly, that Best Management Practices can be continually improved, and that planning and management strategies need to be modified frequently as better information becomes available. Scientific principles and methods are applied to improve management systems incrementally as decision-makers learn from experience, scientific findings, and feedback from the Criteria and Indicators systems and as management goals are adapted to evolving social expectations and demands. Adaptive Forest Management is a process of continuous improvement. It requires ongoing indicator monitoring and analysis of management outcomes which are then reviewed and used to revise management decisions where appropriate. There is an increasing trend for developed countries to adopt an Adaptive Forest Management approach in pursuing sustainability goals.

The certification of environmental management systems developed for forestry operations involves an independent, third-party audit. The audit process for forest management systems requires evaluation of the ability of an organization to manage its forestry operations in an environmentally sound and sustainable manner, and to have management systems in place to achieve goals stated by an organization. Certification has potential for creating a fundamental shift from government-enforced or voluntary forest practice regulation to the adoption of operational performance standards defined by auditors and private organizations such as the American Forest and Paper Association (AF&PA 1995, Lucier and Shepard 1997), the International Standards Organization (e.g. ISO 14001 standards for Environmental Management Systems), or the Forest Stewardship Council (FSC). Forest certification has emerged as part of a larger movement towards increased use of voluntary and market-oriented instruments to pursue environmental objectives. Its principles are similar to those established voluntarily by governments contributing to the development of Criteria and Indicators for sustainable forest management.

Certification is an increasingly important issue, becoming more complex with each successive international discussion. A primary reason for promotion of certification is the reduction of forest loss and degradation while developing forest management practices that are environmentally sensitive, socially acceptable and economically viable. Certification aims to promote forest management through the use of third-party, independent auditing processes. These can be supportive of regional or local standards of forest management shown to be in accordance with world-wide initiatives.

Certification has many advantages. A forest manager may see it as a way to get recognition for improved forest management; an environmentalist may use it to apply market forces to improve forest management. It may be used to reassure forest managers, the public and customers that forests are well-managed, thus improving access to markets and facilitating commercial and trade relations. Forestry workers receive a strong positive message and significant increase in morale. The public and other parties with an interest in forests gain confidence in forest management, which benefits both public and private sectors.

The widespread decision at the corporate level to embrace certification is symptomatic of the effectiveness of 'green' consumerism in the marketplace. The authenticity of environmental claims can be validated and the development of market-based incentives for sustainable forest management encouraged. Certification, by verifying the ability of environmental management systems to achieve a standard, ensures that products are derived from sustainable management. The certification system itself must meet standards of credibility, acceptability, cost-effectiveness and operational usefulness. It will not completely resolve environmental or subjective conflicts, but will help to channel discussions toward rational solutions that are acceptable to all parties.

As yet, there appear to be few demands for forest fuel to be certified. Increased awareness of the need to demonstrate forest sustainability means that certification is likely to be required in the future. Currently there are many green power schemes in which consumers are asked to pay a slightly higher price for energy from green sources such as biomass, solar power, wind power, hydro-electricity or geothermal power. In localized areas, or where bioenergy production is the primary motivation for forest management, there may be pressure for standards and certification.

Under current circumstances, certification of a forestry-energy system should present few problems. Forest biomass is plentiful, and current harvesting and processing techniques mean that more of the harvested wood can be used. More power plants are commercially viable where land is plentiful and extensively managed. Energy production based on sustainable conventional forest management provides widespread benefits at all levels of society, and is an effective way of reducing GHG emissions.

Criteria and Indicators are primarily tools for assessing forest management performance, and are designed to provide information that will help in setting policy directions and strategies for management. On a national or even regional scale, bioenergy is an important part of the overall system of land use and forest management, and contributes to benefits and potential impacts that can be monitored by tools currently available to the forestry and energy sectors.

1.2   WHY BIOMASS FOR ENERGY?

The vision of IEA Bioenergy is to realize the use of environmentally-sound and cost-competitive bioenergy on a sustainable basis in order to provide a substantial contribution to meeting future energy demands. The issues are those of supplying a clean and reliable source of energy as economically as possible (EUREC 1996). All sources of energy have limitations. Radiation risks are associated with generation of nuclear energy; fossil fuel-based energy is not considered to be sustainable; and the harnessing of wind and solar power can be limited by problems associated with site, location, cost and technology. Traditional farming or forestry land use, competition with other energy sources, national energy policies, and local policy/opinion constitute major barriers to increased bioenergy use. Biomass, nevertheless, is a relatively low-risk, clean source of energy, but production is now limited by economic factors. A range of social, economic and political factors will have to change if barriers to its utilization are to be overcome. In a society with increasing levels of technological sophistication and awareness of the importance of environmental considerations, these barriers take on a new light. Many of the paradigms prevalent a generation ago are of increasingly less relevance. If biomass is seen as a useful and necessary source of energy its utilization will interact with these issues in ways not previously encountered.

Fuelwood production could be integrated with conventional forest management systems in order to increase yield and improve the quality of wood raw materials in the future. Demand for wood from sapling stands and commercial thinnings promotes silviculture and good management practices. Integration of silviculture and forest energy recovery requires cooperation of many players and the implementation of Adaptive Forest Management practices. Harvesting, processing and transport of wood-based fuels and energy production can enhance employment and earnings. The choice between highly mechanized and labor intensive methods is based on local needs, energy demands and labor costs. Local economies and national trade balances may benefit from the replacement of imported fuel with locally derived fuel, since revenue from the production of wood-based fuels may be retained by the local community. Greater national self-sufficiency is a policy goal for many countries exposed to uncontrolled increases in the price and availability of fossil fuels.

1.3   CONCLUSIONS

The strategy of bioenergy programs is to assist participants in expanding the use of economical, environmentally-sound bioenergy technologies. IEA Bioenergy was set up to facilitate, coordinate and maintain bioenergy research, development and demonstration through international cooperation and information exchange.

Demands on forests are growing, not only for wood and fiber but also in terms of environmental services, ecological functions, conservation of biodiversity, and economic development. Sustainable development is the balancing of these benefits today and in the future. Assessment of sustainability is based on the use of Criteria and Indicators of forest sustainability, and has led to the implementation of Adaptive Forest Management strategies and to the certification of management systems and forest products.

Recent global events affecting the energy sector are raising public awareness of biomass for energy production, particularly the environmental advantages of bioenergy as applied to the issue of global warming. Bioenergy production relates closely to issues of sustaining forest cover, slowing deforestation, regenerating natural forests, engaging in intensive forest management generally, and improving management of agricultural and rangeland soils. Economic advantages are perceived to be associated with development of biomass energy plantations, use of forest biomass that would otherwise be unmerchantable, and promotion of rural development. A major barrier to the increased use of wood energy is still the poor economic competitiveness of biofuels compared with fossil fuels. When the environmental and social benefits of wood energy are accounted for, utilization of forest biomass for the production of energy becomes environmentally, economically and socially justifiable.

1.4   REFERENCES

AF&PA (American Forest and Paper Association). 1995. Sustainable Forestry Principles and Implementation Guidelines. Washington, D.C.

Anon. 1994. European criteria and most suitable quantitative indicators for sustainable forest management. Adopted by the first expert level follow-up meeting of the Helsinki conference, Geneva, 24 June 1994.

Anon. 1995. Criteria and indicators for the conservation and sustainable management of temperate and boreal forests - The Montreal Process. Canadian Forest Service, Natural Resources Canada, Hull, Quebec, Canada K1A 1G5. 27 p.

Barbier, C. and Schwaiger, H. 1998. Fuelwood in Europe for environment and development strategies (FEEDS). Pp. 61-70 *in* Madlener, R. and Pingoud, K. (eds.). 1998. Proceedings of the workshop "Between COP3 and COP4: The Role of Bioenergy in Achieving the Targets Stipulated in the Kyoto Protocol". September 1998, Nokia, Finland. ISBN 3-9500847-1-1.

Bradley, D. 1998. Potential impact of forestry initiatives on Canada's carbon balances. Pp. 93-106 *in* Madlener, R. and Pingoud, K. (eds.). 1998. Proceedings of the workshop "Between COP3 and COP4: The Role of Bioenergy in Achieving the Targets Stipulated in the Kyoto Protocol". 8-11 September 1998, Nokia, Finland. ISBN 3-9500847-1-1. 130 p.

Brand, D.G. 1997. Criteria and indicators for the conservation and sustainable management of forests: progress to date and future directions. Biomass and Bioenergy 13(4/5): 247-253.

Brand, D.G. 1998. Criteria and Indicators for the Conservation and Sustainable Management of Forests: The special case of biomass and energy from forests. Pp. 1-6 *in* Lowe, A.T. and Smith, C.T. (compilers). 1999. Developing systems for integrating bioenergy into environmentally sustainable forestry. Proceedings of IEA Bioenergy Task 18 Workshop, 7-11 Sept. 1998, Nokia, Finland. Forest Research Bulletin No. 211. New Zealand Forest Research Institute, Rotorua, New Zealand. 128 p.

Brown, A. 1992. Introduction to the IEA Bioenergy Agreement. Biomass and Bioenergy 2(6): 1-7.

Carazo, V.R. 1997. Analysis and Prospects of the Tarapoto Proposal: Criteria and Indicators for the Sustainability of the Amazonian Forest. *In* Proceedings of the XI World Forestry Congress. Volume 6, Topic 37.3. XI World Forestry Congress, Antalya, Turkey, 13-22 October 1997. (*http://www.fao.org/montes/foda/wforcong/PUBLI/V6/T373E/2.HTM#TOP*).

Delucchi, M.A.A. 1999. A model of lifecycle energy use and greenhouse gas emissions of transportation fuels and electricity. IEA Bioenergy Annual Report. ExCO:1999:01. 9 p.

EUREC (European Renewable Energy Centres). 1996. The Future for Renewable Energy. European Renewable Energy Centres Agency, UK. 209 p.

Hall, D.O. and Scrase, J.I. 1998. Will biomass be the environmentally friendly fuel of the future? Biomass and Bioenergy 15(4/5): 357-367.

Hall, J.P., Bowers, W., Hirvonen, H., Hogan, G., Foster, N., Morrison, I., Percy, K., Cox, R. and Arp, P. 1998. Effects of Acidic Deposition on Canada's Forests. Canadian Forest Service, Information Report ST-X-15. 23 p.

Heikkinen, A. 1998. Bioenergy and power production: power company's perspective. Pp. 85-92 *in* Madlener, R. and Pingoud, K. (eds.). 1998. Proceedings of the workshop "Between COP3 and COP4: The Role of Bioenergy in Achieving the Targets Stipulated in the Kyoto Protocol". 8-11 September 1998, Nokia, Finland. ISBN 3-9500847-1-1. 130 p.

Helsinki. 1994. Proceedings of the ministerial conferences and expert meetings. Liaison Office of the Ministerial Conference on the Protection of Forests in Europe. PO Box 232, FIN-00171, Helsinki, Finland.

IPCC (Intergovermental Panel on Climate Change). 1996. Climate Change 1995. Impacts, adaptations and mitigation of climate change: Scientific-technical analysis, Intergovernmental Panel on Climate Change Working Group II Report. Cambridge University Press, Cambridge.

IPCC (Intergovermental Panel on Climate Change). 2000. Land use, land-use change, and forestry - A special report for IPCC. Watson, R.T., Noble, I.R., Bolin, B., Ravindramath, N.H., Verardo, D.J. and Dokken, D.J. (eds.). Cambridge University Press, New York, USA. ISBN 0521-80495.

Jelavic, V., Domac, J. and Juric, Z. 1998. Greenhouse gas emissions and possibilities for reduction using biomass for energy in Croatia. Pp. 53-60 *in* Madlener, R. and Pingoud, K. (eds.). 1998. Proceedings of the workshop "Between COP3 and COP4: The Role of Bioenergy in Achieving the Targets Stipulated in the Kyoto Protocol". 8-11 September 1998, Nokia, Finland. ISBN 3-9500847-1-1. 130 p.

Johansson, T.B., Kelly, H., Reddy, A.K.N. and Williams, R.H. (eds.). 1993. Renewable Energy. Sources for Fuels and Electricity. Washington. 1160 p.

Johnson, D.W. 1992. Effects of forest management on soil carbon. Water, Air and Soil Pollution 64: 83-120.

Lönner, G., Danielsson, B-O., Vikinge, B., Parikka, M., Hektor, B. and Nilsson, P.O. 1998. Kostnader och tillgänglighet för trädbränslen på meddellång sikt. (Costs and availability for forest fuels in medium- and long-term time perspectives). Report No. 51. SIMS, Swedish University of Agricultural Sciences, Uppsala, Sweden. ISBN 0284-379X.

Lucier, A.A. and Shepard, J.P. 1997. Certification and regulation of forestry practices in the United States: implications for intensively managed plantations. Biomass and Bioenergy 13(4/5): 193-200.

Madlener, R. and Pingoud, K. (eds.). 1998. Proceedings of the workshop "Between COP3 and COP4: The Role of Bioenergy in Achieving the Targets Stipulated in the Kyoto Protocol". 8-11 September 98, Nokia, Finland. ISBN 3-9500847-1-1. 130 p.

Maini, J.S. 1990. Sustainable Development and the Canadian Forest Sector. The Forestry Chronicle 66(4): 346-349.

Mann, M.K. and Spath, P.L. 1999. The net CO_2 emissions and energy balances of biomass and coal-fired power systems. Pp. 379-385 *in* Overend, R.P. and Chornet, E. (eds.). Biomass - a Growth Opportunity in Green Energy and Value-Added Products. Proceedings of the Fourth Biomass Conference of the Americas. Volume 1. Pergamon Press, Elsevier Science Limited, Kidlington, Oxford, UK OX5 1GB.

Savolainen, I., Lehtilä, A., Liski, J. and Pingoud, K. 1998. Role of forestry and biomass production for energy in reducing the net GHG emissions in Finland. Assessment concerning the history and future. Pp. 41-52 *in* Madlener, R. and Pingoud, K. (eds.). 1998. Proceedings of the workshop "Between COP3 and COP4: The Role of Bioenergy in Achieving the Targets Stipulated in the Kyoto Protocol". September 1998, Nokia, Finland. ISBN 3-9500847-1-1.

Shell. 1996. The Evolution of the World's Energy Systems. Shell International Ltd. London.

United Nations. 1992a. United Nations Framework Convention on Climate Change. UNEP/WMO, Climate Change Secretariat, Geneva. (*http://www.unfcc.de*).

United Nations. 1992b. Report of the United Nations conference on environment and development. UNDoc. A/CONF.151/26, 14 August 1992. United Nations, New York.

United Nations. 1995. Regional workshop on the definition of criteria and indicators for sustainability of Amazonian forests. UNDoc. E/CN.17/1995/34, 10 April 1995. United Nations, New York.

United Nations. 1997. The Kyoto Protocol to the Convention on Climate Change. Climate Change Secretariat, Bonn. (*http://www.unfcc.de*).

WCED (World Commission on Economic Development). 1987. Our Common Future. Oxford University Press. 400 p.

WEC (World Energy Council). 1993. Energy for Tomorrow's World. St. Martin's Press. New York.

CHAPTER 2

FUEL RESOURCES FROM THE FOREST

P. Hakkila and M. Parikka

Biomass is a widespread source of renewable energy. Potential bioenergy sources include crops, forest residues, process residues from forest industries, agricultural residues, agrofood effluents, manures, as well as the organic fraction of municipal solid waste, separated household waste, methane from landfill areas and sewage sludge.

Forest biomass is defined as the accumulated mass, above- and below-ground, of the wood, bark, and leaves of living and dead woody shrub and tree species (Young 1980). The primary constituents of forest biomass are carbohydrates and lignin, which trees produce from carbon dioxide and water through photosynthesis, thereby simultaneously storing solar energy. The total amount of organic matter produced in the assimilation process is called the *gross primary production* of biomass. Some of the photosynthates are consumed during respiration, and the remainder are incorporated into various tree components and referred to as *net primary production*.

When trees grow, organic matter such as outer bark, foliage and branches, either falls as litter, is sloughed, or is grazed by insects or herbivores. Therefore, in applied forestry, *net biomass increment of a forest stand* is defined as the change in the accumulated biomass between two measurements.

Forest industries use wood raw materials selectively. Even in the most favorable market conditions, industrial use of forest biomass is generally limited to stems that meet given dimensional and quality requirements. During pre-commercial thinnings of young stands, trees are often undersized for industrial utilization and are therefore left intact at the site as *silvicultural residues*. In commercial harvesting operations, low-quality stems and other tree components, such as the crown and stump-root system, are left at the site as *logging residues*. Silvicultural residues and logging residues together are called *forest residues*. In this book, the concept of forest residues is restricted to the above-ground residues from pre-commercial thinnings and harvesting operations.

Richardson, J., Björheden, R., Hakkila, P., Lowe, A.T. and Smith, C.T. (eds.). 2002. Bioenergy from Sustainable Forestry: Guiding Principles and Practice. Kluwer Academic Publishers, The Netherlands.

Forest-based industries also fail to completely use the wood raw material at the mill, thus producing *industrial process residues*. In addition to *primary process residues* such as bark, sawdust, slabs, cores, and lignin-based black liquor from the primary forest industries, downstream industries such as furniture manufacturing produce *secondary process residues* that are typically characterized by low moisture contents.

This chapter describes wood-based fuels derived from conventional forestry as a source of renewable energy. Section 2.1 discusses the consumption and availability of woody biomass for energy. Section 2.2 deals with the fuel properties of woody biomass.

2.1   QUANTITY AND AVAILABILITY OF FUELWOOD

2.1.1   Global consumption of fuelwood

The area covered by forests and other wooded lands of the world is approximately 4 billion ha. FAO (Food and Agriculture Organization of the United Nations) distinguishes three geographical forest regions: developed temperate zone countries (51% of the total area of forest and wooded land), other temperate zone countries (7%), and tropical zone countries (42%). *The developed temperate zone* comprises all the countries of Europe including Turkey and Israel, the Russian Federation, North America, Japan, Australia and New Zealand (United Nations 1992). This chapter deals with conventional forestry in conditions that occur in countries of the developed temperate zone.

The average area of forest and wooded land per inhabitant varies regionally. It ranges between 7.5 ha in Oceania, 3.4 ha in the Nordic countries, 2.7 ha in North America and 0.35 ha in all Europe. It follows that the potential contribution of wood-based fuels to the energy supply of the society also varies from country to country.

Great regional differences occur in the accessibility of forests. It is therefore important to distinguish between *exploitable and unexploitable forestland*, although there is no universal acceptance of this concept. The term exploitable is intended to convey the idea of forest that is in regular and sustainable use as a source of wood, whether managed or not, or is available and accessible for such use, even if it is not yet being so used. Unexploitable forest and other wooded land consists of areas that are deemed to be physically or economically inaccessible in the foreseeable future, or that are protected for various environmental or social reasons. In Europe, the distinction between exploitable and unexploitable forests is fairly clear, but it becomes increasingly difficult in more extensively forested countries such as Canada and Russia (United Nations 1992).

For fuelwood, the cost of transport is high because of the low energy density. Consequently, large-scale production of fuelwood as a by-product of conventional forestry is feasible only in exploitable forests. In the developed temperate zone, the area of exploitable forests is about 900 million ha (Table 2.1-1). If all above-ground woody biomass in exploitable and unexploitable forest land is taken into consideration, the total amounts to 117 billion t of oven dry mass (Figure 2.1-1), half of which is carbon.

Table 2.1-1. The area of exploitable and unexploitable forests, and the stem volume of the growing stock in the exploitable forests of the developed temperate zone (data from United Nations 1992).

Region	Forest and other wooded land	Forest land, million ha			Growing stock, million m³
		Total	Unexploitable	Exploitable	
Europe	194	149	16	133	18 509
Former USSR	941	754	341	414	50 310
North America	749	457	149	308	37 947
Other	177	71	28	43	5 008
Total	2061	1431	534	898	111 774

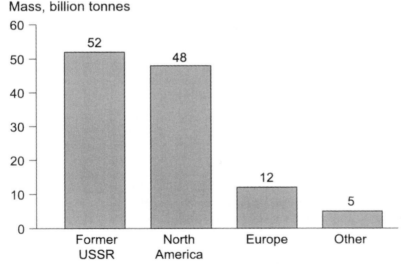

Figure 2.1-1. Above-ground woody biomass in forest and other wooded land in the developed temperate zone of the world (United Nations 1992).

In the developed world wood is used primarily for industrial raw material, and in the developing world, primarily for cooking and space heating. The global use of roundwood was estimated as 3358 million m³ in 1991 (Figure 2.1-2), and consumption increased by 2% annually.

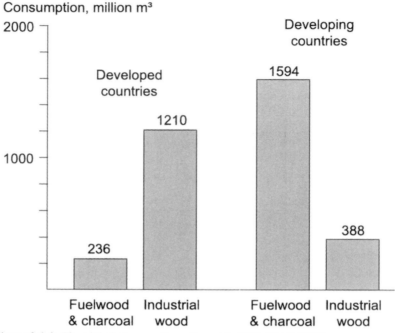

Figure 2.1-2. The global consumption of wood in 1991 (FAO 1999).

The role of fuelwood in the developed world is not very clear, as a considerable part of the raw material consumed by the forest industries becomes process residues, which ultimately have a secondary use in the production of energy. The average proportion of the fuel component in the raw material of various forest industries is shown in Table 2.1.2.

Table 2.1-2. The fuel component in the raw material of forest industries.

Product	Fuel component of raw material	Fuel % by volume
Chemical pulp	Bark, screening residues, black liquor	50-60
Mechanical pulp	Bark, screening residues from chips	10-20
Lumber	Bark, sawdust, slabs, cull logs, etc.	15-60
Plywood	Bark, log ends, waste plies, dust, screening residues, cull logs etc.	40-75

Earlier, these mill residues were burnt to remove harmful waste, without recovering the energy. Today they are a valuable byproduct that is available for the production of heat, steam and electricity. Even in countries with a highly developed forest industry such as Finland and Sweden, almost half of the total consumption of wood may finally end up as fuel (Figure 2.1-3). In 1990, over 200 million m^3 wood equivalent, or 46% of the annual fellings of stemwood were used for energy in the 15 countries of the European Union (Table 2.1-3). Currently the largest single source of industrial process residues is the black liquor from sulfate pulping, representing in Finland 10% and in Sweden 5% of the total consumption of primary energy.

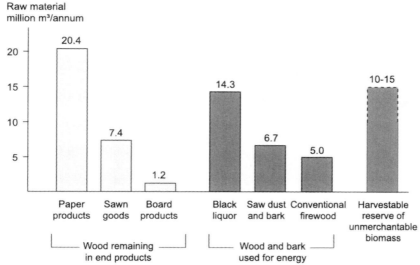

Figure 2.1-3. The destination of wood raw material removed from Finnish forests in the 1990s.

Table 2.1-3. The use of wood-based energy in the 15 countries of the European Union in 1990 (European timber...1994).

Source of fuelwood	Wood equivalent, million m^3	% of fellings
Conventional firewood	92	20.3
Industrial primary residues:		
Wood and bark from mech. industries	24	5.3
Wood and bark from pulp industries	13	2.9
Black liquor from pulp industries	49	10.8
Industrial secondary residues	17	3.7
Recovered wood products	13	2.9
Total wood for energy	208	45.9

While biomass from forestry and agriculture provides as much as one third of the primary energy consumption in developing nations, its role is modest in the industrialized countries. In absolute measures, the USA is by far the largest user of biomass for energy, although only 3-4% of its total energy consumption is derived from biomass. In the European Union, biomass accounts for 3% of the total energy consumption. However, there are large differences between EU countries. Finland derives 19% of its primary energy from wood, while in the densely populated countries of Central Europe the proportion is less than 1%.

Nowhere in the industrialized countries is the potential of wood-based energy fully utilized, and a considerable increase is still possible. An encouraging example is Sweden, where the consumption of wood-based fuels doubled during the last two decades of the 20th century (Figure 2.1-4). This was partly a consequence of an increase in the raw material consumption of the forest industries, but in the district heating sector the rapid development was based partly on the use of chips from logging residues.

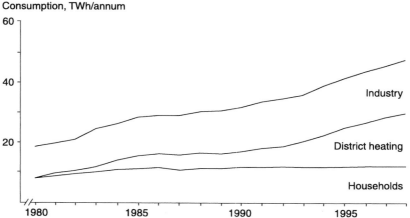

Figure 2.1-4. Consumption of wood-based fuels by households, district heating plants and industry in Sweden (Energy in Sweden 2000).

2.1.2   Biomass components of trees

Against the well-established *merchantable-stem concept*, Young *et al.* (1964) introduced the *complete-tree concept* that embraced the entire tree from root tips to leaf tips. In traditional forest inventories, the stem is measured by volume because of the simplicity and relative accuracy of the scaling. The advantages and accuracy of volume scaling disappear in the face of odd forms, small size and varying basic density of forest residues. Hence, mass becomes a more appealing and additive unit for quantifying all components of a tree, no matter whether the biomass is used for fiber, chemicals, fodder or energy. The result may be given as fresh or dry mass, but the moisture content should be known.

The use of all tree components becomes attractive where the production of energy from conventional forestry is to be expanded. All biomass is then potentially equally important. Unfortunately, some biomass components of a tree are difficult to define and measure. Due to differences in species, tree size, application purpose of data, and the scale of material measurements, the terminology used in literature is neither clear nor consistent. Nevertheless, a uniform and well-defined nomenclature is of utmost importance, particularly for international activities such as the bioenergy research tasks of the IEA. The following classification (Young *et al.* 1964, Keays 1968, Hakkila 1989) is used in this book (Figure 2.1-5).

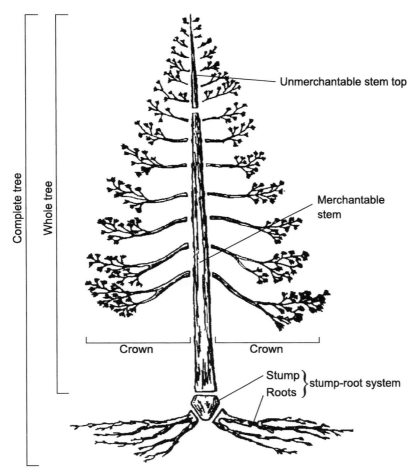

Figure 2.1-5. The biomass components of tree (redrawn from Young et al. 1964).

- *Complete tree* refers to the entire above-ground and below-ground mass of a tree.

- *Whole-tree* (full tree) refers to the mass of a tree above the stump. Stem and crown mass is included, but the stump-root system is excluded.

- *Stem* does not include the stump and its underground continuation. The stem, which includes bark, is divided into merchantable and unmerchantable portions.

- *Unmerchantable top* (sometimes simply top) is the upper section of the stem which is left unutilized in logging operations due to its small diameter and high degree of branching. The size of the unmerchantable top is defined by local logging practice. In industrialized countries, the bottom stem diameter of the unmerchantable top varies between 5 and 10 cm. When there is insufficient demand and wood prices are low, the bottom stem diameter of the unmerchantable top may in some species be 20 cm. In natural tropical forests it may be even greater.

- *Crown* includes all live and dead branches plus all the foliage and reproductive organs. However, in some studies, dead branches are not included in the crown mass.

- *Branches* include wood and bark of live and dead branches but not the leaves, shoots, and reproductive organs.

- *Foliage* refers to all leaves, new shoots and reproductive organs. In some studies, reproductive organs are considered to be a separate biomass component.

- *Stump* is the unutilized above-ground biomass below the bottom of the merchantable stem, and its underground projection including the taproot. Lateral roots are excluded.

- *Stump-root system* is the stump below the merchantable stem, and all roots.

- *Roots* include all side or lateral roots, but not the taproot.

Distribution of biomass within the tree is determined by the *tree architecture*. This is determined by genetic constitution and its interaction with the environment, including competition between individual trees in a stand. Knowledge of the distribution of biomass in an individual tree is the basis for quantitative evaluation of the fuel potential of a stand, by wood procurement area, by region or even globally. Biomass data for individual trees are used to form regression functions based on relationships between various components of standing trees and measurable stem or tree characteristics. In terms of the energy potential of conventional forestry and the ecological consequences of intensive recovery of biomass, crown mass is the most important contributor.

Of all biomass components of the tree the stump-root system is least clearly defined and standardized in the literature. Stump and root wood from mature trees is potentially high-quality raw material for the pulp and paper industry, but the high cost of harvesting, transport, cleaning and processing prevents its utilization for fiber. Since there is no industrial demand for stumps and roots, this massive biomass source can be considered an energy reserve. With some allowance for losses in the harvesting phase, the potentially available biomass from stump and root (greater than 5 cm in diameter) wood in boreal forests is 20% of the stem mass (Hakkila 1989).

Recovering stump and root wood is technically possible, but the costs are high, and the wood is contaminated by grit. Furthermore, the demand for energy in harvesting and processing phases is high, and the removal of stump-root systems may not be desirable for ecological reasons. Consequently, stumps and roots are not discussed further in the present review.

The ratio of crown mass to stem mass, *the branchiness ratio,* is a function of stand density, although it is also affected by site fertility, genetic factors and tree species. In a closed stand, competition for light is severe among tree crowns and even among individual branches of a tree. When a tree grows, older branches at lower locations become shaded. At the base of the crown, the consumption of photosynthate starts to exceed its production, leading to death of the lower branches. Some of the dead branch material falls to the ground as litter, but some remains on the crown and can be recovered together with live branches in conjunction with whole-tree harvesting or the salvage of logging residues.

Tree species differ in the relative quantity and composition of their crown mass. Variability between species includes differences in shade tolerance, thickness of branches and stembark, longevity of foliage, durability of dead branches, stem form, and the basic density of various biomass components. Reactions to aging, size growth, spacing, crown class, and silvicultural treatments are species-specific. Furthermore, the branchiness ratio may vary geographically within the natural range of a species.

Table 2.1-4. Composition of dry whole-tree mass in conifers (averages of 18 species, dbh 20 cm) and hardwoods (averages of 4 species, dbh 16 cm) in British Columbia (Standish et al. 1985).

Tree category	Foliage	Live branches, cm			Dead branches	Crown	Stem	Whole tree
		-0.5	0.5-2.5	2.5+				
			% of whole-tree mass					
Conifers	8.5	3.1	7.8	3.6	1.8	24.8	75.2	100.0
Hardwoods	4.2	2.2	12.7	4.4	1.4	24.9	75.1	100.0

Table 2.1-4 provides an example of the composition of whole-tree mass in the Northern-Hemisphere forests. Stem mass represents three quarters and crown mass one quarter of the whole-tree mass. The most conspicuous difference between conifers and hardwoods is in the foliage component. In British Columbia, conifer foliage amounts to 8.5% and hardwood foliage 4.2% of the dry whole-tree mass (Standish *et al.* 1985). In 22 hardwood species growing on pine sites in the south-eastern United States, the average proportion of foliage is 3.7% (Koch 1985). In boreal forests firs (*Abies* spp.),

followed by spruces, (*Picea* spp.) have the greatest foliage and crown mass, due to shade tolerance. In pines the mass of foliage is substantially smaller.

From the utilization point of view, it is essential to know the mass distribution at the harvesting stage rather than during earlier phases of the tree life cycle. Table 2.1-5 refers to three successive harvests from managed forests in Finland. The proportion of crown mass is largest in the early thinnings from young stands. This is because the *crown ratio*, i.e. the ratio of crown length to tree length, is still high. When a stand matures and lower branches die, the proportion of crown mass decreases.

Table 2.1-5. The composition of dry whole-tree mass from trees removed in selective thinnings and final harvest of Scots pine and Norway spruce in Finland (Hakkila 1991).

Tree species	Treatment	Age, years	Foliage	Live branches	Dead branches	Stem	Whole-tree
				% of whole-tree mass			
Pine	Early thinning	30	7.7	15.0	6.0	71.3	100.0
	Late thinning	55	4.6	9.2	4.1	82.1	100.0
	Final harvest	80	3.5	10.1	1.4	85.0	100.0
Spruce	Early thinning	30	12.3	17.2	2.3	68.2	100.0
	Late thinning	55	10.1	15.5	2.6	71.8	100.0
	Final harvest	80	10.3	17.1	1.1	71.5	100.0

Great differences occur between light-demanding Scots pine (*Pinus sylvestris*) forests and shade-tolerant Norway spruce (*Picea abies*) forests in the amount of biomass residues after harvesting of conventional timber. The collection of logging residues for energy in Sweden and Finland is concentrated largely in the final harvest of Norway spruce stands due to the higher yield of biomass.

2.1.3   The potential availability of forest residues for energy in Europe

In the industrialized world, Sweden has become the pioneer in the production of energy from forest residues. The Swedish energy policy gives a high priority to energy production from renewable resources. An estimate of the potential for wood-based energy in Sweden is given in Table 2.1-6. While the possible annual cut of stemwood is 87 million m³, the theoretical potential of solid fuelwood is estimated as 27 million t of dry-matter or 126 TWh of energy. When the ecological and technical constraints are taken into account, the available supply is reduced to 91 TWh. Using the best techniques of the late 1990s, about 75 TWh of this would be available in the medium term with an average price of 115 SEK (~10-11 USD) per MWh (Hector *et al.* 1995).

Two major sources of fuel from forest operations can be identified. One consists of *forest residues*, defined as all above-ground biomass left on the ground after timber harvesting operations, and pre-commercial thinnings of young stands. The other source includes *stands that should have been thinned* but are left without treatment due to the lack of demand for small-sized wood as industrial raw material. It also includes individual trees that are rejected because of their species, small size, or inferior quality. Unless they are felled, these trees compete with valuable trees for growing space, but do not contribute to timber production.

Table 2.1-6. Potential availability and energy content of solid fuelwood in Sweden in the early 2000s. Black liquor from sulfate pulp industries is excluded (Hector et al. 1995).

Source of fuelwood	Theoretical total	Ecological constraints applied	Ecological and technical constraints applied
	Million t of dry mass yr^{-1}		
Logging residues from integrated harvesting	16.6	14.7	9.9
Direct cutting of fuelwood	4.9	4.9	4.2
Damaged industrial wood	1.0	1.0	1.0
Solid process residues from industry	3.4	3.4	3.4
Recycled wood	0.8	0.8	0.8
Total	26.8	24.7	19.3
	TWh yr^{-1}		
Total	126	116	91

Over-mature, unmanaged forests typically contain large amounts of unmerchantable biomass in the form of undersized, rough, rotten, and dead trees. When clear-cutting these old forests, a great proportion of stemwood is left as residue because of its low quality. Although these residues fulfil the quality requirements of fuelwood, unmanaged forests are often located far from population centers and industrial plants, which makes the recovery of fuel unprofitable.

In the final harvest of *repeatedly thinned managed forests*, there are few dead, defective, and undersized trees. Here most of the logging residues consist of crown mass rather than stem mass. In managed forests, pre-commercial and early commercial thinnings of young plantations contain large volumes of wood in undersized trees (Figure 2.1-6).

In Europe, because of the high degree of utilization, opportunities for increasing energy production from *industrial process residues* are rare, except in conjunction with new production capacity. As Europe is seeking new biomass sources of renewable energy, the primary interest must be in energy crops from plantations and forest residues from conventional forestry.

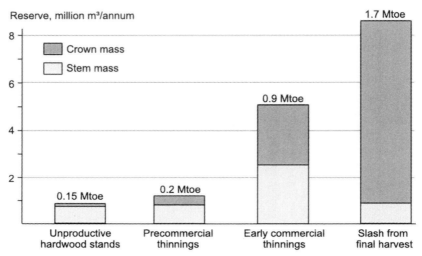

Figure 2.1-6. Technically-harvestable residues from conventional forestry in Finland. The corresponding annual yield of conventional timber assortments is 55 million m³. Mtoe = million tonnes of oil equivalent.

The European Commission White Paper *Community Strategy and Action Plan* considers tripling the present production of biomass for energy in the European Union (EU) by the year 2010 on condition that effective political measures are adopted. The annual use of biomass fuels would grow from the present level of 45 Mtoe to 135 Mtoe, equivalent to 8.5% of the projected total energy consumption in EU in 2010 (Commission of the European Communities 1997) (Table 2.1-7).

Table 2.1-7. Projected additional bioenergy use in the EU.

Additional bioenergy use in EU by 2010	Mtoe yr⁻¹
Biogas exploitation (livestock production, sewage treatment, landfills)	15
Agricultural and forest residues, process residues from forest industries	30
Energy crops	45
Biomass in total	90

Figure 2.1-7 shows the theoretical energy potential of forest residues from logging and tending operations in the EU countries. The above-ground residues from annual fellings are estimated to contain 24 Mtoe, located primarily in Germany, France, Sweden and Finland. This potential cannot be utilized in its entirety, since ecological, technical and economic barriers constrain its recovery. Furthermore, supply and demand do not always coincide geographically. One third of the theoretical potential would be enough to replace 8 Mtoe of fossil fuels, resulting in a 24 million t reduction in CO_2 emissions if fuelwood were to replace oil. Replacement of coal would reduce CO_2 emissions by 30 million t annually. This reduction would be 2-3% of total CO_2 emissions from power generation in the EU countries.

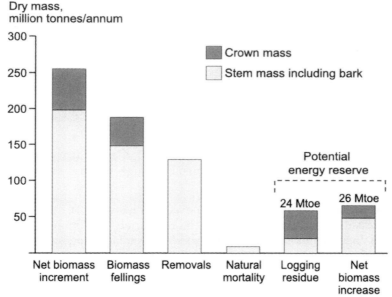

Figure 2.1-7. Forest biomass increment and maximum energy potential in the 15 EU countries in the year 2000. Net biomass increase = net biomass increment - biomass fellings - natural mortality. Logging residue = biomass fellings - removals (data from Karjalainen et al. 1998).

In the year 2000, the increment of stemwood is estimated to be 464 million m^3 in the EU countries. A quarter of it is left growing since it is not needed by the forest industries. Using the low-quality part of this surplus increment for fuel could greatly improve the management of young forests and reduce European CO_2 emissions by another 2-3% through replacement of fossil fuels (Karjalainen *et al.* 1998).

Estimates of the total potential of conventional forestry for reduction of CO_2 emissions from power production in the EU (Table 2.1-8) are of considerable interest, since the target of the Kyoto protocol for the EU is 8% reduction of GHG emissions from the 1990 level by 2008-2012.

Table 2.1-8. The potential of forest fuels for reduction of CO_2 emissions from EU power production.

Source of forest fuel	Potential reduction of CO_2, %
Logging residues	2-3
Unutilized net forest biomass increment	2-3
Unused potential from conventional forestry	4-6

2.2   FUNDAMENTAL FUEL PROPERTIES OF WOOD

2.2.1   Formation and decomposition of woody biomass

Trees *produce woody biomass through the process of photosynthesis* which converts carbon dioxide from the atmosphere and water from the soil to simple sugars or monosaccharides. Biomass is composed of three principal elements: carbon (C), oxygen (O) and hydrogen (H) plus small amounts of nitrogen (N) and mineral elements. Monosaccharides are further converted to cellulose, hemicelluloses and lignin.

Cellulose is the most abundant organic material on earth. It is composed of glucose ($C_6H_{12}O_6$) units which are joined through oxygen atoms to form long-chain molecules or polymers.

While a cellulose molecule is composed exclusively of glucose units, hemicellulose molecules include a large variety of monosaccharides. Softwood hemicelluloses are dominated by galactoglucomannans, and hardwood hemicelluloses are dominated by xylans.

Lignin is an amorphous polymer, the basic structural unit of which is phenylpropane with a phenol ring that is substituted by zero, one or two methoxyl groups. Lignin is responsible for the rigidity and stiffness of tree stems, roots and branches. The presence of lignin thus differentiates wood from other cellulosic materials. In chemical pulping, lignin is removed from fibers and recovered as the primary constituent of black liquor, which in many countries with forest industries is the primary source of wood-based energy.

Forests accumulate carbon from the atmosphere through photosynthesis. In the same process *energy from solar radiation is converted to chemical energy* through photochemical reactions in chlorophyll and other pigments in the chloroplasts. It is stored in carbohydrates in the biomass. Deciduous trees have a higher photosynthetic potential per unit of foliage mass than conifers, although conifers have a larger foliage mass.

Green plants absorb solar radiation from the visual part of the spectrum between the wavelengths 400-700 nm. The photosynthetic conversion efficiency of a forest shows the percentage of the incident solar energy that is stored in various chemical forms during net production of biomass. In temperate forests, the conversion efficiency is 0.6-3.1% of visible solar radiation for long-term storage in woody tissue and 0.4-0.8% for short-term storage in non-woody components such as foliage and fruits. In tropical forests the efficiency is lower but the growing season longer (Jordan 1971).

Forests thus form *a renewable reservoir, or sink, for carbon.* When a tree or part of a tree dies, the dead biomass begins to decompose. Oxidation reactions take place which release both carbon and energy. The release may take place slowly through natural decomposition or rapidly if death is caused by forest fire. In sustainable forestry, the cycle of carbon, oxygen and hydrogen is closed, and the carbon content of the atmosphere remains stable. In the natural cycle, energy released from the biomass is not recovered and used for human needs. When forest biomass is harvested and reduced to chips, and chips are burned in controlled conditions in a boiler system, the released energy can be used for heating or the production of steam or electricity.

The quality of wood-based fuel is dependent on the techniques of storage, handling and combustion. Properties such as high heating value, low moisture content, low ash content, and consistent particle size improve the efficiency and economy of chip combustion. The conventional combustion chamber for fuelwood, *the grate burner*, is also suitable for biomass chips. However, modern techniques based on *the fluidized-bed burner* are especially suitable for multifuel firing of solid material with low-calorific value and variable moisture content and heating values, e.g. wood chips, bark, pulp sludge, peat and municipal waste. Knowledge of fuel properties and careful quality control continue to be of importance for efficient combustion in all boiler systems.

2.2.2   Compounds releasing energy during combustion

The thermal energy content of a fuel depends on its chemical composition and the amount of energy stored in organic molecules. When biomass is combusted, energy is released by the breaking of high-energy bonds between carbon and hydrogen. A high carbon and hydrogen content means a high heating value. Oxygen, nitrogen and inorganic elements are present in wood but do not contribute to heating value (Figure 2.2-1).

The average carbon content of softwoods is 50.7% and of hardwoods 49.0% of the dry mass of wood. The hydrogen content is 6.2% and 6.0% respectively (Kollmann 1951). The lower proportion of combustible compounds in hardwoods results from a lower content of lignin and extractives.

The proportion of combustible compounds varies between tree components. In boreal forests, birch (*Betula* sp.) bark serves as a traditional firelighter. Its excellent fuel properties are due to high levels of suberin and betulin, two carbon- and hydrogen-rich extractives specific to the white outer bark. Table 2.2-1 gives an example of the variation in the carbon and hydrogen contents of different tree components.

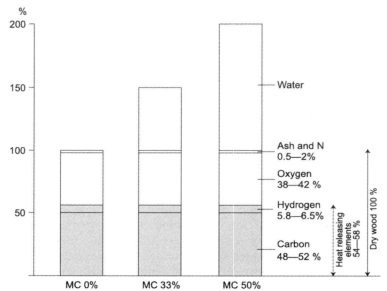

Figure 2.2-1. The chemical composition of bark-free wood in temperate forests (MC = moisture content).

Table 2.2-1. Combustible element content of the stemwood, stembark and foliage of mature Scots pine, Norway spruce and birch trees in Finland (data from Nurmi 1997).

	Stemwood	Inner-bark	Outer-bark	Foliage
		Carbon content, %		
Pine and spruce	52.1	52.1	55.2	51.0
Birch	51.8	51.7	72.5	51.2
		Hydrogen content, %		
Pine and spruce	6.4	5.8	5.6	6.1
Birch	6.2	5.7	9.2	6.0

To mitigate the greenhouse effect, it is desirable for fuel to have a low *specific emission of CO_2*, i.e. a low ratio of CO_2 released to amount of energy produced. The specific emission of CO_2 depends on the ratio of carbon to hydrogen, with higher ratios giving higher specific emissions.

Natural gas, with methane (CH_4) as the main component, has a lower ratio than any other major fuel and is thus more environmentally friendly than other fossil fuels. The carbon:hydrogen ratio of woody biomass, peat and coal is much higher. However, if biomass is a product of sustainable forestry or agriculture, carbon circulates in a closed system without increasing the carbon dioxide content of the atmosphere. Under these conditions, biomass is almost a carbon-neutral fuel, as only a small quantity of fossil fuel is necessary for the harvesting, transport and handling of chips. The input:output ratio of energy is 3-5% for forest biomass chips, 5-10% for coal and about 10% for oil (SLU 1999).

2.2.3   Mineral elements in wood

In addition to carbon, oxygen and hydrogen, trees require small amounts of nitrogen and a number of minerals for their life processes. These elements do not release energy during combustion, and they have a detrimental effect on the heating value of chips.

A tree absorbs minerals from the soil through its root system. These are transported to stem and crown in sap flow. Since the uptake of minerals is partly based on concentration differences between the liquids outside and inside the root tips and hairs, the biomass also contains elements that are not essential to the tree, e.g. lead, arsenic, cadmium and silicon. The presence of these minerals in tree components does not reflect the requirements of the tree, but only their availability from the soil.

The mineral content of tree components varies according to soil origin, site fertility, species, tree size and age, and season. Young trees from pre-commercial thinnings tend to contain a higher concentration of mineral elements than mature trees at final harvest, and trees with a large crown contain more minerals than trees with a small crown. Hardwoods have a higher mineral content than softwoods, and tropical hardwoods contain more minerals than temperate hardwoods. The uptake of minerals is greatest when the growth of the stand is fastest, i.e. normally before and after the first commercial thinning.

The most abundant mineral elements in tree biomass are calcium (Ca) and potassium (K). This is demonstrated in Table 2.2-2, which refers to concentrations of mineral elements and nitrogen in small-sized Finnish softwoods (averages of Scots pine and Norway spruce) and hardwoods (averages of birch, aspen (*Populus tremula*) and grey alder (*Alnus incana*)) felled in first commercial thinnings.

Trace elements occur in woody biomass at such low levels that their concentration is expressed in parts per million, or mg kg^{-1}, of dry-matter. Although the amounts are very small, heavy metals and chlorine may affect the process of combustion and harmfulness of the emissions. Since concentrations differ greatly between tree components, it is important to know the chemical composition of fuelwood in order to control the combustion process, especially when chips are mixed and co-fired with other fuels.

Table 2.2-2. Mineral concentration in the dry mass of small-sized trees from first commercial thinnings in Finland (Hakkila and Kalaja 1983).

Tree component	Primary elements				Trace elements				
	P	K	Ca	Mg	Mn	Fe	Zn	S	B
	Concentration, g 100g^{-1}				Concentration, mg kg^{-1}				
Softwoods:									
Stemwood	0.01	0.06	0.12	0.02	147	41	13	116	3
Stembark	0.08	0.29	0.85	0.08	507	60	75	343	12
Branches	0.04	0.18	0.34	0.05	261	101	44	203	7
Foliage	0.16	0.60	0.50	0.09	748	94	75	673	9
Whole-tree	0.03	0.15	0.28	0.05	296	85	30	236	6
Hardwoods:									
Stemwood	0.02	0.08	0.08	0.02	34	20	16	90	2
Stembark	0.09	0.37	0.85	0.07	190	191	131	341	17
Branches	0.06	0.21	0.41	0.05	120	47	52	218	7
Foliage	0.21	1.17	1.10	0.19	867	135	269	965	21
Whole-tree	0.05	0.21	0.25	0.04	83	27	39	212	6

All fuels contain small amounts of *nitrogen* (N). In the combustion process nitrogen becomes oxidized producing NO_x emissions. As long as these emissions are at a low level, they may have a useful fertilizing effect in forests. They are in fact assumed to contribute to acceleration of the increment of European forests. However, in the long run, excessive amounts of NO_x cause acidification of soil and water, and may be harmful to health.

Nitrogen concentration is typically 0.3-0.4% in heavy fuel oil, 0.01-0.03 in light fuel oil, 0.8-1.2% in coal, and 0.5-2% in peat. Woody biomass has less N: 0.1-0.5% in wood and bark, but as much as 1-2% in foliage (Alakangas *et al.* 1987).

At high combustion temperatures, atmospheric N is oxidized, thus adding combined N to the emissions. Wood has the advantage of burning with a lower flame temperature. This means that NO_x emissions from wood-based fuels are relatively low.

Sulfur (S) is such a harmful element in combustion that fossil fuels are often graded according to their S content. A proportion of the S remains in ash, but the greater part is oxidized and emitted into the atmosphere. Unless separated from flue gases it causes serious acidification in soil and water systems.

In Finland the S content of heavy fuel oil is typically 0.9%, of light fuel oil 0.1%, of coal 0.8%, and of peat 0.05-0.3%. The S content of black liquor is especially high, typically 3-5%, due to the cooking chemicals used for sulfate pulping of wood. In woody biomass, the S content is low: 0.01% in stemwood, 0.02-0.1% in bark, and 0.04-0.2% in foliage. Higher values have been reported in chips harvested from short-rotation willow (*Salix* sp.) plantations (Naturvårdsverket 1983).

Equipment for the removal of S from flue gases arising from the burning of fossil fuels is expensive. No such investment is normally required for wood-based fuels. Some peat-fired operations in Finland and Sweden mix wood chips with their fuel in order to meet the tightening S emission limits and to avoid investment in flue gas cleaning.

Chlorine (Cl) is an essential component of chlorophyll. In forest biomass chips, the primary source of Cl is foliage. Although present in only small concentrations, it may cause problems in boilers, since it forms alkali chlorides with K and Na during combustion. These in turn may form deposits on heat exchanger surfaces, leading to oxidation and corrosion of the tube material (Riedl and Obernberger 1996).

Chlorine concentrations of 0.01% have been reported for whole-tree chips of Scots pine, up to 0.03% for needle-rich chips from logging residue of Norway spruce, and 0.04% for spruce needles in Finland (Orjala and Ingalsuo 2000). Considerably higher concentrations of up to 1% occur in straw. Chlorine contamination may also be caused by salt treatment of the side-walls and floor of a truck to prevent chips from freezing during transport in winter.

The corrosion problem is reduced if chips do not contain foliage. Alternatively, the formation of harmful alkali and Cl compounds on the boiler surface can be prevented if chips are co-fired with fuels richer in S, such as peat or coal. When alkali reacts with S, Cl is released with flue gases in the form of hydrochloric acid, and the formation of the corrosive layer of alkali chlorides on the tube surfaces is substantially reduced (Orjala *et al.* 2000).

Heavy metals are found in all fuels in varying quantities. A proportion vaporizes during combustion and is emitted with flue gas. Higher concentrations in the atmosphere and soil pose a risk to environmental and human health. Part of the heavy metal content remains in ash. This inhibits the recycling of ash in agriculture since heavy metals could be absorbed by plants or leached into ground water.

The concentration of heavy metals is greater in solid fuels than in oil and gas. Among the solid fuels, forest biomass, especially wood, has a lower heavy metal content than coal or peat. Impregnated and painted demolition wood should not be burned in conventional furnaces.

The emission of heavy metals depends on the fuel and also on techniques used in the combustion and cleaning of flue gases. Modern heating and power furnaces are equipped with efficient filters that reduce emission of heavy metals from all fuels to a very low level. However, emission problems remain where older equipment is deployed.

2.2.4   Ash from wood

As biomass is burned, the inorganic components form ash – an alkaline compound of plant nutrients. Depending on temperature, only nitrogen and some of the trace elements escape during combustion. Ash itself is undesirable, and removal from flue gas and furnaces can cause disposal problems (Centre for Biomass Technology 1999).

Hardwoods produce more ash than softwoods, and tropical hardwoods generally produce more ash than temperate hardwoods. The variation between tree species is large. Table 2.2-3 presents the average ash content of stems of softwoods and hardwoods in six countries in different parts of the world. For bark-free softwood it is typically 0.3-0.5%, for hardwoods 0.5-1.6%.

Table 2.2-3. The average ash content of stemwood on a dry mass basis and the standard deviation between species for a large number of softwoods and hardwoods in six countries (Pettersen 1984).

Country	Softwoods	Hardwoods
	Average ash content of stemwood, %	
USA	0.3 ± 0.1	0.5 ± 0.3
USSR	0.5 ± 0.4	0.6 ± 0.4
Japan	0.4 ± 0.4	0.5 ± 0.2
Taiwan		0.9 ± 0.4
The Philippines		1.2 ± 0.7
Mozambique		1.6 ± 1.1

Table 2.2-4. The average content of ash in different biomass components of Scots pine, Norway spruce and birch in Finland on a dry mass basis (Hakkila and Kalaja 1983).

Tree component	Scots pine	Norway spruce	Birch
		Ash content, %	
Stemwood	0.4	0.6	0.4
Stembark	2.6	3.2	2.2
Branch wood with bark	1.0	1.9	1.2
Needles and leaves	2.4	5.1	5.5
Whole-tree incl. foliage	0.9	1.6	1.0
Whole-tree excl. foliage	0.8	1.3	0.8

Ash content of other tree components is higher. Under boreal forest conditions, bark contains 6-7 times and foliage 6-11 times as much ash as bark-free stemwood. On average, the ash content of uncontaminated whole-tree chips from small-sized trees is about 1% (Table 2.2-4), producing 4 to 6 kg of ash per m³ solid of fuel. However, due to impurities and incomplete burning, a power plant always produces more ash than might be suggested by the amount of uncontaminated biomass. The *amount of crude ash* from combustion of chips in a power plant may be substantially higher due to inclusion of sand, charcoal and other substances.

The actual proportion and properties of *ash after combustion* depend on the composition of chips, the occurrence of impurities, furnace controls, and the ash separation apparatus. High proportions of bark and foliage result in higher ash production. The decomposition of organic matter during chip storage increases the concentration of inorganic materials in the remaining biomass. Silica and other impurities in chips decrease the melting temperature of ash and may cause slag formation, resulting in handling and recycling problems.

Ash management is a problem and an opportunity. Removal of ash from the furnace and disposal in landfill areas incurs costs for power plants. If ash is recycled in the forest ecosystem, depletion of plant nutrients (other than nitrogen), and acidification associated with intensive biomass removal, is radically reduced. In Sweden, the National Board of Forestry issued recommendations for good management practice in conjunction with residue removal. According to the recommendations, a maximum of 3 t per hectare of ash may be spread on any given forest area once every hundred years under conditions of conventional forestry (Skogsstyrelsen 1998).

2.2.5   Heating value of woody biomass

The carbon in woody biomass was originally fixed from the atmosphere through photosynthesis. During the complete combustion of biomass oxygen combines with carbon to release CO_2, and with hydrogen to release water. In contrast to the burning of fossil fuels, combustion of forest biomass does not increase the CO_2 content of the atmosphere, since under sustainable forestry practice C is recycled in a closed system. At the same time, thermal energy absorbed from solar radiation through photochemical reactions is released according to the following reactions:

$C + O_2 \rightarrow CO_2 + 32.8 \ MJ \ kg^{-1}$ *of carbon;*

$2H_2 + O_2 \rightarrow 2H_2O + 142.2 \ MJ \ kg^{-1}$ *of hydrogen.*

The heating value of a fuel is a measure of the amount of energy that can be released through combustion. The heating value is expressed in the metric system as MJ kg^{-1} or kWh kg^{-1} of fuel, and in the American system as British

Thermal Units per pound (Btu lb^{-1}). One MJ kg^{-1} is equivalent to 0.2778 kWh kg^{-1} or 429.9 Btu lb^{-1}. A commonly-used unit for comparing types of fuel is the toe, or 'tonnes of oil equivalent'. Toe refers to the heating value of raw oil, one toe corresponding to 11.67 MWh.

The total amount of heat released from fuel is referred to as the *calorimetric or higher heating value*, which is independent of moisture content. In practice, part of the released heat must be consumed during vaporization of water. The water comes from two sources: moisture present in the fuel and water generated when hydrogen and oxygen combine. If the steam does not condense, the heat used in vaporization is lost to the atmosphere. For practical calculations, the energy used for vaporization is deducted from the calorimetric heating value, assuming that the water cools down to its initial temperature while retaining its vapor form. The magnitude of this unit of measurement, the *effective or lower heating value*, is thus affected by the presence of moisture and hydrogen in the fuel.

The formula below can be used for conversions between the calorimetric and effective heating value (W_{ea}) of oven dry wood. In this equation 2.45 MJ kg^{-1} is the amount of heat energy required for vaporizing water at 20°C, and the factor 9 is used because one part of hydrogen and eight parts of oxygen combine to form nine parts of water. For example, if the hydrogen content of dry biomass is 6%, the effective heating value is 1.32 MJ kg^{-1} less than the calorimetric heating value:

$$W_{ea} = W_c - 2.45 \times 0.09 \times H_2$$

Where:

W_{ea} = Effective heating value of dry biomass (MJ kg^{-1} dry mass);

W_c = Calorimetric heating value of dry biomass (MJ kg^{-1} dry mass);

H_2 = Hydrogen content of dry biomass (%).

As a raw material, biomass always contains water. Due to the latent heat of condensation, a further deduction, depending on the moisture content of biomass, is required:

$$W_{em} = W_{ea} - 2.45 \times (MC / (100 - MC))$$

Where:

W_{em} = Effective heating value of biomass with moisture (MJ kg^{-1} dry mass);

MC = Moisture content of biomass on a fresh mass basis (%).

In freshly-felled wood with a typical moisture content of 45-58%, the effective heating value per kg of dry mass is 15-20% lower than the calorimetric heating value. Calculations made on a fresh mass basis naturally show considerably lower heating values (Figure 2.2-2).

Differences in chemical composition between tree species and tree components cause differences in heating value. Lignin, resin, terpenes and waxes have higher heating values than cellulose and hemicelluloses. The mineral elements do not have any heating value. Softwoods have higher heating values than hardwoods, and branches have higher heating values than stemwood. Bark and foliage have higher heating values than wood in spite of their higher ash content (Table 2.2-5).

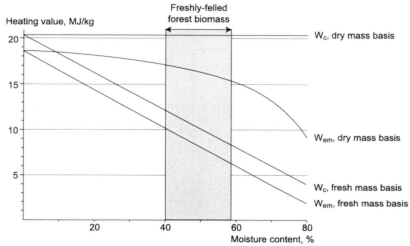

Figure 2.2-2. The calorimetric and effective heating values of bark-free wood on dry and fresh mass basis as a function of the moisture content of the fuel.

Table 2.2-5. Effective heating values of oven dry biomass components of young Scots pine, Norway spruce and birch trees in Finland (data from Nurmi 1993).

		Pine	Spruce	Birch
		Effective heating value, MJ kg^{-1}		
Stem mass	Wood	19.3	19.0	18.6
	All bark	19.5	19.7	22.7
	Whole stem	19.3	19.0	19.2
Crown mass	Wood	20.0	19.7	18.7
	All bark	20.7	19.8	22.3
	Foliage	21.0	19.2	19.8
Whole-tree		19.5	19.3	19.3

Three consecutive stages can be identified in the combustion process. First, the moisture is evaporated to dry the wood. Second, the volatile compounds are driven off and burned. Third, the non-volatile carbon compounds are burned. The volatiles burn in the gaseous phase, whereas non-volatile carbon burns in the solid phase.

Proximate analysis reports the proportions of water, ash, volatile matter, and non-volatile carbon compounds in the fuel. The moisture content is used to predict the amount of heat required to evaporate and to superheat the water. The ash content indicates the ash handling requirements. Relative amounts of volatile matter and non-volatile carbon compounds can be used to predict requirements for the division of combustion airflow between over-fire and under-fire air. Careful division of airflow is useful for smoke and air pollution control.

Compared to other solid fuels, wood has a low carbon content, high hydrogen content, and a low heating value (Figure 2.2-3). The proportion of volatile substances is high which makes wood burn with a long flame. This should be taken into account in the boiler design. The proportion of volatile matter is somewhat lower in bark (Table 2.2-6).

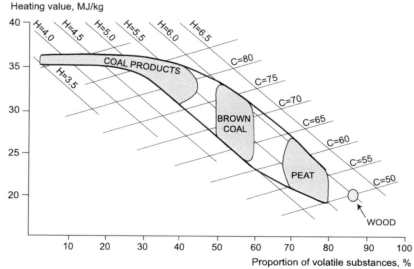

Figure 2.2-3. The heating value and proportions of carbon, hydrogen and volatile matter in solid fuels (adapted from Ryti 1962).

Table 2.2-6. The average proportions of volatile matter, carbon and ash in five softwoods from the western United States (Mingle and Boubel 1968).

	Volatile matter	Non-volatile carbon compounds	Ash
	%		
Wood	85.1	14.6	0.3
Bark	72.6	25.8	1.6

2.2.6   Energy density of forest chips

Heating value is determined per unit mass of dry or fresh fuel. As woody biomass is frequently bought and measured by volume, and transport and storage facilities are dimensioned for volume rather than mass, it is also important to know the effective heating value per unit volume. This is the *energy density* of a fuel. The basic density, i.e. the oven dry mass per green volume in kg m^{-3}, serves as a conversion factor from mass to solid volume.

Variation in the basic density of tree species and biomass components is considerably greater than the variation in the effective heating value of dry mass. Therefore, the differences are larger when calculations are made on a volume basis. Energy density is highest in chips from high-density species such as oaks (*Quercus* spp.). In the Nordic forests, a solid cubic metre of birch bark has a heating value equal to 0.30 toe. The corresponding figure for Scots pine bark is only 0.13 toe (Table 2.2-7).

Table 2.2-7. The energy density of forest biomass chips and crushed bark in Finland at 40% moisture content of biomass (data from Hakkila 1989).

Source	Basic density,	Energy density		
	kg m^{-3}	MJ m^{-3}	kWh m^{-3}	toe m^{-3}
Whole-tree:				
Scots pine	395	7100	1970	0.169
Norway spruce	400	7020	1950	0.167
Birch	475	8270	2300	0.197
Bark:				
Scots pine	280	5460	1520	0.130
Norway spruce	360	7090	1970	0.169
Birch	550	12490	3470	0.297
Crown without foliage:				
Scots pine	405	7780	2160	0.185
Norway spruce	465	8400	2330	0.200
Birch	500	9040	2510	0.215
Crown with foliage:				
Scots pine	405	7660	2130	0.183
Norway spruce	425	7730	2150	0.184

In forestry the primary unit of volume is m³ solidwood. For biomass chips, although m³ solid is a practical unit for comparison between timber assortments with varying bulk density, m³ loose is a more commonly used unit (although less accurate) unit. Therefore, the *bulk density* or *solid content of chips*, i.e. the ratio of solid volume to loose volume of chips, must also be known. The solid content of chips is affected by the following factors:

- *Particle shape.* The greater the diagonal-to-thickness ratio in chip particles, the lower the solid content (Edberg *et al.* 1973).
- *Particle size distribution.* Material with a heterogeneous size distribution has smaller spaces between particles. Chipped fuel from whole-trees or logging residue contains more fine material than uniform chips from pulpwood logs, and tends to have a higher solid content. The dry mass of a load can be increased by mixing sawdust with the chips.

- *Tree species*. Fuel chips from brittle low-density material contain more fine particles and have a higher solid content.
- *Branch content*. Fresh branches and pliable twigs tend to produce long particles which reduce the solid content of chips.
- *Storage*. Stored biomass tends to contain more fine material and fewer long particles than fresh material. The solid content is slightly higher than that of chips from fresh biomass.
- *Season*. Frozen biomass produces more fine material during comminution due to brittleness. This results in a higher solid content.
- *Loading method*. Blowing chips through the discharge spout of a chipper into a truck increases the solid content per unit volume to a greater extent than freefall from a conveyor, tractor bin, loader or silo. Blowing from above gives a higher solid content than lateral blowing. The stronger the fan pressure, the greater the compaction of the particles.
- *Settling*. The solid content of a chip load increases during transport due to vibration and settling. Factors contributing to settling are the initial solid content of the load, length of haul, evenness of the road and possible freezing of the chips. Settling takes place rapidly at first but slows down after the first 10-20 km of travel. From the standpoint of transport efficiency, the solid content before haulage is more important than the content after haulage.

The solid content of fuel chips varies between 0.38 and 0.44 (Nylinder and Törnmarck 1986), depending on the factors listed above. A commonly-used conversion factor is 0.4. A solid m^3 of wood thus produces approximately 2.5 m^3 loose of chips.

Low energy density is a problem associated with biomass chips. The space required for transporting and storing chips is 11-15 times greater than that needed for oil and 3-4 times that required for coal, resulting in higher costs. For this reason fuelwood is traditionally and ideally used close to source. If woody biomass is ground, dried and pressed into pellets, its energy density is increased significantly (Figure 2.2-4).

As industrial demand for biomass chips increases, the average distance between utilization point and source will increase, as will the cost of transportation. In contrast to fossil fuels, economy of scale becomes negative for wood-based fuels, although bulk transportation (train or ship) can make the operation less dependent on distance. To control the cost of fuel in large installations forest biomass chips are often co-fired with bark, sawdust, peat, coal or municipal waste.

Fuel is needed to produce fuel. Despite its low energy density, production of wood-based fuel from conventional forestry does not necessarily require a high input of additional energy. As long as the forest fuel is a byproduct, inputs are limited to harvesting and transportation, none being needed for cultivation. Consequently, the *energy balance ratio* or energy return of investment (i.e. the ratio of energy output to energy input) is high compared with energy crops such as rape, alfalfa, canary reed, pasture plants, or short-rotation willow (Figure 2.2-5).

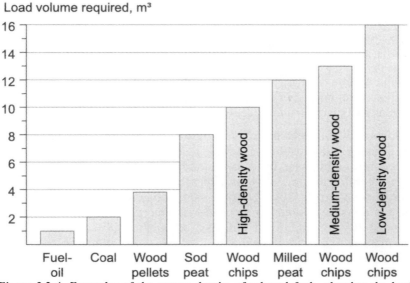

Figure 2.2-4. *Examples of the energy density of selected fuels, showing the load volume required for one toe.*

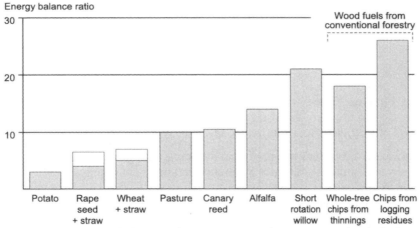

Figure 2.2-5. *Energy balance ratio (output energy:input energy) for fuels from conventional forestry and other selected cultivated crops. The unshaded sections of rape and wheat bars refer to straw. The energy input for each system includes cultivation, harvesting and transport over a distance of 50 km (adapted from Blümer 1997).*

2.3   CONCLUSIONS

The global use of wood approaches 4 billion m^3 per annum. About 55% is used directly for fuel, mainly in developing countries. The remaining 45% is used as industrial raw material, but some 40% of this ends up as primary or secondary process residues, suitable only for energy production. Thus 70-75% of the global wood harvest is either used or potentially available as a renewable source of energy. This estimate does not include the large amount of logging slash and other biomass left at site after silvicultural and logging operations associated with conventional forestry.

To mitigate climatic change caused by GHG emissions, the industrialized world struggles to substitute renewable energy sources for fossil fuels. In many countries, biomass from stems and tree crowns left unutilized after forestry operations is an obvious alternative to fossil fuels. In developed temperate zone forests, 10-20% of the stemwood cut annually is left intact at the site after silvicultural tending and commercial logging. Tree crowns contain 20-30% of additional biomass. Altogether, low-quality stems and crown mass left as residues comprise 25-45% of all fellings in conventional forestry. For environmental, technical and economic reasons only a part of this potential biomass reserve is recoverable for fuel, but it is nevertheless of great interest in countries with a large *per capita* forested area.

Under sustainable forestry practice, woody biomass is an almost carbon-neutral fuel. The carbon content of wood is about 50% and hydrogen content 6%. When biomass is substituted for heavy fuel oil CO_2 emission is reduced by 0.5-0.6 t m^{-3} of wood, depending on the wood basic density. If coal is replaced instead of oil, the reduction in CO_2 emission is 0.6-0.7 t m^{-3} wood.

In addition to elements of primary importance in combustion, i.e. carbon, oxygen and hydrogen, mineral elements and nitrogen are also present in woody biomass. Since they are especially abundant in foliage and other parts of the crown, the harvest of all above-ground components greatly increases the loss of nutrients from a forest ecosystem.

As biomass is burned, most of the inorganic components form ash. Only nitrogen and some of the trace elements escape during combustion. The ash content is typically 0.3-0.5% of dry bark-free softwood. Under boreal forest conditions, bark contains 6-7 times and foliage 6-11 times as much ash as bark-free wood. On average, the ash content of uncontaminated whole-tree chips from small-sized trees is about 1%. However, due to impurities and incomplete burning, power installations always produce more crude ash than theoretical considerations might suggest. Ash management during power generation and recycling of ash in the forest ecosystem become important issues when utilization of forest chips increases.

Low energy density of fuel chips is a problem. Energy density is affected by the basic density of biomass components; the bulk density of chips; and moisture content. The amount of space required for transport and storage of biomass chips is 11-15 times that needed for oil and 3-4 times that required for coal. The resulting higher cost of transport means that wood-based fuels are typically used close to source.

As industrial demand for forest biomass chips increases, the average distance and cost of transport will also increase. To maximise advantages from large-scale operations and simultaneously control the cost of fuel, forest biomass chips will in the future be co-fired with other fuels such as bark and sawdust residues from timber milling, peat, coal or municipal waste. Co-firing affects fuel quality requirements and the quality of the residual ash.

2.4   REFERENCES

Alakangas, E., Kanervirta, M.-L. and Kallio, M. 1987. Kotimaisten polttoaineiden ominaisuudet. Technical Research Centre of Finland. Research Notes 762. 125 p.

Blümer, M. 1997. Energieffektivitet i bioenergisystemet. Rapport från Vattenfall Utveckling AB. Projekt Bioenergi.

Centre for Biomass Technology. 1999. Wood for energy production. Technology-Environment-Economy. Second Edition. Trøjborg Bogtryk.

Commission of the European Communities. 1997. Energy for the future: Renewable sources of energy. Communication from the Commission. Brussels.

Edberg, U., Engström, L. and Hartler, N. 1973. The influence of chip dimensions on chip bulk density. Svensk Papperstidn. 76: 529-533.

ECE/FAO. 1994. European timber trends and prospects in the 21st century.

Energy in Sweden. 2000. The Swedish Energy Administration. Stockholm.

FAO. 1999. State of the World's Forests. Rome.

Hakkila, P. 1989. Utilization of residual forest biomass. Springer Series in Wood Science. Springer. Heidelberg, New York. 568 p.

Hakkila, P. 1991. Hakkuupoistuman latvusmassa. Folia Forestalia 773. 24 p.

Hakkila, P. and Kalaja, H. 1983. The technique of recycling wood and bark ash. Folia Forestalia 552. 37 p.

Hector, B., Lönner, G. and Parikka, M. 1995. Trädbränslepotential i Sverige på 2000-talet (Wood fuel potential in Sweden). Sveriges Lantbruksuniversitet, Institutionen för Skog-Industri-Marknad Studier (SIMS). Utredningar Nr. 17.

Jordan, C.F. 1971. Productivity of tropical forest and its relation to a world pattern of energy storage. Journal of Ecology 59: 127-142.

Karjalainen, T., Kuusela, K., Hakkila, P. and Päivinen, R. 1998. Logging residues and unutilized increment as a source of renewable energy in the EU countries. ETM/97/501294.

Keays, J.L. 1968. Whole-tree utilization studies: selection of tree components for pulping research. Canadian Department for Rural Development. For Prod Lab. Info. Report VP-X-69.

Koch, P. 1985. Utilization of hardwoods growing on southern pine sites: the raw material. U.S. Department of Agriculture. Forest Service, Agricultural Handbook 605(3): 2543-3710.

Kollmann, F. 1951. Technologie des Holzes und der Holzwerstoffe I. Springer Berlin Heidelberg. 1050 p.

Mingle, J.C. and Boubel, R.W. 1968. Proximate analysis of some western wood and bark. Wood Science 1: 29-36.

Naturvårdsverket. 1983. Miljöeffekter av ved- och torvförbränning. Meddelande SNV pm 1708. 232 p.

Nurmi, J. 1993. Heating values of the above ground biomass of small-sized trees. Acta Forestalia Fennica 236. 30 p.

Nurmi, J. 1997. Heating values of mature trees. AFF 256. 28 p.

Nylinder, M. and Törnmarck, J. 1986. Scaling of fuel chips, sawdust and bark. The Swedish University of Agricultural Sciences, Department of Forest Products, Report 173.

Orjala, M. and Ingalsuo, R. 2000. Metsähakkeen poltto: toinen väliraportti. ENE3/T0063/2000. VTT Energia, Jyväskylä. 31 p.

Orjala, M., Ingalsuo, R., Patrikainen, T. and Hämäläinen, J. 2000. Combusting of wood chips produced by different harvesting methods in fluidised bed boilers. The 1st world Conference and Exhibition on biomass for Energy and Industry. Sevilla. 6 p.

Pettersen, R-C. 1984. The chemical composition of wood. *In* Rowell, R.M. (ed.). The chemistry of solid wood. Am. Chem. Soc., Adv. Chem. Ser. 207. 614 p.

Riedl, R. and Obernberger, I. 1996. Corrosion and fouling in boilers of biomass combustion plants. *In* Chartier, P., Ferrero, G.L., Henius, U.M., Hultberg, S., Sachau, J. and Wiinblad, M. (eds.). Biomass for energy and environment. Proceedings of the 9th European Bioenergy Conference. Volume 2: 1123-1129.

Ryti, H. 1962. Termodynamiikka. Teknillisen korkeakoulun moniste nro 151. Helsinki. 282 p.

Skogsstyrelsen. 1998. Rekommendationer vid uttag av skogsbränsle och kompensations-gödsling. 16.11.1998. 6 p.

SLU. 1999. Energi från skogen. SLU Kontakt 9.

Standish, J.T., Manning, G.H. and Demaerschalk, J.P. 1985. Development of biomass equations for British Columbia tree species. Can. For. Serv. Pac. For. Res. Cent. Infornation Report BC-X-264. 47 p.

United Nations. 1992. The forest resources of the temperate zones. Main findings of the UN-ECE/FAO 1990 forest resource assessment. ECE/TIM/60. 32 p.

Young, H.E. 1980. Biomass utilization and management implications. *In* Weyerhaeuser Science Symposium 3, Forest-to-Mill Challenges of the Future: 65-80.

Young, H.E., Strand, L. and Altenberger, R. 1964. Preliminary fresh and dry weight tables for seven tree species in Maine. Maine Agricultural Experiment Station, Tech. Bulletin 12. 76 p.

CHAPTER 3

PRODUCTION OF FOREST ENERGY

G. Andersson, A. Asikainen, R. Björheden, P. W. Hall, J. B. Hudson,
R. Jirjis, D. J. Mead, J. Nurmi, G. F. Weetman

While the production of fuelwood from forest residues in industrial forestry in developed countries is the focus of this chapter, we have included some mention of production under other conditions. In the developing world, fuelwood is often collected from natural forests and shrublands, or grown under agroforestry practices. It provides local people with a greater proportion of their energy requirements than in the developed countries described in Chapter 2. Brief reference is made to the growing of short-rotation bioenergy crops, which, although of interest in some developed countries, does not greatly affect their energy supplies.

Forests have always been a primary source of energy for mankind. Peasants in Western Europe once fought to maintain the right to 'assart' (i.e. clear forests for agricultural crops) and collect branches, dead wood and litter in the wooded preserves of kings and nobles (James 1981). Early forest laws and the struggles of individuals for their 'rights' were centered on forest and energy-use issues. Today the same issues are important in developing countries where millions of poor people depend on wood or other biomass for energy. About half of the global utilization of wood is accounted for in this way. This traditional form of tree biomass use is simple and small-scale in application. Over the past few decades, however, we have witnessed the development of large-scale operations that enable tree biomass utilization to be integrated with industrial energy systems. The main incentives for this development are concerns about rising oil prices, reduced oil supplies and the effects of the use of fossil fuels on the environment through GHG emissions.

The production of fuelwood from industrial forests follows a logical progression from the forest to the industrial complex in which it will be used:

Richardson, J., Björheden, R., Hakkila, P., Lowe, A.T. and Smith, C.T. (eds.). 2002. Bioenergy from Sustainable Forestry: Guiding Principles and Practice. Kluwer Academic Publishers, The Netherlands.

Forest	**Forest & transport operations**	**Industry**
Primary production $\Rightarrow$	Secondary production $\Rightarrow$	Tertiary production
Silviculture	Harvest and transport	Drying and storage

This chapter follows a similar sequence beginning with a consideration of silvicultural practice and ending with the handling of the fuel prior to use by the industry. This approach is developed in further detail in Section 3.2.

3.1   SILVICULTURE

The nature of *forest management* is determined by specific human requirements for tree stands, forests and landscapes. Thus optimum silviculture for a particular situation is dependent on integrating or balancing the following factors:

- Objective of the owner.
- Forest type and species (ecological factors).
- Site characteristics.
- Human resources and equipment.
- Environmental and sustainability issues.
- Economics, including risk and markets.
- Social issues.

Most of these factors are interrelated. The scope for production of bioenergy using trees is wide, and a range of silvicultural systems and practices is possible. There are six broad categories that need to be considered:

1. Natural immature and mature forest stands with high timber value but low potential for bioenergy production except from logging residues.
2. Naturally regenerated dense immature and mature stands with low timber value (often due to high-grading in the past) and high potential for bioenergy production. Includes some natural coppice forests.
3. Plantations grown for high-value timber or fiber. Bioenergy production is a secondary objective.
4. Forests for which the major objective is not timber production but often conservation, recreation etc. Bioenergy production is incidental.
5. New plantations with production of biomass for energy as the primary goal. These are usually grown in ultra-short rotations.
6. Developing countries where fuelwood is often grown in agroforestry systems.

Only one of these categories (No. 5) has growth of biomass for energy as the major objective. In the others, bioenergy production is secondary and utilization will often depend on local conditions or the perceived needs of society. For example, in British Columbia where 25 million ha of public commercial land is held under license, bioenergy is of little interest since oil, natural gas, coal and hydroelectricity are cheap and readily-available. Forest biomass is used only on a very local scale in the form of firewood or industrial process residues.

In California, interest in bioenergy is increasing as a result of recent energy shortages. Similarly many European countries have developed policies to encourage bioenergy use and have been actively researching and implementing its production (see Chapter 7).

Silvicultural systems designed for bioenergy production are historical and current products of situations in which wood is a valuable energy source. Before the industrial revolution in Europe when transport was limited, communities had to rely on wood for fuel and construction. *Coppice* and *coppice-with-standards* are two very ancient silvicultural systems (National Academy of Sciences 1980) which supplied short rotation fuelwood, charcoal and longer rotation construction timbers within horse-cartable distances. When iron was first made, using charcoal, the ironmakers fostered coppice production. Records of coppicing in Europe go back to Roman times. Some tree stands in Britain have been coppiced on a 10-11 year cycle for over 300 years (National Academy of Sciences 1980). Coppicing was the first silvicultural system in the US and is still favored for energy plantations.

Silvicultural systems (clearcutting, seed tree, shelterwood, selection, and coppice) were developed over a 200 year period and were originally used to upgrade devastated and high-graded forests in Europe. Recently conversion of old coppice to more valuable high forest has slowed to preserve the biodiversity and recreation values of the coppice.

Historically, and still today, forests are successively *high-graded* time and time again. Loggers take the most valuable and accessible stands and trees and leave the rest. They return only when residual stands have recovered sufficient volume or when markets for poorer species and grades have improved. This high-grading results in extensive areas of naturally-regenerated, poorly- to fully-stocked forest having low commercial value. In both the US and Canada, some states and provinces are characterized by these low value forests, especially on private land where there has been no regulation. High-grading has occurred elsewhere in the world including the tropics.

During the OPEC oil crisis in the 1970s, when oil prices up to USD50/ barrel were forecast, calculations suggested that biomass harvesting would be feasible in these low-grade forests. Wood-burning stoves and heating plants became fashionable. In Sweden, central heating plants were powered by logging residues. The Chernobyl nuclear disaster also focused attention on short rotation energy plantations. However, in countries where oil, natural gas and coal are still cheap and abundant the economic efficiency of bioenergy is only a reality at wood processing plants where harvesting and transportation costs are already covered. A few industrialized countries such as Denmark have instituted a fossil fuel carbon tax which has made bioenergy harvesting more cost effective.

In developing countries where fuelwood is a major forest product, and fossil fuels are unaffordable, forests are often degraded by fuel harvesting. Short rotation coppice plantations and agroforestry coupled with social or community forestry are being actively pursued with varying degrees of success in many of these countries.

Silvicultural methods are quite different in developing and developed parts of the world. This review of silvicultural systems associated with bioenergy production will distinguish between industrial and non-industrial forestry.

3.1.1   Biomass production, stand dynamics and silvicultural systems

In even-aged tree stands, maximum growth rates occur after crown closure and achievement of 'full stocking'. Soon after canopy closure, leaf weight per hectare becomes relatively constant and as stands grow, self-thinning takes place. As a consequence 20-30% of the total above ground volume dies and falls to the ground. Theoretically this volume could be recovered by repeated light thinning but in practice it is not. In naturally regenerated untended forests, high tree density results in small average stand diameters. Harvesting is delayed until the trees are large enough to harvest economically. Much of the biomass is unused for the following reasons:

- Natural mortality occurs due to self-thinning.
- Many trees are too small to harvest with machines.
- Tops of trees, branches and stumps are left behind.
- There are many unlogged residual, partially rotten and crooked trees.
- Snags, i.e. dead standing trees, are common.
- Stands contain non-utilizable tree species and understorey shrubs.

In addition, natural forests incur huge losses of biomass through damage by fire and insects. For example in Canada these losses usually exceed the annual harvested area of about 1 million ha – sometimes up to 3 million ha per year have burned or been killed.

The rate at which *maximum biomass production* is reached during stand development depends on the density of regeneration and the site quality (height growth rate).

Different silvicultural systems have different lag times between successive rotations:

- For clearcut systems, including the seed tree method, there is a long lag. Rotation length depends on growth rate and product type being grown.

- For shelterwood, there is less lag since rotations overlap.

- For selection, there is no lag although the cycle of cutting in any one area is periodic rather than annual.

- For coppice, recovery from a clearcut is rapid because root systems are already established. Rotations tend to be short as the product is mainly small-diameter wood.

3.1.2   Industrial forestry

In developed countries, large industrial companies, sometimes state enterprises, operate at the upper end of the size scale for managed forests. They often practise forestry to obtain industrial wood but may use lower grade logs and process residues to meet their energy needs and improve profits. These companies are likely to have good equipment and access to highly skilled manpower. They often have to meet clear environmental standards and long-term sustainability is likely to be an important objective. Their choice of silvicultural system is likely to be driven by a need to produce specific types of wood for primary processing. However, they will be interested in fuelwood recovery during logging operations if this will lead to greater efficiencies and profits (see Section 3.2). This is often the case in the Nordic countries.

Immature and mature natural stands with high timber/low bioenergy value

As markets and timber values improve, greater areas of natural forest become economically operable and stumpage rates allow more formal regeneration and tending treatments. This situation is illustrated in Canada where huge areas of forest are currently coming to the economic margin. Bioenergy production is secondary to more valuable sawn timber, pulp and fiberboard production. Extensive use of feller-bunchers for whole-tree harvesting has resulted in residue accumulation at roadsides, where removal could be relatively cheap. As yet, there is little interest in bioenergy although wood is sometimes chipped in the forest for pulp production (Araki 1999). In the Nordic countries and recently in northern California, where there is more interest in bioenergy, integrated harvesting systems, including in-forest chipping, have been developed to recover logging residues (see Section 3.3).

In Canada clearcutting, often with some stand structure retained to preserve biodiversity, is normally prescribed for stands of natural origin that are well past technical maturity. In temperate countries where the forests have been managed for longer periods, clearfelling or seed-tree systems tend to predominate, although shelterwood or selection systems are sometimes used. In humid tropical countries there is increasing use of the selection system.

In many Canadian forests there has been little or no *thinning* to recover competition-induced mortality. This 'waste' of productivity is not usually considered in stand regulation based on net merchantable volumes in untended stands. Recently, computerised stand simulators have included quantitative estimates and timing of competition-induced mortality, and the effects of thinning regimes on recoverable volumes. Diagrammatic representations of the effect of stand density management in even-aged pure stands have focused attention on stand dynamics and thinning.

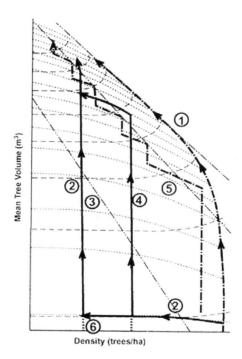

Figure 3.1-1. A stand density management diagram showing trajectories for stands of different densities, with decline in density due to competition-induced mortality. Classic options for crop planning are indicated and are explained in the text (from Farnden 1996).

Figure 3.1-1 is a stand density management diagram for natural stands of lodgepole pine (*Pinus contorta* var *latifolia*) showing classic options for crop planning. These include: (1) Dense natural regeneration carried through to rotation for high total volume production at low cost. (2) Pre-commercial thinning of dense natural regeneration to a final crop density. (3) Thinning to final crop density coupled with pruning to increase product values. (4) Pre-commercial thinning to a density which allows for a single commercial thinning. (5) Establishment of moderately dense stands followed by frequent light commercial thinning to maximise volume production. Establishment can also be by planting to the desired density rather than through pre-commercial thinning of natural stands (6).

Currently *pre-commercial thinning* is widely used in the regulation of stand density to accelerate stand operability. Usually these small trees are not used. Commercial thinning must be done late in the life of untended stands if trees are to be of economic size for mechanized logging. These late thinnings do not recover mortality, and do not improve total stand yield. If a market for small trees for bioenergy existed, repeated light thinnings could recover mortality and increase yield by 15-30%. This occurs in a few European plantations (see Section 3.3).

At present it seems unlikely that use will be made of competition-based mortality in natural forests, particularly in industrialized countries where fossil fuel energy is cheap. In Canada, salvage of competition-induced mortality on leased forestlands could increase annual allowable cuts by about 25 million m^3.

Naturally regenerated dense immature and mature stands of low timber value but of potential high bioenergy value

Throughout the world there are extensive forests that have been repeatedly high-graded and often abandoned. Low commercial log values often preclude pre-commercial thinning or stand improvement cuts. These stands are often degraded to the point where economic silvicultural intervention is not feasible. The despair of silviculturists would have a bright side if bioenergy harvesting were to become feasible on a large scale. Occasionally, for instance in the drier areas of California, small trees and undergrowth are removed for bioenergy use if prevention of forest fire is critical (Figure 3.1-2). Many of these stands are fully stocked, growing at maximum potential for the present species composition. One option is to clear-cut and plant valuable species if the owner can bear the cost and there are no other constraints. A notable example in Canada has been the replacement of degraded stands in New Brunswick with spruce plantations.

A recent estimate of total biomass in Maine, USA, gives an example of the biomass production potential (Wharton and Griffith 1998). Biomass energy values will have to increase further before such stands and forests can be formally committed to energy production. Short rotation coppice may also be biologically feasible.

Figure 3.1-2. A ponderosa pine stand in inland California. Shrubs and small trees have been removed, primarily for fire protection (Photo: D.J. Mead).

Harvesting of low value biomass in degraded natural forests with mixed species composition does not currently provide enough revenue to pay for regeneration. Clearcutting results in even-aged stands of less tolerant species which some people consider undesirable. In the Appalachians, clearcutting for fuelwood or pulpwood once stimulated more valuable second growth tolerant hardwood stands, many of which were repeatedly picked over as markets developed for different species, grades and sizes (Smith 1994). Improved biomass markets may create opportunities for placing huge areas of low-value hardwood forest under more formal silvicultural regulations, i.e. for changing to selection and shelterwood systems that are sometimes seen to be more ecologically or aesthetically desirable.

Fiber production using natural coppice after clearfelling mixed conifer/hardwood stands

Aspen (*Populus tremuloides*) which reproduces by suckers, and balsam poplar (*Populus balsamifera*) which reproduces by seeds, suckers, stump sprouts and rooting of buried branches, both cover huge areas of North America (Figure 3.1-3). The stands typically arise through replacement of poplar/white spruce (*Picea glauca*) mixtures (Samoil 1988) and constitute one of the most extensive silviculture problems in Canada. Recently, use of poplar for oriented strand board and pulp has increased utilization, and these former 'weed' species are now regarded as 'acceptable species' under legal regeneration requirements for commercially clear-cut lands. Petersen and Petersen (1992) have reviewed their ecology, biomass productivity and management and the Aspen Managers' Handbook (Petersen and Petersen 1995) gives practical advice. Today fewer 'heroic' silvicultural attempts are made to control poplar regeneration on many cutovers.

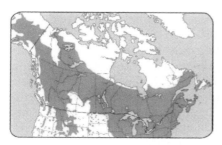 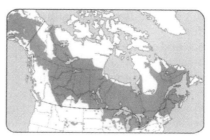

Figure 3.1-3. Current natural distribution of aspen (left) and balsam poplar (right) in North America (from Farrar 1995).

Biomass productivity in unmanaged stands of aspen at age 40 is 83,000 and 165,000 kg ha^{-1} on Site Indexes of 16 m and 24 m respectively (Petersen and Petersen 1992). There are no growing costs and much of the production is from clones. Coppicing is thus possible on a grand scale. Over most of the geographical poplar range, fossil fuel energy is cheap and abundant and there has been little interest in poplar bioenergy since the OPEC oil crisis in the 1970s. However, regeneration may be uneven and disease is a problem. Silvicultural decision guides have been prepared for aspen stemwood production and integrated management of wildlife, livestock and biodiversity (Petersen and Petersen 1992). The ability of aspen and birch to reproduce with age is shown in Figure 3.1-4.

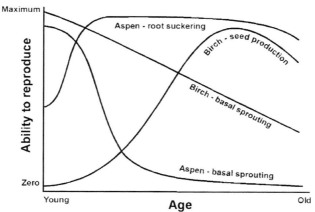

Figure 3.1-4. As aspen stands mature they develop and maintain a high capacity for producing root suckers, but rapidly lose the ability to produce basal sprouts from cut stumps. In birch stands, seed production increases and basal sprouting declines with age (from Petersen and Petersen 1995).

If the need should ever arise Canada could produce large amounts of bioenergy from boreal aspen and balsam poplar with little silvicultural effort or cost.

Industrial plantations

The area covered by *industrial wood plantations* worldwide in 1995 was about 123.7 million ha. Another 26 million ha is planted with traditionally non-forestry species such as rubber, oil palm and coconut (FAO 2001, Pandey and Ball 1998). While representing only 3.5% of the total forest area of the world, plantations provide 27-35% of the industrial roundwood supply and this proportion is expanding (FAO 2001). Of the total plantation area about 83% is classified as industrial. Approximately 55% of all plantations occur in temperate and boreal regions and these are largely composed of conifers. In the tropics about 57% of plantations contain hardwoods. Two examples of large-scale industrial plantation programs – one from temperate areas and the other from the tropics – are discussed below.

Radiata pine plantations

In the temperate areas of the Southern Hemisphere, large-scale industries have developed around plantations of radiata pine (*Pinus radiata*) grown for sawlog and/or chipwood production. These include pulpmills, fiberboard factories, veneer plants and sawmills. In many cases process residues are important for energy production. Clearcutting is the most appropriate silvicultural system for radiata pine plantations.

The intensity of silviculture used varies widely and depends on many of the factors listed at the beginning of this chapter. The New Zealand direct sawlog schedule for clearwood is generally very intensive: location, site selection, plant production, establishment, pruning, thinning and rotation length are closely integrated with the objective of producing high value timber and high returns to the owner (Burdon and Miller 1992, Maclaren 1993, Hammond 1995). Current practice is to use the direct sawlog schedule on the best sites, planting highly improved stock from specific genotype crosses. Because of high seed cost, stock is multiplied by use of cuttings from juvenile plants or by tissue culture. These highly improved and expensive plants can be planted at lower densities than was traditional for the species – 600-800 stems ha^{-1} are now common. Intensive establishment practices include mechanical site preparation and soil amelioration, weed control, fertilizer application on nutrient-poor sites, use of cover crops, and careful hand planting (Maclaren 1993). Stands are thinned to 250-350 stems ha^{-1} final crop (sometimes as low as 200 stems ha^{-1}) in one or two stages. Usually the best trees are pruned in three or four lifts to a final pruning height of either 5.5-6 m or to 8.5 m. Rotation length is 20-30 years. The objective is production of a large, valuable butt log and diameters over 60 cm at breast height are common. However, considerable quantities of lower grade logs are found above the pruned logs (Figure 3.1-5). These, together with thinnings and logging and industrial process residues could be used for bioenergy. For this purpose, extraction must be economic and for some assortments there would be competition with other uses.

Figure 3.1-5. The large pruned butt logs of these mature radiata pine trees growing on a fertile site are very valuable. Potential use of the upper, heavily branched logs for bioenergy would depend on relative product values and demand (Photo: D.J. Mead).

Radiata pine crops grown primarily for lower value products such as knotty timber for structural uses, pallets, chipwood etc. usually receive less intensive silviculture. Planting stock is likely to be cheaper and derived from less improved seed. Stocking rates are often higher, e.g. 1250 stems ha^{-1}. Stands may be unpruned or pruned to 2.5 m for access, and perhaps thinned once. Final tree size tends to be smaller, e.g. 45 cm dbh. Rotation length varies with site and will be shorter if chipwood is the primary objective. However, few owners grow radiata pine solely for chipwood and none specifically for bioenergy.

Growers of radiata pine use specially-designed computer software when evaluating silvicultural options and controlling operations (Maclaren 1993). These decision support systems could be adapted to include bioenergy options. Estimates of different types of biomass, costs and economic evaluations could be provided on a stand or forest estate basis. Similar decision making systems are available for other species and in other countries.

New Zealand and Australian growers of radiata pine are concerned about long-term sustainability and a considerable amount of research is done on this topic. To date there is little evidence that production of radiata pine and other plantation species will decline, provided that good practices are followed, e.g. avoiding soil compaction during logging, and using mulching techniques rather than burning between rotations (Evans 1998). On nutrient-deficient sites the application of fertilizers containing phosphorus, boron and zinc may be required and in some cases this will lead to long-term enhancement of site productivity (Mead and Gadgil 1978, Will 1984, Burdon and Miller 1992). Nitrogen fertilizer is occasionally applied to boost growth of established stands but more frequent use is made of N-fixing understorey plants to enhance the supply of this nutrient (West 1995). Tree breeding is also an avenue for increasing productivity. In New Zealand there are comprehensive strategies and systems in place to prevent and control the impacts of disease, insects or other disasters (Burdon and Miller 1992).

If branches and tops were to be extracted for bioenergy then combustion ash or other fertilizers should be applied to the site to ensure sustainability, especially where the soil is less fertile. Ash application is common in Scandinavia (see Section 4.1.11).

Short-rotation eucalypt plantations

Some large industrial companies, primarily in the Southern Hemisphere, produce short-fiber pulp from eucalypt plantations (Pandey and Ball 1998, Jenkins and Smith 1999, Turnbull 1999). Process residues are often used to supply their bioenergy. In Brazil about 4 million t of eucalypt wood is harvested annually for making charcoal that is primarily used in the steel and cement industries (Turnbull 1999). For this purpose 5-12 year rotation coppice systems are appropriate and *Eucalyptus grandis, E. saligna, E. urophylla* or hybrids are widely grown (Figure 3.1-6). Initial stocking rates range from 900 to 1500 stems^{-1}. Mean annual increment can exceed 45 m^3 ha^{-1} yr^{-1} (Jenkins and Smith 1999, FAO 2001) but is usually less than half this rate (Pandey and Ball 1998).

Figure 3.1-6. Fast-growing Eucalyptus urograndis cuttings at 6 years of age in Brazil (Photo: J. Richardson).

Silviculture in these large pulpwood plantations can be very intensive, the aim being production of uniform, high volume crops. The following practices are frequently included:

- Active breeding programs, e.g. clonal forestry and hybrid crosses.
- The use of advanced nursery techniques, occasionally including tissue culture.
- Intensive establishment practices such as soil cultivation, chemical weed control and fertilizer application at time of planting.
- Control of pests and diseases.
- Mechanized harvesting. Where sustainability is a priority, nutrient-rich parts of the biomass are left and woody biomass is removed (see Turnbull 1999).

In these coppice systems trees are felled cleanly with saws. Stumps are cut on an angle at a maximum height of 120 mm as this improves the wind-firmness of the sprouts (Mathews 1989). The number of sprouts is reduced to two or three per stool. Thinning is labor-intensive and costly (Evans 1992). As a general rule 3 or 4 coppice rotations are allowed before replanting.

Where intensive breeding is being undertaken, crops of lower genetic productivity may be replaced earlier. In Brazil, Aracruz Forestal replaces stands after two rotations (Jenkins and Smith 1999). This plantation system is suitable for production of industrial fuelwood, but widespread application would be dependent on costs and competing markets for the products.

Natural and artificial multiple-use forests – a changing scene

In some richer industrialized countries there is a trend away from wood production towards *multiple use.* Conservation, recreation, biodiversity, habitat protection and aesthetics may be the main objectives. In Germany, for example, pure plantations of introduced trees are sometimes replaced with mixtures of native species, often with the objective of producing uneven-aged stands. In the USA and many other temperate and tropical countries similar objectives are sought for natural forests; society is clearly looking to forest ecosystems for a variety of products and functions. The USDA Forest Service which manages the national forests in the USA has reduced its cut to a third of the historical maximum achieved between 1960 and 1980. It is now well below annual rates of growth. The rise of green certification is a reflection of the concern of society for different forest values. Forest managers often have to abide by strict regulations on where and how to harvest and what silvicultural practices are allowed. Often it is necessary to work closely with other government agencies and stakeholders.

Figure 3.1-7. 'Green tree retention' as practised in the redwood forests of northern California (Photo: D.J. Mead).

There is a distinct trend towards the use of selection silvicultural systems or the use of small clearcut coupes to maintain forest cover. There are many other changes from traditional systems. These range from practices such as 'green tree retention' (Figure 3.1-7) to the harvest of only a proportion of mature trees as they approach death. Such practices are not compatible with cost-effective, large-scale bioenergy production, although harvesting for local fuelwood consumption may occur.

Very short rotation tree crops for energy production

Research and development of very short rotation tree crops has also been undertaken (Nyland 1996, Tuskan 1998). In the 1960s there was an intensive research effort in eastern USA to develop coppice systems with a 1-4 year cutting cycle to provide a pulpwood resource. This was often based on sycamore *(Platanus occidentalis)*. A similar approach has been applied to bioenergy plantations of shrub-willow, alder or poplar (Christersson *et al.* 1993, Makeschin 1999). Maximizing coppice production is quite complex and calls for an understanding of whole-plant physiology and growth responses to silvicultural practices. Mitchell *et al.* 1992 has described the ecophysiology of short rotation crops in detail.

For these energy crops, very intensive management has been envisaged, usually on deep, fertile, well-drained soils on flat land having a history of farming (Mitchell *et al.*1999). This can be costly. Poorly drained and waterlogged sites are not recommended because of low productivity and susceptibility to soil compaction. Nutrients, principally N, P and K, and even water should be applied as required. Because of the high rate of nutrient removal due to harvest of all above-ground biomass at frequent intervals, fertilizer may be required in later rotations. It is not always necessary on fertile ex-farmland during the first 10 or more years (Mitchell *et al.* 1999). Cover crops of legumes and use of nutrient-efficient clones may reduce the demand for fertilizer-N.

Planting densities between 5000 and 20,000 trees ha^{-1} are often used for willows (Mitchell *et al.* 1992, Willebrand and Verwijst 1993, Nyland 1996). These represent a compromise between cost, rate of growth and frequency of cutting. Current commercial schemes often involve planting willows in double rows 1.5 m apart at 0.9 m spacing within and 0.75 m between rows (Mitchell *et al.* 1999). Headlands are left unplanted and may be sown with grass or game-bird cover mix. Such planting arrangements facilitate harvesting. Typical cropping cycles for shrubby willows are 3-5 years, for alders 5-15 years and for poplars 6-15 years (Mitchell *et al.*1992).

Clonal cuttings, selected for different regions and for disease or insect resistance are usually planted after thorough soil cultivation and pre-plant chemical weed control (Makeschin 1999, Mitchell *et al.*1999). Semi-automated planters are commonly employed. Further weed control is often required, the aim being to keep weed competition to a minimum.

Mechanical harvesting can lead to soil compaction and damage to stools; these adverse effects must be minimized. Harvesting is best undertaken in the dormant season when the foliage has dropped. This reduces nutrient removal and ensures good sprouting in the spring.

Insects and diseases pose problems in willow and poplar plantations in Europe (Makeschin 1999). The ecological effects and prospects for long-term sustainability of this type of plantation have also been studied (see Makeschin 1999). The effects are mainly positive when compared to those of previous agricultural practices. There may be some negative impacts during conversion of grassland to bioenergy plantations.

This agronomic approach to bioenergy production is costly, involves risk, and requires fertile soil. In North America risks and costs have outweighed the bioenergy value, even when very high yields are achieved. Greater potential is seen in Europe.

3.1.3   Non-industrial forestry

Farm forestry – agroforestry

There is often a need to provide fuelwood and other forest products to poor rural communities. In such situations a wide range of *agroforestry practices* can be adopted. Where owners use small woodlots for bioenergy, coppice systems would be appropriate if sprouting is reliable. Many nitrogen-fixing trees such as *Acacia, Prosopis* and *Casuarina* species, *Leucaena leucocephala* and other hardwoods, coppice readily. Rotations are likely to be short; between 3 and 12 years. If the owner wishes to grow a proportion of larger-sized logs, the coppice-with-standards system can be ideal. In this system, part of the crop is grown as a normal coppice but some scattered trees (e.g. 100 out of 900-1500 trees ha^{-1}) are left to grow on a longer rotation, to be cut down when they reach a suitable size or maturity. Often these larger 'standards' are cut on a multiple of the normal coppice rotation length.

The following criteria are used when selecting species for fuelwood in developing countries:

- Species should adapt well to the site conditions, establish easily and require minimum care, as villagers do most of the work.
- Seed or planting stock should be readily available.
- Trees should grow rapidly and have an early culmination of current annual increment.
- Nitrogen-fixing ability is advantageous.
- Trees should coppice readily.

- The wood should have high calorific value and burn without sparks or toxic smoke. It should split easily and dry quickly. Usually this implies moderate to high wood density.

- Resistance to goat and wildlife damage is advantageous, unless trees are also grown for fodder.

- Multi-product species which are a source of fruit, medicines, fodder, poles or timber as well as fuelwood are especially valuable.

Stem straightness is not a criterion for fuelwood; nor are large-sized trees always desirable since they can make handling difficult. A wide range of tree species meet many of the above criteria (National Academy of Sciences, 1980, 1983, Nair 1993). Some 1200 have been identified of which 700 are highly suitable. Examples of multiple-purpose species which make excellent fuelwood are various acacias, alders, eucalypts, *Casuarina, Leucaena,* and *Prosopis* species, *Albizia lebbeck, Azadirachta indica, Calliandra calathyrsus, Cassia siamea, Gmelina arborea, Melia azedarach, Robinia pseudoacacia, Sesbania sesban, Tamarinus indica.* Some potentially useful species are aggressive pioneers and can become weeds, e.g. *Prosopis juliflora* and some acacias. Often it is best to use species that already grow in the area and are sold as fuelwood rather than newly-introduced ones. Difficulty has been encountered in the marketing of fuelwood from exotic species such as *Leucaena, Casuarina* and *Eucalyptus* by state agencies in India.

Silviculture will vary with species and site, but it is usually simple and readily adopted by local people. The most important considerations are species, seed or plant availability, nursery facilities, spacing and layout, planting, initial weed control and animal control. Where block plantings are used for firewood, typical spacings range from 1 to 3 m in a square pattern. Spacings of 1 to 2 m will produce the greatest amount of biomass with smaller piece size in the shortest possible time. Wider spacing (3 m) will produce larger piece sizes and give more flexibility in rotation length without risking stool suppression. Woodlots grown for fuelwood only are more common in developed countries.

Agroforestry practices, which may include bioenergy as one of several products, vary widely. Nair (1993) identified 18 different practices within the three main agroforestry systems:

- *Agrosilvicultural systems* – crops plus trees. These range from improved fallow to home gardens and shelterbelts associated with crops.

- *Silvopastoral systems* – trees plus pasture and/or animals. These can range from trees widely-spaced over pasture to trees used as protein banks.

- *Agrosilvopastoral systems* – trees plus crops plus pasture and/or animals. Included are home gardens with animals, multipurpose woody hedgerows and woodlots.

Many agroforestry practices are designed for multiple functions such as shelter, erosion control, food, fiber and firewood production. Nair (1989, 1993) emphasized that farmers and communities in developing countries rarely plant for fuelwood alone or for fuelwood as a primary product, even where wood is in short supply. Wealthy farmers may use intensive silviculture with the goal of maximum profit from their investment, but it is unlikely that the highest value will be derived from fuelwood. Resource-poor farmers are likely to use low-input approaches which aim at small but attractive returns with minimum investment. A good example of integrated agroforestry is provided by small farmers in southern African tropical savannas who grow *Sesbania sesban* sequentially with maize. The N-fixing trees supply N to the maize crop, and also yield fodder and fuelwood (Sanchez 1995).

Sometimes agroforestry is based around natural forests or forests regenerating after shifting cultivation. In Nepal the state is allocating patches of native forest to village control, provided that an agreed management plan is put into practice.

In tropical developing countries, land ownership and social patterns vary widely. They can range from individual farms to communally-owned village land or state-owned land on which local people may have some rights such as the collection of non-timber forest products and fuelwood. Community or social forestry addressing the full needs of the people, through their participation, is important even where the focus is on promotion of fuelwood crops (Nair 1993).

3.1.4   Adaptation of silviculture to fuelwood production

Silvicultural methods vary widely. If fuelwood production is the primary management objective it would be difficult to go beyond the coppice system. There are many opportunities for increasing fuelwood production under other systems if socio-economic conditions are favorable. But are there ways in which silviculture could be adapted to bioenergy production without changing socio-economic conditions if the primary objective is a high value product?

The answer is likely to depend on specific site factors and may involve risks. It will be governed by demand, economics, stand type, stand condition and existing management requirements. One option might be the alteration of stand density and thinning schedules to make harvesting more cost-effective. It is usually cheaper to harvest greater volumes of trees with larger piece size, so delay of thinning could swing the economics in favor of bioenergy production. This could be risky, because the stand would become overstocked if the market did not eventuate, and the primary objective would be jeopardized. Another option is utilization of trees or shrubs that are seen as detrimental and would otherwise be controlled by other means. Other longer-term strategies could include a change of species or product mixtures.

Well-developed and validated decision support systems, such as stand and forest simulation models coupled with economic packages, would be useful for analyzing the effects of different options. Unfortunately few exist at present and without them forest managers find it difficult to justify changes to silvicultural schedules. If major changes in bioenergy economics or in other factors affecting forest management occur, silvicultural systems could easily be adapted to allow greater fuelwood productivity.

3.2   INTEGRATION OF ENERGY PRODUCTION INTO FOREST MANAGEMENT

In this section, an attempt is made to describe strategies for integrating forest fuel procurement with the harvesting of industrial roundwood. Tactical and operational aspects of woodfuel harvesting are covered in Section 3.3, which may provide further food for thought for the practitioner. But a basic understanding of the forestry sector is also needed. Clear definition of fundamentals serves as a foundation for the integration process. By contrast, lack of understanding about the role of forest management and the basic functions of the forestry sector may impede development and lead to faulty or counterproductive decisions.

The general approach to the topic of integration elaborated in this section is based on common *systems theory*. Organizations are systems in which people and things are purposefully arranged in order to effectively carry out specified tasks. Forestry is an organized effort to derive benefit from the forest ecosystem. Forestry, accordingly, is seen here as an organization[1] engaged in the supply of materials, energy and other outputs from forests under human supervision.

According to the systems school, organizations in general (Arbnor and Bjerke 1994) including forestry and forest energy organizations (Vikinge 1999) are regarded as *'open systems'*, meaning that they interact with the environment in which they operate. An organization consists of 'dynamically intertwined and interconnected elements, including its inputs, processes, outputs, feedback loops… A change in any element of the system inevitably causes changes in its other elements.' (Shafritz and Ott 1987). The implication of this is that the effects of changes are not limited to the organization itself, but provide leverage between the organization and its environment. An organization must therefore adapt to environmental change.

---

[1] In reality, of course, 'forestry' is the sum of activities of a multitude of separate but cooperating, competing or complementing organizations.

Natural starting points for studying the integration of energy production and forest management may be:

1. Definition of the principal aim of forest management.
2. Description of the forestry sector as a production system in generic terms, placing special emphasis on structural analysis of the forest fuel and bioenergy components.

From this basis we move on to strategic considerations to be made in the integration process, including:

1. Product assortment definition and quality demands for wood-based fuel.
2. General logistical layout of the physical product flows.

3.2.1   The objectives of forest management

Forest management is concerned with activities aimed at fulfilling specific human needs through forest utilization. One of the most important goals of management is to design these activities in the most rational and efficient way, so that identified needs are fulfilled with a minimum of resource depletion. In generic terms, the managerial task may thus be described as a process of constantly striving to maximize operational efficiency, defined by the ratio of useful output per unit input:

f_{max} *for f = output/input*

Given an ideal market situation, monetary values contain all relevant information needed to compare different alternatives. The managerial problem may then be formulated in an extremely tangible way as *fulfilling market demands with minimized costs*.

Unfortunately, market mechanisms are far from ideal. Björheden (2000) discusses criticism of pure monetary evaluation criteria in greater detail and gives references in support of this criticism. He concludes that criteria other than monetary value have to be added in order to produce a more complete analysis. Such criteria are formulated as constraints, limiting allowable sets of solutions of a monetary efficiency function:

f_{max} *for f* $= \sum x_{i-n}\text{-}y_{i-n}$, $x_{i-n} \in \Omega$

> Where:
>
> x_{i-n} *is the revenue, x, of every activity, i-n;*
>
> y_{i-n} *is the corresponding cost, y, of each activity, i-n;*
>
> $x_{i-n} \in \Omega$ *is all activities, x_{i-n} within the allowed set of constraints Ω.*

The constraints include environmental and social restrictions supported by legislation as well as ethically-founded considerations laid down by more or less variable moral standards reflected in company policy. Many of these are discussed in Chapters 5 and 6. In addition, there are a number of 'hard' technical restrictions that limit the number of possible solutions to the efficiency function.

Although many of the constraints are difficult to assess in monetary terms, they must be included in the managerial evaluation process. There are a number of techniques for such evaluations (cf. Bulmer 1977, Caldwell 1992, and Emory 1985). One approach is to analyze the hierarchical structure, putting more weight on high level constraints than on constraints of a lower order. Hierarchical analysis of constraints makes evaluation easier and more structured. When broken into detail, many criteria may be linked to economic reasoning. In an example of hierarchical analysis shown in Figure 3.2-1, the assessment of yield losses for different levels of nutrient removal is indicated. In accordance with the saying *'What gets measured gets done'*, it is better to use measurable criteria as delimiters of the allowable solutions.

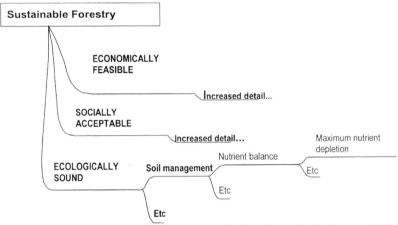

Figure 3.2-1. An example of hierarchical analysis. The method aims at dividing complex constraints into measurable criteria, usable in follow-up of operations. Eventually, many of the criteria may be estimated in monetary terms. (From Björheden 2000).

The appearance of modern forest energy harvesting on an industrial scale adds a new dimension to forestry operations. The introduction of a new product assortment adds complexity to the system but also increases the opportunity for forest management to achieve its goals.

To be feasible, integrated harvesting must add efficiency to forestry without violating social, environmental, technical and policy constraints. The managerial task is to maximize efficiency gain in this process. Thus, logging sites serving as sources of biomass to be salvaged for fuel need to be chosen on the basis of harvesting costs and expected revenues. In addition, ecological sustainability, opportunities for concentrating operations, and seasonal variation in consumption must be taken into account. The wishes of landowners and other stakeholders also affect the choice of harvesting sites.

3.2.2   Description of the forestry sector from a logistical perspective

Fuelwood procurement strategy is affected by the general layout of the forest fuel recovery system, and also by interaction with the parallel supply chains connecting logging areas with other wood-consuming industries, notably sawmills and pulp mills.

The forestry sector can be described by the following terms:

- *Primary production*, activities related to the biological processes yielding forest biomass.

- *Secondary production*, activities connected with harvesting and transporting forest products to various industries.

- *Tertiary production*, processes within the wood-consuming industries.

From a systems analysis point of view the industrial forestry sector can be described as a set of supply chains, binding forests together with industries that utilize forest biomass as raw material in their production (Figure 3.2-2). Forest harvesting and transport supplies the logistic solutions for all industries dealing with primary conversion of forest biomass.

Each of the three main sector components plays a specific role in the system (Figure 3.2-2). The spatial dimensions of each component are important for understanding the workings of the industrial forestry sector. In geometric terms, forests represent areas, forest harvesting operations and transport can be described as a line network, and industries as points.

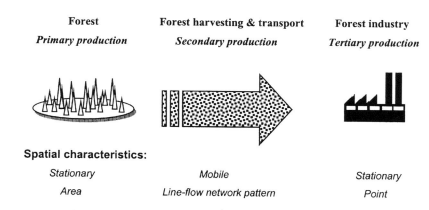

Forest	**Forest harvesting & transport**	**Forest industry**
Primary production	*Secondary production*	*Tertiary production*

Spatial characteristics:

Stationary	*Mobile*	*Stationary*
Area	*Line-flow network pattern*	*Point*

Figure 3.2-2. Characterization of components of the forestry sector.

Forests

Forests provide raw material through a process that is basically endogenous and natural, although it can be manipulated through silviculture. The process is driven by incoming solar energy, and it is area-dependent. Although the spatial distribution of forests can be altered through active management, forest resources must be regarded as stationary at any given time. In general, forestry has a limited economy of scale. As a result it is normally fragmented into many companies and public and private holdings. Major forest enterprises and other large forest owners, such as states or municipalities, even adopt policies leading to a diseconomy of scale. They may fragment their forestry holdings through geographical division and by leasing or contracting activities to an array of smaller companies.

Forest harvesting operations and transport

Harvesting and transportation of primary forest products from forest to industrial site are operations controlled and maintained entirely by man. The main objectives are:

- Collection of forest biomass according to demand.
- Preparation of this biomass for transportation.
- Transfer of the biomass to the industrial site.

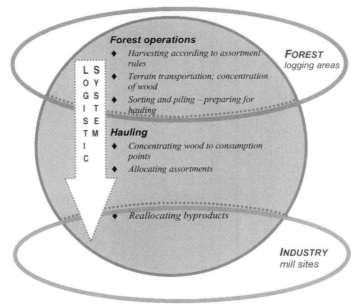

Figure 3.2-3. Harvesting operations, including hauling, produce a logistic system for the forest sector by pre-processing and concentrating biomass products and linking the forest areas with industrial sites.

Figure 3.2-3 illustrates the logistic system for forest harvesting and hauling, which links the forest to industrial sites. In this diagram, secondary production is described as a simple process of concentration and transportation from 'forest' to industrial site. In reality the process is extremely complex, because at each point there are a number of decisions to be made which will affect the flow pattern.

Part of this complexity is due to the nature of forest biomass as a variable commodity. Its properties vary with species, phenotype, tree size and age, felling season and tree component. As a result of this variability, different industries demand different types of forest biomass of which fuelwood is only one. Species, diameter, stem length, shape and properties such as decay and structural characteristics normally decide assortment composition.

Thus the logistic system is not normally a simple supply chain from raw material source to consumer. It resembles a set of parallel and sometimes interlocked chains. A supply network is probably a better description. An increased degree of integration may lead to increased efficiency, better precision and lower procurement costs for the forestry sector (Hektor 1998). Integration of the planning and implementation of wood-based fuel procurement with production of roundwood assortments for other industries is probably a feasible undertaking (Ling 1999).

Harvesting activities include the cutting and collection of biomass at the logging site. They also include preparation of harvested material for transport and pre-processing to make the material more suitable or acceptable to the user. Silvicultural treatment of the stand or site may be an additional task combined with harvesting. Harvesting activities generally show limited economies of scale, since the optimum scale of operations is determined by crop density together with cost of concentrating the crop through harvesting. High crop density coupled with efficient harvesting techniques will increase the economy of scale. However, it is easy to exceed the optimum size of an operation, with resulting decreased economy.

In spite of the importance of other harvesting activities such as sorting, pre-processing for industry, thinning or preparing the site for regeneration, the key activity in secondary production is *transportation*. This begins when biomass is moved from the stump. Transportation includes several steps of concentration of the harvested material. At each step transportation conditions are affected by increased concentration which makes it economically possible to invest in further machines or new technologies in the following transport chain:

Harvester head → Forwarder grapple → Forwarder-loader → Truck → Industrial stride lifter → Conveyor.

Transportation thus displays extreme economy of scale (Sundberg and Silversides 1988) but is nevertheless fragmented as a result of unavoidable inefficiencies of interaction with the forest resource. This is mainly due to the need for concentration of material before economies of scale can be exploited. The fragmentation of forestry itself and legislation controlling maximum allowed gross vehicle weight increase the fragmentation of transportation in spite of the obvious economies of scale (Björheden and Axelsson 1991).

The total product flow resembles a complex web of interacting supply channels. Different products are transported in different patterns within this web, the basic layout being decided by the composition and spatial distribution of forests (fixed) and the location and demands of industry (fixed in the short- to medium-term). The web-like structure increases complexity and the need for co-ordination. In addition, there is a need to divide the harvested material into assortments corresponding to specific demands from industry.

Decisions about when, where and how the division into different assortments is made is a strategic one which will influence logistic solutions as well as the cost structures of all forest products.

Forest industries

The role of forest industry is the processing of forest biomass into merchantable commodities. In order to understand this role from a logistic perspective, forest industries may be divided into groups based on the type of processed product produced. For our purpose three main groups of processing industries can be distinguished. The first consists of the *solidwood (SW) industries* such as sawmills and veneer plants that have high quality requirements for structural and sometimes aesthetic properties of each log procured. The second group consists of the *pulpwood (PW) industries* such as pulp mills, fiberboard, chipboard and oriented strand board plants etc. These demand a lower level of log quality since disintegration of solidwood, which is their basic process, provides opportunities for removing undesirable material. The third group is the *energywood (EW) industries* such as bio-fuelled heating and power plants. They have no interest in structural and aesthetic properties, demand being based on calorific value, ash content and suitability for storing, processing and handling.

As mentioned earlier, forest industries are points of demand usually located close to forest areas. Location of industrial sites in relation to the forest and the integration of industrial processes determine flow patterns. Thus industries may be located apart from each other may share a joint location. Integrated installations may contain any combination of *SW, PW* or *EW* industries. Figure 3.2-4 shows transport flow patterns for separate industries, *PW+EW* and *PW+SW* combinations and a fully-integrated complex *(EW+PW+SW)*. The figure shows a very simple situation. Frequently, integrated complexes and several separate industries are present within a forest region. These generate complex flows from harvesting to industrial sites, as well as between industries of different types. Increased industrial integration would simplify transportation planning and allow the postponement of division of harvested material into industrial assortments.

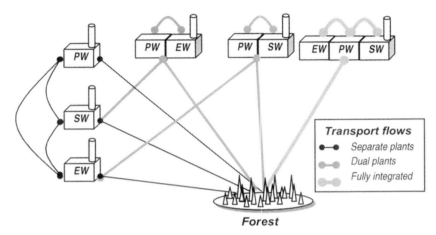

Figure 3.2-4. Transport flow patterns for separate, dual and full combinations of industries. In reality, industry layout is often a mix of separate and integrated industries, and transport flows form extremely complex, web-like patterns.

Forest processing industries generally display strong economy of scale, but with increased size, wood acquisition costs will also rise due to longer hauling distances (Figure 3.2-5). The recent trend towards decreasing transportation costs is pushing the optimum size towards larger and larger units. This sets the limit for optimal concentration of a forest industry. Furthermore, the economy of scale is variable for the different parts making up the forest commodity supply chain. It will thus be impossible to optimize the system by optimizing the individual parts separately. Optimization must be addressed at the system level.

The implicit area dependency of all production based on raw materials from forests makes logistics a key element. Fixed costs of production are generally harder to influence than variable costs (operating costs). Logistic costs are a major part of the variable costs. Even small cuts in these costs make large volumes of raw material available and also give access to large new markets. The reason is that for every cut in logistic costs, the area of economic dominance will increase by the square of the cost reduction (Figure 3.2-5). Thus, decreasing logistic costs will strongly affect the optimum scale of operations. This phenomenon is known as 'Lardners law of transport and trade'.

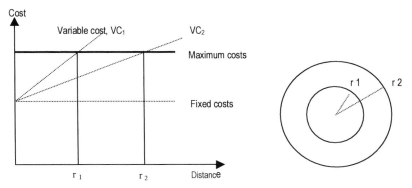

Figure 3.2-5. Decreasing costs of logistics (transport) from VC_1 to VC_2 make it possible to procure raw material at distance r_2 instead of r_1. The area of dominance will increase, as shown on the right, by the square of the distance and so strongly affect the possible scale of operations.

3.2.3   Balancing demand and supply

A common problem in forestry is that supply and demand for forest products are not evenly balanced over time. This adds to the complexity of forest management, increasing demands for elaborate planning and investment to create buffering capacity. Balance of demand and supply is an important element of procurement strategy. There are three principal ways of solving the balance problem (Læstadius 1990, Johansson 1994):

- Storing produce during periods of low demand.
- Storing production capacity during periods of low demand.
- Storing alternative operations during periods of high demand.

In boreal regions the problem is especially pronounced when forest fuel is a product of modern, mechanized harvesting that operates year round. Biofuels are particularly suited to low-grade energy production such as heating. But while the need for heating follows seasonal temperature changes the intensity of logging does not. Figure 3.2-6 shows a typical demand curve for industrial-sized heating plants in Scandinavia. The corresponding supply curve represents production throughout the year. The Scandinavian solution to the balance problem is storage of fuelwood harvested during periods of excess production and delivery of this wood when demand exceeds production. The advantage of this solution is that it does not strain harvesting/transport logistics and associated planning, although it operates at high buffer levels. Disadvantages are that large amounts of capital are tied up in storage facilities, also that degradation may occur during storage. The current trend in Scandinavia towards decreased storage levels of pulpwood and solidwood and consistent logging intensity throughout the year paradoxically leads to higher average forest fuel storage levels dependent on demand and supply, as illustrated in Figure 3.2-6.

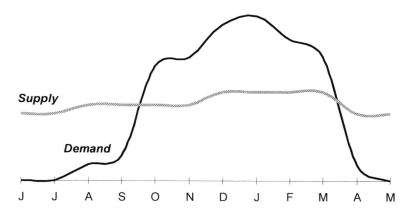

Figure 3.2-6. Relationships between typical demand and supply curves for fuelwood in Sweden. (After Björheden 1986).

The second alternative, storage of production capacity in periods of low demand, pre-supposes that extra production units are readily available when demand is high. This has been the case in the forwarding component of Finnish and Swedish logging operations, where some temporary recruiting of contractors for slash forwarding has occurred during the season of high demand. The advantage of this solution is that capital commitment is minimized. A disadvantage is that operational planning is seasonally intensive, thus varying the workload. Another negative feature is that the system inhibits development by failing to promote lasting relationships between contractors and the principal.

An example of the third alternative is seasonal use of basic machinery used elsewhere, e.g. in agriculture. Flexible operational planning in terms of stand selection and bucking rules may also be used to adjust work content. Another interesting solution is the exploitation of opportunities offered by versatile, multi-task machinery. Johansson (1994) reports that excavators and backhoe loaders normally used in ditching, road construction and maintenance can be equipped with harvesting heads and used to boost wood production during periods of high demand. This combines low capital commitment in storage facilities and machinery with flexibility and incentive for technical development. It also increases advantages associated with employment of full-time professional contractors.

3.2.4   Harvesting operations

Degree of integration

Technology and methods of forest fuel procurement are described further in Section 3.3. This section deals with the relationship between biofuel harvesting and conventional logging. One of the key issues is the degree of integration between forest fuel recovery and other harvesting operations (Björheden 1989). Systems originally used had a low level of integration, and were characterized by the use of existing technology, sometimes with simple modifications. The organizational relationship with traditional harvesting was loose, fuelwood recovery being carried out as a separate, subordinate operation. The first extensive Scandinavian fuelwood recovery systems based on collection and forwarding of slash from shortwood logging sites are good examples. They provided a good opportunity for testing and design of operations with low economic risk and little need for organizational and technical change. However, these two-pass systems did not fully utilize the economic potential of integration.

Higher levels of integration incorporate methods and technology associated with parallel and coordinated one-pass harvesting operations. Full integration of fuelwood and industrial assortment harvesting allows exploitation of all the positive interactions. On the other hand, risk and degree of commitment are increased. In highly integrated systems, fuelwood recovery is an integral part of operational planning; decisions on assortment range are based on their net contribution, and technology and methods are adapted to the task of integrated harvesting. Technology may include multifunction machinery designed to handle several assortments and work tasks, or single-function machines performing one task on a single assortment. When viewed as an integrated system, the logging operation includes purposeful technologies for the complete range of assortments.

An example of a highly integrated harvesting system, suitable for short haul operations to transport terminals or industry sites, is a whole-tree system using feller-skidders/forwarders and whole-tree-carrying trucks. In remote areas, where the industry sites are not close to each other, it is often more efficient to separate some or all of the components at the logging site. Highly integrated operations include tree-section systems yielding sawlogs and non-delimbed energy/pulpwood; systems based on whole-tree chipping and sorting into fuel and fiber components at the mill; and systems for harvesting and treating different types of trees with different machines.

Transport systems

Although forest harvesting operations sometimes include initial processing steps, the principal aim is concentration of dispersed raw material into a form that is suitable for subsequent handling, processing and consumption. Usually forest resources are distributed over an area while conversion into commodities occurs at specific points such as sawmills, pulp and paper mills, fiberboard plants, and heating or co-generation plants. The required quantities and qualities of wood are accumulated and delivered to the points of demand. Transport is thus a key element of forest activities, and the way that it is organized can have implications for the production system as a whole. Since fuelwood is a low-value commodity, transportation costs constitute a relatively large part of the total production cost. Under Nordic conditions and with an average transportation distance of 60 km, more than half the cost of residues delivered at heating plants is incurred by transportation. Eriksson and Björheden (1989) noted that 'optimizing forest-fuel production essentially means minimizing transportation costs'.

Choice of transport technology

Efficient *transport technology* minimizes the cost of moving goods. When the commodity is forest fuel, the transported product is actually energy. The goal should be to transport energy as efficiently as possible. This is not necessarily the same as optimizing transport of a physical load.

Figure 3.2-7. Proportion of solids in uncompacted logging residues and tree-sections, wood chips and conventional pulpwood. All loads have the same solid content. (After Nilsson 1983).

A basic problem of forest energy transportation is illustrated in Figure 3.2-7. Slash, un-delimbed small trees and tree-sections are typical forest bioenergy products (Hakkila 1989). Their low bulk density increases the cost of transportation. Water is a major constituent of the transported mass and the complex texture of the material makes handling technically difficult. Bulk density may be increased by compaction or by chipping. Processing into chips will decrease durability under storage. Green chips are highly vulnerable to exothermic microbiological, physical and chemical degradation which can cause health hazards, loss of substance, loss of energy content and the risk of self-ignition (Björheden and Eriksson 1990, Kofman 1994, see also Section 3.4.2). Chipping can only be recommended if it takes place shortly before consumption.

To ensure cost-efficient transportation without loss of quality, an array of methods for compacting unprocessed fuel materials has been tested. Up to now they have been too time-consuming, too capital-expensive, or have reduced overall system performance. An interesting new technical solution is the FibrePac system, described in more detail in Section 3.3.2. A machine bundles forest residues into composite residue logs (CRLs). Current technology does not lend itself to integration with the logging operation proper, although there is a possibility of using a conventional forwarder for CRL transportation. Later in the chain CRLs may offer high integration opportunities since hauling can be carried out by conventional roundwood rigs. This would allow more efficient use of the transport fleet than the present situation in which several highly specialized vehicles are used. With the Nordic tradition of development of increasingly multifunctional machines, it seems probable that efforts will be made to mount the CRL bundler on the chassis of a harvesting machine. This would lead to complete integration with the shortwood system and permit harvesting as a one-pass operation, thus greatly reducing the cost.

Selection of transport systems is affected by the quality and structure of the forest road network and by conditions at the landings. Under a given set of circumstances, transport technologies will differ with respect to terminal cost (costs connected with loading and unloading) and the direct cost of transportation. A comparison between two hypothetical systems is presented in Figure 3.2-8. Terminal costs are represented by interception of the line with the y-axis, and direct transport costs by the slope of the line. A system with high terminal costs and low direct transportation costs (System A in Figure 3.2-8) could be bulk cargo haulage with a simple loading system. System B in Figure 3.2-8 might represent a container truck or railway system. System A with low terminal costs is competitive when terminal work makes up a large part of the transport activities and where there are short hauls. Over longer distances the direct transport cost becomes more important and System B is preferable. Provided that there is sufficient volume, both could be part of an efficient overall system.

Transport systems may be selected on other than purely economic grounds. In Sweden, containers are sometimes used for chip transportation in order to prevent littering of roadsides with chips. System flexibility and applicability for a variety of goods is an important consideration when dealing with marginal quantities or when the amount of transport capacity needed is uncertain.

Cost/ton

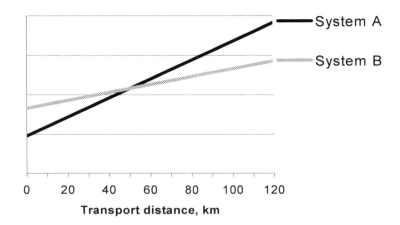

Figure 3.2-8. Cost functions for two transportation systems. The intercept of cost function with the value axis represents direct cost at the terminal; increment with distance represents the variable direct cost of transportation.

Transport flow layout

The choice of supply chains or production flows influences the choice of harvesting and processing technology. This justifies brief discussion of the logistic layout of the fuelwood recovery system. In a study of marginally-integrated forest fuel production (Eriksson and Björheden 1989), five different theoretically available production flows were identified and evaluated using linear programming techniques. The alternative production flows, shown in Figure 3.2-9, were:

1. Raw material transported directly from source (logging site) to consumer and processed at the heating plant.

2. Raw material transported via a terminal and processed at the heating plant.

3. Raw material transported via a terminal equipped with processing and fuel storage facilities. Processed material transported to heating plant.

4. Raw material processed at source, and transported via terminal to heating plant.

5. Raw material processed at source and transported directly to heating plant.

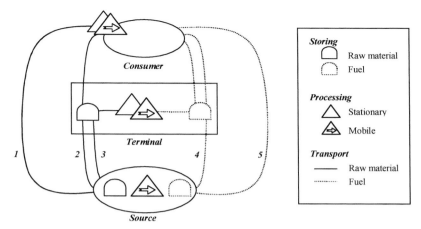

Figure 3.2-9. Alternative production flows from raw material source to consumer used to study fuelwood transport and processing systems for a heating plant. The five numbered systems are described in the text.

Some important logistic characteristics were:

- No transloading (Flows 1 and 5).
- Optimized hauling distance (1, 5).
- Minimized storing cost (5).
- Opportunity for use of cost-effective stationary processing units (1, 2, 3).
- Opportunity for use of buffer storage for increased reliability etc. (2, 3, 4).
- Potential for optimizing fuel quality (3, 4).

Based on assumptions relating to equipment, costs etc. used in the study, the most straightforward solutions were also the most economic ones (Eriksson and Björheden 1989). These were Flows 1 and 5, where material goes directly from logging site to heating plant. This had not been expected. Previously the leading hypothesis suggested that intermediate terminals with storing, mixing and processing facilities would improve the profitability of forest energy activities. The reasons for the new results are discussed below. They are still applicable in Scandinavia where the basic assumptions, such as a 55 km average transportation distance, are valid.

Flow 1 was the most favorable solution for short hauling distances. In order to become the exclusive transport flow, comminution at the heating plant had to be 30-45 percent cheaper than chipping at the landing, in order to compensate for the high cost of slash transport. Since 1989, transport of unprocessed material has become more efficient and better compaction devices have been developed. The competitive strength of Flow 1 is probably greater today than it was over a decade ago. It may be noted that compaction and chipping have a similar effect on transportability. Transport of compacted material should therefore be compared with Flow 5 rather than Flow 1.

Flows 2, 3 and 4 all include transport to a terminal where various activities take place before transport to the heating plant. An advantage of the terminal is the opportunity for mixing raw materials of varying quality into a homogeneous fuel. A closely related benefit is the opportunity for producing an acceptable fuel by mixing cheap sub-standard material, such as wet bark from sawmills, with higher-grade materials. Furthermore, stationary comminution equipment at a terminal is very cost-efficient compared to mobile processing. Solutions based on a terminal also give higher security to transport and delivery during times of limited forest road access. The most important disadvantages of a terminal are the high capital and maintenance costs, the increased direct and indirect transport work and very high storage costs. Use of a terminal will increase the number of internal transactions for the forest fuel supplier, thus adding complexity to the operations. In the 1989 case study, use of a terminal had to add value corresponding to at least 50-75% of the cost in order to break even – a goal that was not achieved. In recent times fuelwood volumes delivered through large terminals in Sweden have decreased, mainly because of quality management problems (Andersson 2000).

It should be pointed out that the terminal used in the study of Eriksson and Björheden was equipped for fuelwood alone. If the level of integration were to be increased so that the terminal handled assortments for several industries, it might become more viable. Large centralized log-making and sorting terminals (not necessarily including fuelwood) are used in some other developed countries where the advantages, including increased rationalization through integrated planning, improved value recovery, traffic flows, etc. lead to increased overall efficiency. The viability of a terminal is very sensitive to the relative location of customer and forest.

3.2.5   Quality and forest fuel

Quality is a relative term. Andersson (1996) defines *wood quality* as the 'suitability of a certain type of wood for a specific purpose'. High-grade lumber is not used for fuel, but this does not mean that it is unsuitable, rather that the sawmilling industry outbids the energy industry. Large, straight trees with clear stems, expected to yield high quality timber, might also yield excellent fuelwood for very reasonable extraction costs. Rejects from conventional industry, often used for energy production, are expensive to harvest, difficult to handle and process, and variable in quality, but they are not entirely unsuited to their purpose.

An important group of constraints affecting procurement strategy is relted to technical minimum requirements stipulated by consumers. For fuelwood, requirements are based on particle size, dry-matter content, contamination and pollutants. Procurement strategy must be based on customer needs. In an ideal situation the combustion plant would be built to consume easily-harvested fuelwood. The supplier must meet the consumer's minimum requirements, but a clear distinction has to be made between properties that are required and those that are negotiable, such as quality gradients. As with other commodities, differences in quality should be reflected in the price. It is recommended that prices be based on retrievable energy value, rather than the amount of energy actually retrieved, as this will stimulate efficient combustion.

It is not unusual for different consumers to have different demands on the fuelwood fraction depending on the size and type of heating plant, storage and processing facilities. Whenever a contract is being drawn up for a new fuelwood consumer it is essential to find out whether quantities and qualities can be supplied at the required times.

The definition of assortments

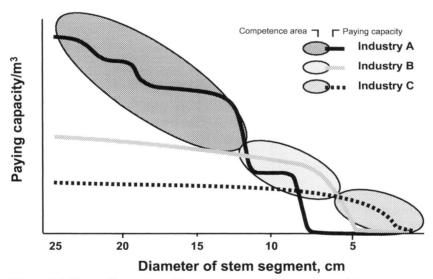

Figure 3.2-10. An illustration of how paying capacity may vary with average tree diameter in different types of industries (A = sawmill type; B = pulpmill type; C = forest energy type).

Except where forests are being grown or harvested solely for fuelwood, the bioenergy component is material rejected by other industries. Fuelwood consists of the tops, branches, twigs, foliage, stumps and small trees that remain after extraction of high-value logs and pulpwood. If forest energy is considered as one of a number of assortments marketed from forests, the distinction may not always be so obvious. The paying capacity of different industries varies with stem diameter (Figure 3.2-10). Interception points between the lines for different industries indicate that the difference in paying capability is moderate. These are areas in which an altered definition of assortments may increase economic efficiency.

Defining fuelwood assortments through calculation and analysis often leads to solutions which differ from decisions guided by current markets. If calculations show that there may be cause for redefining assortments, this is an indicator of a potential efficiency improvement. Under Swedish conditions it has been shown that a general increase in pulpwood minimum diameter would lead to lower total costs and higher quality for both the pulp and forest energy industries. The efficiency of the forest enterprise would also increase (Björheden 1989). In stands with large proportions of unmerchantable wood, especially small diameter material, it may even be economically feasible to simplify operations by harvesting only energy assortments (Eickhoff 1988, Parikka and Vikinge 1994).

The spatial dimension of wood procurement

As stated above, the collection of geographically-dispersed materials and accumulation of these at specific points is the quintessence of forest harvesting. Strategies for wood procurement therefore always include a spatial dimension. If, as with forest fuels, the cost of transportation constitutes a large part of the total acquisition cost, the spatial dimension becomes a dominant factor. A simple way of optimizing or highlighting this aspect is to divide transport costs into distance classes. This supports intensified procurement of materials close to the point of demand. The strength of this stimulus can vary, depending on the market situation (Björheden 1993). The competitiveness of different wood-consuming industries is affected by differences in transport cost, especially in the case of industries that are not located in the same place. In Figure 3.2-10, this could be illustrated by raising or lowering the paying capacity lines, thus changing the intercept points. The cost effects of the spatial dimension are illustrated in Section 4.2.

3.3   FUELWOOD HARVESTING - TECHNOLOGY AND METHODS

3.3.1   Conditions affecting fuelwood harvesting

Many bioenergy fuels are derived from the forest (Figure 3.3-1). This section deals with technical aspects of the harvesting of fuels in conjunction with logging, together with the secondary haulage of the fuels to the energy producer.

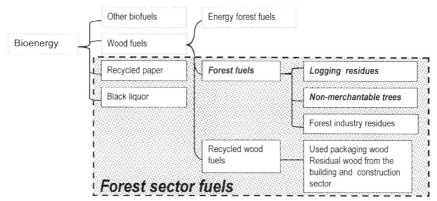

Figure 3.3-1. Bioenergy fuels derived from the forest.

The procurement of fuelwood in conjunction with logging consists of a number of operations and sub-operations. In principle, these make up a supply chain:

Collection → Off-road hauling → Secondary hauling.

Various forms of *processing*, such as comminution and measures to facilitate handling of the material, as well as *storage*, can take place at different points in the chain (cf. Björheden and Eriksson 1989). Fuelwood can consist of conventional logging residue, whole-trees from thinning or premature clearfelling operations, or non-merchantable trees from an exclusive fuel-harvesting system (Vikinge 1999).

The only type of fuel comminution discussed here is the simple transformation into *chips* or *hog* that takes place in conjunction with harvesting and extraction. The most usual operation is comminution by chipper or hammer mill. These fuels are mainly used for heating or for combined heat and power production. Hakkila (1989) notes that 'comminution has a central role' in the overall design of the harvesting system layout. The choice of comminution device and the location of comminution in the forest fuel supply chain are among the most important components of a supply system for fuelwood. Factors affecting choices include:

- Customer requirements for the raw material.

- Total volume of fuelwood in the system.

- Stand characteristics and nature of the road network.

- Conditions at the end-user reception facility, and feasibility of creating terminals for efficient handling and storage without incurring excessive additional haul distances.

Quality parameters are always important for the forest fuel supplier. Cost-effective handling of fuelwood requires careful harvesting, avoidance of contaminants, and promotion of drying (Brunberg *et al.* 1998). Qualitative demands arise from the point of view of transportation and handling. Parameters such as friction and bridging are affected by particle shape and size distribution, which in turn are the result of interaction between a specific raw material and the comminution device (cf. Mattsson 1988).

It is also essential to select the most favorable logging sites. Larger areas with high residue density are usually more economic to harvest. For small sites with little fuelwood, machine moving and setup costs will substantially increase production costs (see also Chapter 4).

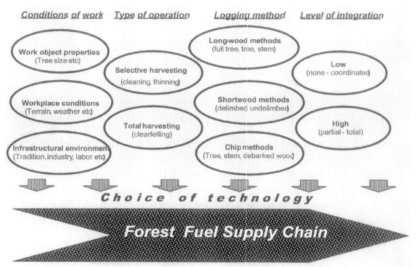

Figure 3.3-2. Factors influencing the choice of technology and methods for harvesting forest-derived fuel.

The technology to be employed is determined by the conditions under which the fuel is harvested (Harstela 1993) and by the scale of operations. The nature of the forest site, forestry traditions, infrastructure, and the desired level of integration into conventional logging systems - all these influence the choice of technology and methods (Figure 3.3-2). Developments in technology and methods for fuelwood harvesting should be viewed against the background set by individual countries.

Interest in forests as a source of industrial energy is international, but most of the following examples come from the Nordic countries where industrial forest fuel procurement systems were initiated earlier than elsewhere. Countries without a history of large-scale use of forest residues for bioenergy should look at the experience of the Nordic countries and select parts of their systems that suit local conditions.

In Finland, Norway and Sweden, factors influencing the development of fuel harvesting techniques include:

- A high demand for timber, even of small dimensions. Most of the bioenergy is produced from branches, tops, small trees and other unmerchantable wood.

- Dominant use of a shortwood or cut-to-length logging system. This is a consequence of the small size of forest holdings, the importance of thinnings and the scattered locations of the mills. From a logistic point of view this makes it advantageous to separate assortments in the forest.

- Large timber volumes are harvested by clearcutting at the end of each rotation. Two to four improvement cuttings (pre-commercial thinning and thinning from below) are carried out during the rotation, each removing 20-70 $m^3 ha^{-1}$.

- Forests contain only a few tree species with relatively uniform growth. Small-branched, straight-stemmed conifers predominate and account for 85% of the standing stock. Average tree volume varies between 0.2 and 0.8 m^3 at time of clearcutting, to a few cubic decimetres in cleaning and early thinnings.

- Forest soils are geologically young and most of the nutrients are in the mineral soil. The predominant forest ecosystem has a natural succession pattern which includes recovery from forest fires.

In Denmark the factors include:

- Lack of a strong home market for small roundwood arising from thinning operations. This has resulted in the development of a market for small roundwood for energy use.

- A high proportion of row-spaced spruce plantations on even-surfaced ground.

- Relatively high prices paid by energy producers for fuelwood as a result of fossil fuel taxation.

In the United Kingdom the factors include:

- A high demand for timber, even of small dimensions. Most potential fuelwood would be produced from logging residues. However, assortments are cut at stump using tree tops and branches as 'brash' mats to support harvesting machinery on wet sites. This reduces the potential volume of fuelwood.

- Substantial potential for use of fuelwood from first and second thinnings if cost-effective harvesting methods can be found.

- Use of whole-tree harvesting systems, particularly on steep terrain where cable logging brings whole-trees to the roadside for processing. Residues are therefore available 'free of charge' at the roadside.

- Potential for fuelwood utilization to act as an incentive for bringing under-managed broadleaf woodland into production using small-scale harvesting equipment.

Whilst north European, and especially the Nordic countries, lead the way in the use of forest residues for bioenergy, other countries have very large forest residue resources which could be used if the economics were favorable. In many of these countries terrain and tree crops are different from those found in the Nordic countries and different harvesting systems will be required.

On the West Coast of Canada, in the Pacific Northwest of the USA and in New Zealand, stand management strategies differ markedly from the northern European pattern. Most residue harvesting is linked to the final felling phase. In these regions factors influencing the choice of technology include:

- A large percentage of tree crops on steep terrain. These sites are not suited to use of ground-based harvesting systems and whole-trees or delimbed stems are moved to the roadside or landing by cable hauling.

- Trees have an average size of 2-3 m^3, and in some cases are much larger.

- Logging systems produce residues at two locations within the forest:
  - at stump (stem breakage during felling),
  - at the point where stems are cut into logs (this may be at roadside, landing or increasingly at central processing yards).

- Cutover residues on steep terrain cannot be recovered for economic and environmental reasons.

- Over 50% of logging residues consist of short sections of stemwood cut to waste during the log-making process (Figure 3.3-3); the remainder is bark and large branches. Volumetric weight of this material is comparatively high and it is not suitable for baling.

- Despite the large piece size, logging residues dry out if stored for 3-4 months in exposed summer conditions. Storage occurs in the forest, at a central processing yard or at point of use. Storage in the forest is preferable as this maximizes delivered energy.

- A competing demand for fiber often exists. Some of the short stemwood material can be used for pulp or other reconstituted fiber products if this component can be separated economically.

- In many situations logging residues are left to decay in the forest. However, there is potential for utilization in bioenergy production. The probable supply chain is simple:

  Residue at landing→ Pile → Store (dry) → Transport → Comminute.

Due to variability in terrain, tree crops, infrastructure and transport regulations between and within countries, description of any one harvesting system is not useful. Each situation will need to be assessed individually and a system designed to suit specific combinations of conditions.

Figure 3.3-3. Logging residues in the form of discarded stem sections at a New Zealand logging site (Photo: R. Björheden).

3.3.2   Harvesting fuelwood in conjunction with final felling

Two-pass harvesting systems

Apart from the bark and woody rejects from industrial processing, logging residues from final felling are the most important source of forest fuel in the Nordic countries.

Neither whole-tree logging, whereby un-delimbed trees are extracted to the landing, nor whole-tree chipping of final-harvest stands is practised in the Nordic countries. However, both are relatively common in North America (Richardson 1986) and other parts of the world. Recovery of fuelwood in this way is simple, but if the chips are to be used in the manufacture of paper, they need to be cleaned.

In Canada, recovery of logging residues at the roadside following whole-tree harvesting is localized to supply power-generating plants within 200 km of the harvest operation site. Residues are usually chipped in trailer-mounted drum or disk chippers, blown directly into chip vans and transported to the point of utilization for outside storage.

Totally non-integrated harvesting of residues is rare in the Nordic countries and North America. Low-level harvesting integration, on the other hand, is widespread. Typically, fuelwood is removed in a separate operation after final felling. The final-harvest operation is adapted to facilitate removal of logging residues through accumulation of branches and tops in larger piles. These are positioned so that the harvester or forwarder will not disturb them. This prevents contamination and compaction, which would impede the circulation of air needed for effective drying.

A variety of harvesting methods can be used, choice being determined by the type of harvester and the type of stand in which it will operate.

The *two-grip harvester* has a chassis-mounted processing unit (Figure 3.3-4). Here, the machine is positioned at an angle to the main direction of travel. The roundwood is deposited closest to the operating strip, and the fuelwood piles behind it.

Figure 3.3-4. Roundwood and for fuelwood harvesting using a two-grip harvester.

Figure 3.3-5. Roundwood and fuelwood harvesting using a single-grip harvester.

The *single-grip harvester* has a harvester head mounted on the end of the boom (Figure 3.3-5). Adaptation for fuelwood harvesting is somewhat easier because the machine does not have to be positioned at an angle to the operating strip. In stands with a high proportion of wood and long stems felling can be done from one side, at right angles to the main direction of travel. Fuelwood piles are sited behind or in front of the roundwood, depending on which side of the strip the machine is operating. Felling can also be done parallel to the operating strip, in which case fuelwood and roundwood will be deposited in alternate piles along the strip.

Storage of logging residue in piles allows litter, mainly needles, to fall off. This reduces the ash content and raises the calorific value (see also Section 3.4). Residues are often stored on the logging site so that more nutrients will be retained in the forest area. Since storage affects fuel properties, this operation can be regarded as a form of processing. Prudent storage can increase the value of the raw material. Conversely, careless storage may lead to significant losses in fuel value (Björheden and Eriksson 1989, see also Section 3.4).

Pile size is determined by the amount of residue per unit area, by requirements for economical loading and extraction methods and also by the need for drying and litter drop. To promote drying, the piles should not be too high. Normally, they are mound-shaped, up to 3 m wide at the base and up to 2 m in height. After drying from initial moisture content (50-55%) to 30-40%, pile height will have shrunk to about 1.5 m. In the Nordic countries, piles should not be left at the site after early autumn, because the moisture content will increase again (see Section 3.4).

The most widely used system for fuelwood procurement in Sweden and Finland incorporates forwarder extraction of residues piled at the site after adapted logging, to a landing located at or close to the roadside (Figure 3.3-6). Forwarding should take place at the end of summer or in early autumn, after residues have dried and litter has fallen, but before reabsorption of moisture during the colder, damper months.

Figure 3.3-6. Extraction of fuelwood by forwarder.

The forwarder should be equipped with an enlarged load space. This can be achieved by laying some logs across the bunks at the bottom or by extending and widening the load space with additional stakes. More advanced solutions for increasing the payload are telescopic bunks or the installation of compaction equipment on the forwarder. To prevent excessive contamination, open rake-type grapples are preferred.

The landing for storage of residue piles should be located in an open position so that wind and sun can help to dry the material. Residues should be stacked at right angles to the direction of the prevailing wind and covered with reinforced paperboard to keep off rain and snow. An efficient covering method involves use of a roll of paperboard that can be lifted and manipulated by the loader and grapple of the forwarder. The paperboard is several meters in width and can be cut to the required length for each pile. The paperboard is fixed in position by placing small trees etc. on top of the covered pile.

Special bulk carriers were once used for secondary haulage of unprocessed residues, but chipping of the logging residues, either at the landing or at the logging site, soon became widespread. Chippers operating off-road are mounted on forwarders, while chippers mounted on trucks are often used at the landings. Hammer mills, which are more tolerant of contaminants, are also used. Due to their heavy weight, they are normally used at landings, terminals and industrial sites. Secondary haulage by bulk carriers or container rigs is not integrated with other forestry haulage.

Figure 3.3-7. Bundling of residues using the FiberPac machine. (Courtesy of Timberjack Ltd.).

In the Nordic countries, the development of fuelwood harvesting and final felling operations is moving towards greater system integration. A new procurement concept that has recently been introduced involves compaction of logging residues at final felling into cylindrical bales known as *composite residue logs* (CRLs). Typically, CRLs have a diameter of approximately 0.60-0.75 m and a length of 3 m (Figure 3.3-7). Each weighs between 400 and 600 kg (fresh weight) and has an energy content exceeding 1 MWh. The compaction units, which are mounted on medium-duty forwarders, weigh 6-7 t.

Systems for CRL production are still under development by two independent companies, Fiberpac and Wood Pac (Figures 3.3-7 and 3.3-8). Performance studies carried out during 1999 on Fiberpac 370 and Wood Pac units have indicated a level of productivity of 20-25 CRLs per $G_{15}h$ (PMH including downtime <15 min per occasion) and approximately 15 CRLs per $G_{15}h$, respectively.

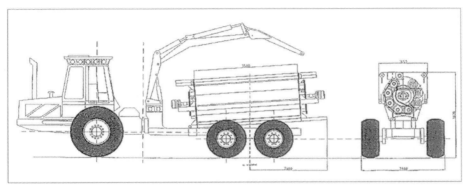

Figure 3.3-8. The Wood Pac mounted on a Rottne forwarder. The compaction chamber has eight cylinders and is fed from above.

The technology is attractive because CRLs can be handled as roundwood throughout the supply chain and conventional forwarders and roundwood haul rigs can be used (Figure 3.3-9). This allows for full integration of transportation which is not possible with unprocessed residues. Payloads are high in comparison with other residue types and forwarders can even load CRLs crosswise to increase the payload. The cost of off-road transport is about half of that for loose, uncompacted residues. Conventional roundwood rigs can be used for secondary haulage of CRLs, although problems with leaves and twigs falling from the load may require containment of the load with netting and/or thin side-boards. The CRLs are suited to efficient haulage by rail. This could be an important consideration in the future if a substantial rise in demand for fuelwood leads to increased transportation distances.

Use of a large mobile or stationary chipper or crusher at the heating plant could achieve a cost saving of two-thirds compared with conventional off-road chipping at the logging site. Large drum chippers (Figure 3.3-10) can achieve high levels of productivity. This machinery, however, is highly sensitive to contaminants, which reduce their productive time and increase knife and repair costs. A solution is the use of hammer mills or crushers, which are more tolerant to contaminants, but less productive. Development of robust machinery that can deliver high productivity at reasonable cost is an important area for development.

Figure 3.3-9. A conventional roundwood rig can accommodate approximately 66 CRLs - 15 per stack on the trailer and 10 or 11 per stack on the truck. A pulpwood bolt is placed on either side at the bottom of each stack to prevent the load from overhanging the load-securing devices.

Figure 3.3-10. A Bruks 1004 CT drum chipper mounted on a dumper-truck chassis.

Integrated one-pass harvesting systems

In one-pass, or fully-integrated harvesting systems, all products are harvested in one operation. Sometimes the whole-tree method is used and the trees are transported to the road, landing or central processing plant for separation and processing of components. One-pass operations may be based on processing into assortments at the stump.

One-pass systems are used throughout Europe and North America (Richardson 1986) where they produce forest fuel commercially in addition to conventional roundwood. The most common system in North America uses chain flail delimbing/debarking combined with off-road chipping, white wood chips being used for pulping and the remainder for hog fuel (Watson and Twaddle 1990).

One-pass harvesting systems take a number of forms and utilize equipment in a variety of combinations. Supply of fuelwood to a relatively insecure market involves a high risk for any investment. Initial development is therefore usually based on the adaptation of equipment used for conventional operations. Specially-designed machinery is not normally produced until a steady market has been established (Björheden 1989). One non-Nordic example of a 'first wave' system used increasingly on steep terrain, is whole-tree extraction by cable crane. In this operation trees are felled and extracted to the roadside where they are processed into conventional roundwood assortments. Large piles of residues are left at the roadside, and therefore have a high potential for use as fuelwood.

Greater integration of roundwood and fuelwood harvesting offers opportunities for further development of cost-effective systems within the Nordic shortwood tradition (Glöde 2000). A highly-integrated system could include a single-grip harvester equipped with a compaction unit and fed with logging residues during delimbing. Cutting of timber and collection of residues into CRLs would be carried out in one operation (Figure 3.3-11). There would be no need for special machinery to handle residues. Simultaneous harvesting of timber and logging residues would reduce the cost of fuelwood. If production of conventional timber assortments could be maintained, the cost of fuelwood would be more than 20% lower than that from the current, non-integrated system. Such cost reductions would make it economically possible to harvest smaller or more distant sites and increase the economically available volume of logging residues.

In New Zealand, considerable potential exists for fuelwood harvesting to be integrated with the recently-introduced 'Superskid' operation. Here a single centralized processing site is positioned to service a number of clearcut harvesting sites within one large forest area (Figure 3.3-12). With processing operations concentrated at a single site, the potential for recovery of logging residues (in this case large-dimension stemwood from crosscutting and large limbs) has considerable scope providing that transport distance to the end user can be minimized.

Figure 3.3-11. Simultaneous harvesting of roundwood and logging residue - efficient and cost-effective?

Figure 3.3-12. The 'Superskid' operation in New Zealand. Large volumes of residues are concentrated at a central processing site (Photo: J. Ford-Robertson).

3.3.3   Harvesting fuelwood in thinning operations

Single tree felling, delimbing and handling of small-diameter trees are labor-intensive operations which result in high roundwood costs. Rationalisation of multi-tree handling methods can reduce logging costs. Small trees can be extracted more efficiently with accumulating felling heads, bunch bucking (crosscutting), and multi-tree delimbing if intact limbs are extracted.

Because tops and branches constitute a high proportion of the biomass, small trees are not only harvested more cost-effectively but more fuelwood can be recovered.

These advantages may explain the early development of highly-integrated systems for harvesting fuelwood from young stands. One Nordic example is the *tree-section method*, which is, in principle, a shortwood version of whole-tree harvesting. Trees are felled and bucked (crosscut) without delimbing, then hauled to the landing, where a mobile bunch-delimber separates the wood into pulpwood and fuel. A more common practice involves an integrated secondary haulage system in which conventional roundwood rigs are equipped with removable steel or aluminum side-covers. Pulpwood and fuel are separated either at the mill or at a nearby terminal. Storage of tree-sections is minimized, because the mills require fresh pulpwood.

In whole-tree harvesting systems for fuelwood, either the whole-tree or the branches and tops are used for fuel, the rest of the tree being utilized as industrial wood. In Sweden and Finland, the harvesting of fuelwood in the form of whole-trees or tree-sections is less common than the removal of logging residue from final fellings. In Denmark, fuel chips are produced mainly from early thinnings. Whole-tree and tree-section methods can also be employed for clearcutting, when harvesting under-sized trees handled as tree-sections in diameter-graded harvesting.

Chipping or delimbing can be carried out efficiently through bunch processing at the landing, mill or heating plant. Specially-equipped haul vehicles are used for carrying tree-sections.

Until 1990, the tree-section method was used for integrated pulpwood and fuelwood harvesting from small-diameter thinnings in some parts of Sweden. During the first half of the 1990s, extraction of tree-sections in a pulpwood assortment almost died out. Only a few mills now use tree-sections. This can be explained as follows:

- The introduction of the single-grip harvester proved to be highly competitive, causing a marked reduction in the cost of delimbing (Brunberg *et al*. 1994, 1998). Detailed analysis of the working patterns of a single-grip harvester revealed that piling/bucking without delimbing was often at least as time-consuming as delimbing and production of conventional roundwood. This was because delimbing is a highly efficient process when compared to crane handling which increases the manipulation of un-delimbed wood (Björheden 1997).

- Stricter requirements for freshness and bark content in the raw material were imposed by the mills.

- The method was not suited to existing procurement logistics (Andersson 1985).

- Ability to pay was reduced when fuelwood took the form of tree-sections (Hillring 1995).

In Denmark, harvesting of whole-trees as fuelwood in first thinnings of conifers is common, largely because of the absence of a pulp industry and the low price for pulpwood. A comparison of the net revenue has revealed a profit for fuelwood and a clear loss for pulpwood (Kofman 1998).

In Sweden and Finland, the extraction of whole-trees or tree-sections for fuelwood only has been at a low level for a long time. However, interest in this option has developed in recent years, especially for early thinning of dense, small-diameter stands with mixed species composition. One reason for this is the marked increase in the backlog of pre-commercial thinning (Swedish Forest Statistics Yearbook 2000). Under the present circumstances of slightly declining profitability of forestry, forest fuel harvesting is seen as a way of subsidizing silvicultural costs. Conditions are similar in Finland. The number of installations using fuelwood has increased over the same period, and new technologies have been developed for harvesting small trees for bioenergy.

Tree-section harvesting is used for fuelwood-only production from pre-commercial and commercial thinning. The method is especially advantageous if the pulpwood option yields only small volumes spread over multiple assortments. In this case, the harvesting of fuelwood concentrates biomass into a single product, and lower processing and logistic costs result (Vikinge 1999).

Chipping of whole-trees offers similar savings in logistic costs, and makes handling easier and transport cheaper. However, chipping makes storage more difficult and expensive. It is therefore mainly used for 'hot' systems, in which comminution takes place shortly before combustion. In the Nordic countries, whole-tree chipping is used mainly in pre-commercial thinning and in removal of underbrush at final felling. For tree lengths greater than 6-7 m, tree-section harvesting is used almost exclusively because conventional technology can be employed for both forwarding and secondary haulage. Harvesting of forest fuels from later commercial thinnings has become less common in the Nordic countries since the 1980s. This indicates that none of the other available systems is really cost-competitive with harvesting residues from final felling. Chipping of whole-trees is conducted in North America but mainly for pulp chip production from small stems, using chain flail delimber-debarker-chippers. Occasionally, residues are recovered and used as biofuels. Experiments with stems from early thinnings have shown costs to be prohibitive.

In many parts of the world, there is substantial potential for small-scale utilization of fuelwood from under-managed woodlands, former farmland or non-industrial forestry plantations. The scale of such operations calls for different approaches that are often labor intensive and use *small-scale harvesting equipment*. Harvesting equipment is often developed for use with an agricultural tractor. Felling is motor-manual; extraction is by winch or trailer forwarder; and comminution by a tractor-mounted chipper. Small-scale operations are better suited to the owner/user than to commercial contractors, but can make important contributions to local energy supply. All small-scale chipping operations are unique in some respect (McCallum 1997). They will not be dealt with in greater detail here.

Examples of thinning operations for fuelwood harvesting

The layout of harvesting systems can be varied according to circumstances. The following examples serve to illustrate this. The large number of layouts indicates that system development is still in its initial phase.

Whole-tree chipping in first and second commercial thinnings

Logging is carried out by motor-manual methods or by machines such as feller-bunchers, feller-preskidders or tree-section harvesters. A chipper operating on a strip road (temporary logging road) comminutes raw material, and chips are extracted either by the chipper itself or a special chip-forwarder, also known as a chip-shuttle. Alternatively, unprocessed raw material is extracted to the landing by a forwarder equipped with a grapple-saw for cutting trees into sections.

In Denmark the commercial first thinning is carried out when stands are 15-20 years old. Thinning takes place between December and April in two phases separated by one year. Trees are felled with chainsaws, every sixth or seventh row being removed to reduce the number of stems by 14-17% to 1500-2000 ha^{-1}. Chipping starts in August, giving the tree crowns time to dry through transpiration during the summer. An off-road chipping unit is employed, the operator using the integral loader and grapple to feed trees into the hopper at the front of the machine. Chips are blown into a collection bin mounted at the rear of the machine. When the bin is full the contents are transferred to a forwarder fitted with a chip bin. The forwarder transports the chips to a larger detachable container (big hook skip) at the roadside. This is collected by truck and transferred to the plant/storage facility. Problems associated with this system are as follows:

- It is only applicable on dry mineral soils.
- High levels of specialist investment are required for both harvesting and transport equipment.
- Investment requires a secure guaranteed annual output.

An advantage of this system is the minimal contamination of the fuel.

Harvesting at the subsequent second thinning is mechanized, using a feller-buncher and a mobile off-road chipper (Kofman 1998).

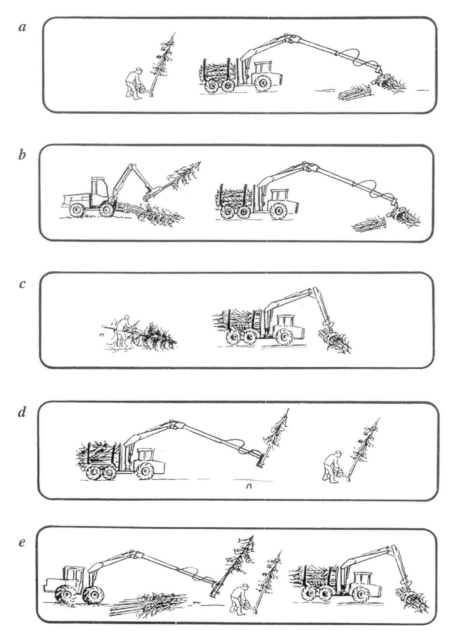

Figure 3.3-13. Schematic diagrams of tree-section logging as practised in Sweden in the 1980s. a) motor-manual felling plus forwarder with grapple-saw; b) feller-preskidder and forwarder equipped with grapple-saw; c) motor-manual felling and pre-limbing plus conventional forwarder; d) harwarder; e) tree-section harvester and conventional forwarder.

Motor-manual felling plus forwarder with grapple-saw (Figure 3.3-13a). Trees are felled with chainsaws away from or towards the strip road by the 'top-butt method' i.e. at right angles to the strip road and with tops and butts lying alternately closer to the road. The felled trees are then bucked using the grapple-saw of the forwarder in conjunction with bunching and loading (Eickhoff 1989). Björheden (1997) notes that the tree-sections should be as long as possible in order to optimize payload and productivity.

Motor-manual felling and delimbing plus conventional forwarder (Figure 3.3-13c). After felling, chainsaws are also used to carry out a rough delimbing. This leaves approximately 40% of the green biomass in the stand. The trees are bucked into transport lengths and if possible, bunching is carried out. Tops are cut off if this does not incur too much extra work. A conventional forwarder extracts wood to the roadside. This system is designed to increase the payload during forwarding and more importantly, secondary haulage. Similar 'rough delimbing' systems based on single grip harvesters are uneconomical in comparison with conventional shortwood harvesting.

Feller-preskidder plus forwarder equipped with grapple-saw (Figure 3.3-13b). Trees are felled, bunched and extracted to the strip road with a feller-preskidder. This is a small four-wheeled machine with a parallel-operating knuckle-boom loader and a felling head mounted on the end of the boom. It operates off strips running at right angles to the road. A forwarder equipped with a grapple-saw is then used to buck the bunches and extract them to the landing.

Tree-section harvester and conventional forwarder (Figure 3.3-13e). The tree-section harvester is a machine equipped with a felling grapple-saw mounted on a long boom. Most versions consist of a converted forwarder chassis with the cab repositioned and the load section removed. Trees within the operating zone of the boom are felled, bunched and bucked. Trees beyond boom reach are felled motor-manually, normally after completion of thinning within the boom-reach zone. Finally, the harvester bunches and bucks the motor-manually-felled trees. Sometimes small trees in the boom zone are felled motor-manually, which increases the productivity of the harvester. A conventional forwarder is used for extraction.

Tree-section 'harwarder' (harvester-forwarder) (Figure 3.3-13d). A single machine performs the entire operation. The 'harwarder' is either a forwarder equipped with a long boom and grapple-saw or a purpose-built combined harvester-forwarder. Trees are felled in the strip road and boom zone, bunched and bucked, then loaded and extracted. Trees out of reach of the machine are motor-manually felled into the boom-reach zone.

Felling plus strip-road-operating chipper. The trees can be comminuted on-site by means of a mobile chipper operating from the strip road (Figure 3.3-14). Felled trees are usually aligned at right angles to the strip road with butt-ends nearest to the road. The design of the chipper intake may require a herring-bone pattern. Trees are fed into the chipper butt-first.

Figure 3.3-14. Self-propelled Chipset terrain chipper.

3.3.4   Delimbing of trees and tree-sections

Haulage of tree-sections over long distances is expensive, owing to the low bulk density of the load. It can therefore be advantageous to process tree-sections at a landing or a nearby terminal if long secondary-haul distances are involved. Delimbing of trees and tree-sections is thus carried out for two reasons: separation of assortments, or simplification of transportation (Watson and Twaddle 1990).

As mentioned in Section 3.3.3, the high cost of processing small-diameter trees is one of the reasons for producing tree-sections rather then roundwood. Dahlin (1989) analyzed the relationship between productivity associated with delimbing and tree diameter. Assuming a constant feed speed, he showed that this relationship could take three forms:

1. Production capacity is proportional to the square of the tree diameter:

$P=a_i d^2$

Where:

P = *production capacity;*

a_1 = *a constant;*

d = *diameter.*

This relationship is valid for trees fed lengthwise through a delimbing device such as a conventional harvester. It is also valid if a fixed number of trees is delimbed simultaneously, and for transverse feeding of single trees or tree-sections over a delimbing device.

2. Production capacity is proportional to tree diameter:

$P=a_j d$

The relationship holds for both lengthwise and transverse feeding during simultaneous delimbing of a layer of trees or tree-sections, providing that the distance between the trees or tree-sections is not fixed but determined by their respective diameters.

3. Production capacity is not influenced by tree diameter:

$P=a_k$

This is true for lengthwise and transversely-fed delimbers used for simultaneous delimbing of a bunch of trees or tree-sections, provided that the number of trees or tree-sections in the bundle is not fixed but determined by their respective diameters.

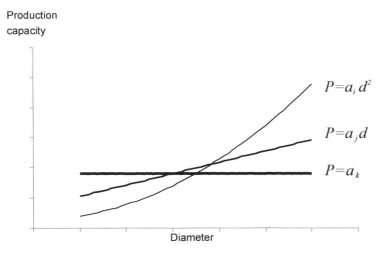

Figure 3.3-15. Three possible relationships between tree diameter (d) and the production capacity of a delimber (P) at constant feeding speed: a_{i-k}= constants. After Dahlin (1989).

The most efficient techniques for delimbing small-sized trees are based on relationships shown in Figure 3.3-15. It follows that bunching is the key to efficient processing. Bunch-delimbing equipment varies but the largest volume of tree-section material is processed in standard debarking machines at pulpmills. Purpose-built machines such as static or mobile drum delimbers and flail delimbers are also available (Dahlin 1991).

The drum delimber is smaller than the conventional debarking drum at a pulp mill and has a lower capacity. It is usually located at a terminal or mill. The inside of the drum is equipped with scraper bars, which help to remove the branches. Billets, branches and tops are separated from the roundwood and reduced to fuel in a hammer mill. Due to the different action of drum delimbers and flail delimbers, and to different pulpwood recovery rates, it would be desirable to combine different bunch delimbing methods. Hakkila *et al.* (1998) describes a Finnish installation which combines chain flail delimbing with subsequent drum debarking of first thinning stems at the mill site.

A compact mobile delimber is useful if roundwood and fuel chips are destined for different industrial locations. One example in use today is a flail delimber mounted on a semi-trailer. Although productivity is lower than that of a static delimbing unit, lower capital cost is a compensating factor. The flail delimber removes the branches by means of chains attached to rotors. Branches, tops and undersized wood are chipped.

3.3.5   Comminution

Reduction of the size of fuelwood components should be delayed as long as possible to shorten chip storage time and minimize the risk of biomass loss and deterioration. Reduction of logging residues to usable fuel size is achieved with a drum chipper or a hammer mill.

The main feature of a *drum chipper* is a rotary drum with knives mounted along its length parallel to the axis. Behind the knives are chip breakers designed to operate at right angles to the cutting action of the knives (Figure 3.3-16). This equipment reduces forest residues to relatively small, uniformly-sized chips. Regular maintenance of the knives is required. Residues must be free of contaminants that would damage or cause excessive wear to the knives.

A *hammer mill* comminutes forest residues by the crushing and tearing action of fixed or pivoting hammers mounted on a rotor. Hammer mills produce larger and less uniform chips than drum chippers but they are more robust and less vulnerable to contaminants. Hammer mills can be mounted on forwarders for off-road operation at the logging site. Due to their considerable weight they are more commonly static or mounted on a truck located at the roadside, terminal or heating plant.

Figure 3.3-16. Mobile drum chipper (courtesy of LHM Hakkuri Oy), and a schematic illustration of its operating principles (courtesy of Jan-Erik Liss).

Figure 3.3-17. A drum chipper mounted on a forwarder chassis empties a bin of chips at the landing after chipping residues on the logging site.

The most commonly-used comminuter in Sweden and Finland is a mobile chipper mounted on a forwarder or truck (Figure 3.3-17). Chips are discharged into an exchangeable tip-up container or blown directly into a road-haulage rig.

Chipping at the landing and direct-loading of custom chip vehicles is a 'hot' or highly sensitive system dependent on the smooth functioning of all components of the productivity chain. If a hold-up occurs at chipper, truck or reception facility, downtime will be costly. This problem is greatest when a container system or custom chip vehicle is being used.

Off-road chipping is a less sensitive system. Piles of logging residues left by the harvester can be chipped in late summer or early autumn at the logging site. This system can be used to slow down the flow of chips during periods of low demand. As well as keeping the contractors working, it can produce high-quality fuelwood at a competitive cost. The need for forwarding uncomminuted fuel and storing it in piles covered with paperboard is eliminated.

In the Nordic countries, most chipping is done off-road or at the landing. Growth of end-user plant capacity is associated with a trend towards centralized comminution at the plant. This offers economy of scale, and robust equipment can be used because maneuverability is not an issue. Costs are therefore considerably reduced. Offsetting the advantages are increased noise levels at the plant, and the need for storage of uncomminuted fuelwood.

Regardless of the type and location of the comminution device, the quality of the material produced is of paramount importance to the operation of the energy plant. Poor quality fuelwood contaminated with soil, stones or harvesting debris, or consisting of over- or under-sized particles, can reduce the efficiency of handling and combustion equipment at the energy plant (see Section 3.4).

3.3.6   Secondary haulage

The transport component of the supply chain can account for up to 30% of the total cost of low-value products such as small roundwood (Jaakko Pöyry 1998). It has been estimated the costs incurred in four hours' transport are equivalent to those involved in growing a 17 year-old pine tree (Gardner 1996).

A great number of techniques have been offered for transportation of small trees, tree-sections and forest residues (Anon. 1990, Axelsson and Björheden 1991).

Discarded stem sections that, as in New Zealand, may be a major constituent of logging residues, are currently transported in large-volume bin trucks or set-out containers. A separate knuckle-boom loader is often used, although self-loading crane trucks are an option in areas with a scattered roadside resource. Material can be transported to point of use and comminuted by chipper or hog as required.

In the Nordic countries, the dominant secondary haulage method for wood chips usually involves rigs equipped with three containers, each with a capacity of 35 m³, giving a total payload of 35 t. It takes about 45 min to load and 25 min to unload a three-container rig.

If comminution takes place at a terminal, purpose-built large-volume truck-trailer rigs, each with a capacity of 145 m³, are normally used for hauling unchipped logging residue. Truck and trailer have solid side-boards and a steel deck. Some rigs are equipped with a heavy-duty loader having double-acting lift cylinders, which can be used to compact the load. Separate, robust mobile loaders are used for the same purpose. Tree-sections are not normally compacted.

Where comminution is centralized and biomass is transported in the form of CRLs, integration into the conventional round timber supply chain is possible without additional investment for haulage companies. Systems based on CRLs have the following advantages:

- Economic payloads can be accommodated on conventional flat-bed non-specialist trailers, or (more importantly) on specialist timber trailers.

- If drying is to take place prior to comminution in the forest, economic payloads can be formed by mixing conventional roundwood products and CRLs.

- The procurement system has increased flexibility.

Because of their uniform cylindrical shape, CRLs can be transported in rail wagons. This allows the incorporation of greater haulage distances into the supply chain whilst minimizing costs.

3.4   DRYING AND STORAGE OF FUELWOOD

Logging and industrial process residues, which are the raw material for fuelwood, are produced all year round although logging sometimes may be seasonal. Because fuelwood in temperate developed countries is still mainly utilized for heat, consumption is concentrated in the winter season (Figure 3.2-6). In warmer developed countries the position may be reversed since more energy is used for cooling than for heating. This difference between time of production and time of consumption means that considerable volumes of fuel have to be stored during periods of low demand. The choice of storage location and method is usually influenced by biological, economic and logistic considerations (see also Section 3.2.4).

After extraction, forest biomass usually passes through a number of processes, including storage, comminution and transport. The sequence of these processes depends on the type of material, industrial facilities, harvest season, and economic considerations (see Section 3.2.4). The most common production flows are shown in Figure 3.2-9.

Maintenance of fuelwood quality during processing is essential. It is especially important in the case of green material stored prior to handling or utilization. Fuelwood is stored either uncomminuted or in comminuted form.

3.4.1   Drying and storage of uncomminuted fuelwood

On-site transpiration drying of small-sized trees and logging residues

Water is present in a bound form in cell walls and in its free state in the cell lumens of wood and bark. Moisture content varies from one part of the tree to another. It is often lowest in the butt of the stem and increases towards the roots and the crown. It is higher in sapwood than in heartwood. The moisture content of live trees varies with season. In temperate broadleaved species it is highest in spring, just before budbreak, and lowest in summer. In temperate softwood species, moisture content is highest during winter and lowest during summer (Hakkila 1962). In tropical countries it is lowest in the dry season.

After felling, moisture content is dependent on post-harvest handling and storage conditions. Moisture may be lost by *transpiration drying,* through foliage, or from open wood surfaces. Transpiration drying is an old, but often forgotten, method of drying whole-trees. The rate of drying depends on many factors including ambient temperature, relative humidity, wind speed, season, rainfall pattern, tree species, and tree size. The best season for drying is when the vapor pressure deficit of the ambient air is low, usually summer in temperate climates and the dry season in the tropics. Transpiration drying can be successfully used with most species. Moisture contents as low as 30% may be reached under favorable conditions (Figure 3.4-1).

Transpiration drying is also possible for logging residues and tree-sections. The moisture content of fresh logging residue in the Nordic countries is 55-60% on a green weight basis. When left by a harvester in small heaps at the logging site or in larger windrows at the roadside, the moisture content of residues can decrease to 20-30% (Nurmi 1999a,b). Small residue piles dry faster than large ones, but they also regain moisture more rapidly (Figure 3.4-2).

Excessive precipitation or low temperatures may hinder the efficiency of transpiration drying (Lehtikangas and Jirjis 1993a). If on-site storage is extended to late fall or even until winter in temperate countries, the advantage gained by transpiration drying will be lost due to absorption of moisture directly from the air and from precipitation (Lehtikangas 1991). An additional benefit of on-site storage in small heaps is that leaves may be left in the forest, resulting in reduced nutrient loss from the site and better fuel quality. In the Nordic countries the weight of needles on detached *Pinus sylvestris* and *Picea abies* branches, expressed as a percentage of the dry matter, may decrease from 25% to 30% in one summer.

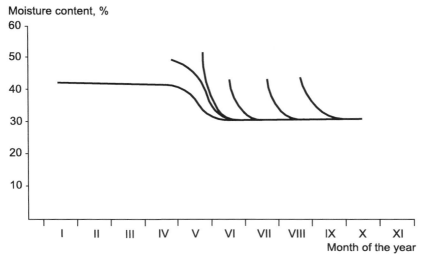

Figure 3.4-1. Transpiration drying rates of birch (whole-trees) felled at different times of the year (Hakkila 1962).

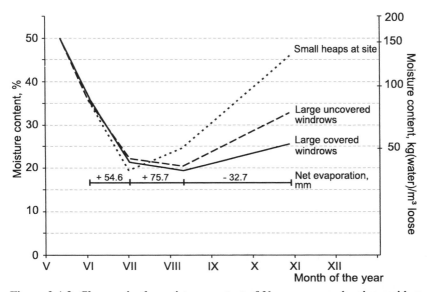

Figure 3.4-2. Changes in the moisture content of Norway spruce logging residues left in small heaps at the logging site and in large windrows at the roadside in 1999 (Hillebrand et al. 2000).

Storage of residues in roadside windrows

Transpiration drying is usually more efficient in small heaps than in roadside windrows. Under favorable weather conditions uncomminuted fuelwood, including whole-trees, whole stems, split firewood or logging residues will slowly lose moisture during storage, although the results can be highly variable (Uusvaara and Verkasalo 1987). Covering fresh or seasoned fuelwood with impregnated paper or other material further reduces the moisture content (Uusvaara 1984, Lehtikangas and Jirjis 1993b, Jirjis and Lehtikangas 1993). Moisture variation within a covered pile is smaller.

Storage of logging residues in large windrows does not allow the needles to fall to the ground and this has several effects. Fuel yield from roadside windrows can be up to 30% higher than from smaller piles left on the logging site (Figure 3.4-3). With defoliation the proportion of wood increases and ash content is reduced (Thörnqvist 1984a). Ash content of logging residues is highly dependent on the proportion of foliage and small-diameter twigs since leaves and bark have a higher mineral content than wood. The ash content of material from roadside windrows is usually higher than the ash content of residues stored in small piles at the logging site.

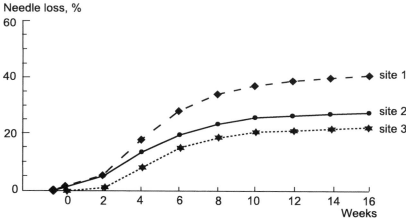

Figure 3.4-3. Needle loss as a percentage of initial needle weight at three clearcut sites in Sweden (Lehtikangas 1991).

Weight losses associated with fungal decay are smaller during storage in windrows than during storage as chipped material (Jirjis and Lehtikangas, 1993). Heat development, a problem with chip storage, is also eliminated.

Storage of compacted materials

Handling and transportation of uncomminuted material can be greatly facilitated by compacting it into bales or bundles. This does not hinder drying, especially when the bundles are stored under cover. In Scandinavia dry-matter loss from bundles due to shedding of needles varied between 12% and 21% after nine months' storage, the proportion depending on whether or not the bundles were covered (Lehtikangas and Jirjis 1993b). When bundles contained summer-dried logging residues, most of the foliage had been lost already and no further dry-matter loss occurred.

Bundling fuelwood into relatively high-density bales may result in lower dry-matter loss associated with microbial activity than is found under chip storage. In a Scandinavian study, up to 18% dry-matter loss, largely due to microbial activity, was recorded in baled forest residues after five months storage (Lehtikangas and Jirjis 1998). Despite the high bulk density, a substantial reduction in the moisture content of baled material is possible. Better drying results are obtained if residues are allowed to dry during the summer prior to bundling.

3.4.2   Storage of comminuted fuelwood

Self-heating in fuel chips

Comminution, usually by chipping, has resulted in major changes in the properties of fuelwoods. It is an essential part of procurement because it facilitates transportation, handling, feeding and combustion of the material.

Most forest residues chipped for heat and energy production are utilized directly after size reduction. To ensure a continuous supply of chips, heating or electricity generating plants are often forced to maintain a store of chips in the plant yard. Storage of large quantities of comminuted material is much more complex than storage of uncomminuted material because it can initiate both biotic and abiotic heat-generating processes (Björheden and Eriksson 1989).

Parenchyma cells in sapwood, cambium, inner bark and foliage can survive and continue respiration for a variable time after comminution. This can cause initial heat development (Hatton 1970, Nylinder and Thörnqvist 1980, Gjølsjø 1995). Microbial activity, which begins as soon as the chip pile is built, is also considered to be a major cause of heat generation. Depending on the nature of the biomass components and their moisture content, both fungi and bacteria can, even at fairly low temperatures, colonize the material and contribute to heat generation.

The rate at which heat develops depends greatly on the season (White *et al.* 1986) and the composition of the material (Thörnqvist 1983). Eventually the composition of the material changes so that the proportion of bark and foliage decreases. The percentage of wood and fine particles increases simultaneously.

Pile shape affects heat development by determining the ventilating chimney effect within the pile. This effect is important as ventilation provides the oxygen required for metabolic activity. It also cools the interior of the pile by convection. According to Kubler (1982), self-heating is affected more by pile width than by pile height.

Once the temperature reaches 42°C, respiration of the living cells in tree material declines, and at 60°C the production of CO_2 ceases as cells die (Springer *et al.* 1971). However, the activity of fungi and bacteria continues and eventually leads to dry-matter loss. Some fungi can tolerate temperatures as high as 60°C and bacteria survive at 80°C (Rothbaum 1963, Assarsson *et al.* 1970). When moisture content drops below 20%, water is found only in the cell walls and microbes cannot utilize it. Hence decreased moisture content has a direct influence on microbial activity and the lowering of pile temperature. Unlike the wood cells, fungi and bacteria can survive in a dormant state, and rewetting will revive them. Lack of oxygen and waterlogging are detrimental to their survival (Kubler 1987). Rapid rise in temperature is often followed by a long period of slow temperature decrease (Figure 3.4-4). In small piles, internal temperature is strongly affected by ambient air temperature (Gislerud 1974).

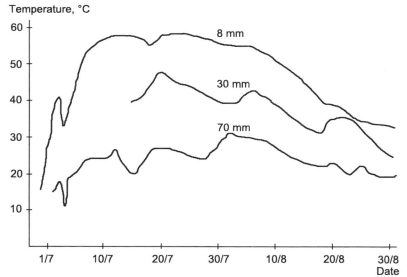

Figure 3.4-4. Effect of nominal particle size on temperature variation at the center of three chip piles during the first two months of storage (Björklund 1983).

The quality of fuel chips is reduced not only by high moisture content but also by uneven moisture distribution within the pile. Over time the interior of the pile warms up and moisture condenses in the cooler outer and upper parts of the pile. As a result, moisture content of chips at the center can be 20-25% of total weight while that of chips in the upper and outer parts is 65-70%. This has been demonstrated for pulp chips as well as fuelwood chips

(Thörnqvist 1983, 1984b, White *et. al.* 1983). Especially noticeable is an over-saturated 'cap' or 'lens' at the top of the pile (Hatton 1970). This causes problems in boreal winters when the wet outer parts freeze and have to be rejected as unusable. Fortunately the heterogeneity can often be overcome by mixing the material.

Abiotic processes such as moisture adsorption, hydrolysis, pyrolysis, chemical oxidation and charring cause further dry-matter loss and generate heat. Where the temperature reaches high levels (300°C), degradation of lignin and carbohydrates occurs, especially in needles and bark. Breakdown of carbohydrates takes place more readily than the breakdown of lignin components (Jirjis and Theander 1990). If these processes are allowed to continue, they may result in combustion and open fire (Assarsson *et al.* 1970).

Self-heating with risk of self-ignition is more rapid in piles of material with low thermal conductivity such as wood chips, sawdust and bark. Generated heat accumulates and only a small proportion flows to the surroundings. In piles composed of larger particles with wider air passages and better heat dissipation, the self-heating process is slow and only moderate heating occurs. Piles containing material with a wide range of particle sizes such as bark, can self-heat relatively quickly due in part to slow air convection.

Concentration of oxygen in the pile is one of the most important factors influencing self-ignition. At reduced oxygen concentrations the minimum temperature required for self-ignition of wood is increased. However, self-heating may occur at oxygen concentrations as low as 4% (Feist *et al.* 1971).

Dry-matter loss in fuel chips

Storage of wet fuelwood in any form will inevitably entail some dry-matter loss. The types of microorganisms that inhabit the material influence the extent of these losses. Microorganisms found in chip piles include fungi (mold, stain and rot fungi) and bacteria (especially thermophilic actinomycetes). These organisms need a relatively high moisture content for growth and can tolerate a wide range of temperatures. Their metabolic activity and the consequent development of heat affects the chemical composition of woody material. Elevated temperature plays a major role in accelerating oxidation and other chemical processes that result in dry-matter loss.

Chipping of fuelwood contributes to dry-matter loss in a number of ways. Firstly, it increases the area of exposed surfaces on which microbial activity can take place. Secondly, it releases the soluble contents of plant cells, providing wood-colonizing microorganisms with easy access to vital nutrients. Finally, small particles restrict air movement and prevent heat dissipation. For all these reasons, stored chips are subject to greater dry-matter loss than uncomminuted fuelwood.

Initial rapid colonization of chips is mainly by mold fungi and does not involve high dry-matter loss since soluble nutrients available on the surface of the chips are utilized without serious degradation of cell wall components. Later development of rot fungi can lead to high losses due to their ability to degrade lignin, cellulose and hemicellulose.

Biomass composition affects the rate of fungal colonization and consequently the extent of weight loss. Easily-degradable components such as needles usually incur high losses within a very short storage period. Stemwood chips, with a higher lignin content and low nitrogen concentration, can be stored for a longer period without serious reduction of dry-matter content.

Microbial processes depend greatly on original moisture content. Forest residue chips with moisture content below the fiber saturation point will lose less dry matter than chips with higher moisture contents (Nilsson 1987). Chips made from fresh material will therefore generate more heat and lose more dry matter than chips made of seasoned material (Kubler 1987). Larger piles and longer storage times will increase dry-matter loss (Thörnqvist 1984b).

Dry weight reduction during storage of fuelwood has been observed in many trials. Fresh logging residue chips stored in a large pile for seven months lost approximately 12% of their dry matter, mainly during the first few weeks (Thörnqvist and Jirjis 1990). Greater losses are observed when bark is stored. A 25.8% dry-matter decrease was observed in a 4 m-high bark pile after six months of storage. This resulted in 20% reduction in the energy content of the fuel (Fredholm and Jirjis 1988).

Wet material, including industrial process residues such as bark, is often useless as a fuel unless mixed with drier products, e.g. shavings. Mixing increases costs but lowers the moisture content and reduces loss of dry matter. Mixed fuels are usually utilized over a short period of time to avoid storage problems that increase with fuel heterogeneity, especially self-ignition.

Increase in ash content is an indirect effect of dry-matter loss. During storage the ash content of logging residue chips can be increased by 50% in one year as a result of decomposition processes. Whole-tree chips and larger pieces have not shown such a high increase in ash content. The type of storage facility (pile, bin or silo) has not been found to affect the ash content of whole-tree chips of conifers or broad-leaved species in the Nordic countries (Nilsson 1987, Fredrikson and Rutegård 1985).

Health risks

Growth of fungi and bacteria usually increases soon after a pile of fuel chips is built. Numbers of species and growth rates depend on many circumstances. The most important internal factors are moisture content, composition, and particle size. External factors include size and form of the pile and duration of storage. The majority of microorganisms colonizing chips are molds and actinomycetes (bacteria with a growth pattern similar to fungi). These have a limited effect on dry-matter loss but during their short life cycle extremely large numbers of microspores (<5 μm in diameter) are produced. Handling of wood chips, e.g. shoveling or unloading, causes hyphal fragments and microspores to be released into the air. The degree of emission is affected by the amount of disturbance, location (indoors or outdoors), and weather.

Exposure to high concentrations of airborne spores and fragments constitutes a health hazard. Upon inhalation they enter the respiratory system and can cause allergic reactions in workers who handle the material. Allergic alveolitis, an immunological lung disease, can be caused by the inhalation of mold dust (Rask-Andersen 1988). This disease, with symptoms of dyspnoea, cough and fever, can lead to permanent impairment of lung function. Another disease associated with exposure to airborne fungal spores is a transient febrile condition known as organic dust toxic syndrome (ODTS). Allergic alveolitis appears to result from repeated exposure while ODTS is associated with occasional heavy exposure to mold dust (Malmberg *et al.* 1988).

Microbial growth, particularly of molds on freshly harvested biomass is minimal. Shortly after felling, fungal colonization occurs especially on damaged surfaces where nutrients are readily available. Here the ambient temperature is the growth-limiting factor. At freezing temperatures hardly any microbial activity is detectable. At higher temperatures the growth rate of fungi and bacteria can be considerable.

Under conditions associated with development of symptoms of alveolitis, the number of spores in the respiratory zone during the handling of moldy material was reported to have exceeded 10^9 microspores m^{-3} air. Lower spore concentrations caused no symptoms (Malmberg *et al.* 1986). Several studies show that specific fungi are associated with allergic alveolitis. Those most frequently reported include *Penicillium* species (Van Assendelft *et al.* 1985), *Rhizopus microsporus* and *Aspergillus fumigatus* (Belin 1987). The dominant bacteria in stored chips are thermophilic actinomycetes, e.g. *Thermoactinomyces vulgaris* (Terho *et al.* 1980).

High concentrations of fungal spores (over 10^{10} spores kg^{-1} dry mass) have been observed during large-scale storage of wood chips. Handling of heavily-infected material can lead to emission of large numbers of spores. Personal protection can be achieved through use of a ventilated helmet fitted with a filter effective against particles smaller than 5 μm in diameter.

If biofuel is stored uncomminuted, fungal colonization will be less than in chipped fuel. Handling will therefore cause a negligible health risk. Once green biomass is chipped, colonization by various microorganisms is inevitable, and will be greater if the material is stored in a large pile. During the first few months of storage, the number of viable spores and other fungal components increases rapidly. After a while, growth rates of many fungi decrease, particularly at the center of the pile, due to high temperatures and low moisture content. The presence of fungal biomass in the remainder of the pile means that the health risk is still present.

3.5   CONCLUSIONS

Fuelwood derived from industrial forests can be viewed as the product of a process that begins in the forest and ends at the point of consumption. In most parts of the world fuelwood is a low-value product that is residual to logging. Its production currently has little influence on the silviculture prescribed for tree crops. In some northern European countries the use of logging residues for energy production is common in heating plants. Industrial process residues are widely used as an energy source for the forest industry, the more valuable products covering the costs of extraction and transport.

Except in the case of relatively small bioenergy plantations of short-rotation coppice species, the industrial forest manager seldom regards fuelwood production as a major goal. Even in developing countries where firewood is a basic necessity of life, trees are seldom grown primarily for fuelwood. Nevertheless there are situations in which a huge potential for bioenergy production is currently not being exploited. Firstly large areas of naturally-regenerated dense immature and mature stands exist throughout the world. These have low timber value, often due to high-grading in the past. There are very large areas of naturally-occurring coppicing species, particularly in North America. The main constraints on utilization of these potential fuel sources are the high cost of bioenergy production compared with fossil fuels and sometimes the priority of values other than commercial exploitation. Secondly, there is unused potential in forests where natural mortality occurs as stands mature. Repeated light thinnings could recover this material for energy before the trees die. Insect- or disease-induced mortality could also be exploited. If fossil fuel became more expensive, silviculture could easily be adapted for bioenergy production and in many cases this would overcome current economic difficulties. In Denmark, for example, it is currently more economic to use pre-commercial and commercial thinnings for fuelwood than for pulpwood.

Extraction of forest residues for bioenergy adds to the complexity of forestry practice but also contributes to possibilities for increased efficiency. The economic potential is especially great if residue removal is integrated with other harvesting operations. Since development of highly integrated systems is associated with higher levels of risk and commitment, it is unlikely that more sophisticated systems will be developed where prices for other energy sources are low. The projected financial performance of fuelwood recovery will probably be too poor to justify the risks.

The Nordic countries currently lead the way in development of integrated harvesting, transport, storage and comminution systems and practices for recovery of fuelwood. The most widely used systems associated with final felling involve two-pass procedures, where harvesting of forest fuel follows the main logging activity. The major operation is often modified to facilitate extraction of logging residues. These may be stored in piles on the logging site or at the roadside, where they lose moisture and leaf material before being transported to a terminal or industrial plant.

In integrated one-pass systems all products including fuelwood are harvested in a single operation. One-pass systems are still being developed, the more advanced forms requiring new or adapted equipment. The recent introduction of composite residue log technology shows considerable promise for improving harvesting and transport systems.

A number of fuelwood harvesting systems based on the recovery of thinnings have been developed, but they are not widely used because of economic considerations.

Comminution of forest biomass for use as fuel is carried out with a drum chipper or hammer mill and may take place in the forest, at terminals or at the industrial plant. Because of problems with storage this process should be delayed as long as possible. Piles of comminuted fuelwood are subject to self-heating and dry-matter loss and may pose health risks due to the buildup of populations of fungi and bacteria in the stored material.

3.6   REFERENCES

Andersson, G. 2000. Technology of fuel chip production in Sweden. International wood energy technology seminar. Nordic treasure hunt: Extracting energy from forest residues. 30th August, 2000. Jyväskylä, Finland. 10 p.

Andersson, M. 1996. Aptering och virkeskännedom (Quality grading and wood knowledge). Swedish University of Agricultural Sciences InfoSkog, Swedish University of Agricultural Sciences, Garpenberg.

Andersson, S. 1985. Visionära möjligheter och verkliga problem inom skogstekniken (Visions and problems for forest operations and techniques, in Swedish). Research Stencil No. 38, Swedish University of Agricultural Sciences, Department of Operational Efficiency, Garpenberg.

Anon. 1990. A databank of equipment for transportation of small trees and forest residues. FERIC/IEA Task VI Activity 3 Report, Spring 1990.

Araki, D. 1999. Recovery of quality chips from unconventional sources. Forest Engineering Institute of Canada Special Report: SR-131. Vancouver, BC. 19p.

Arbnor, I. and Bjerke, B. 1994. Företagsekonomisk metodlära (Methodology of managerial economics). Second edition, Studentlitteratur, Lund.

Assarsson, A., Croon, I. and Frisk, E. 1970. Outside chip storage. Svensk Papperstidning 73(16): 493-501.

Axelsson, J. and Björheden, R. 1991. Truck systems for transportation of small trees and forest residues. Proceedings of the IEA/BA Task VI workshop in Oregon & California, USA, June 3-8 1991. Aberdeen University Forestry Research Paper 1991:2.

Belin, L. 1987. Sawmill alveolitis in Sweden. International Archives of Allergy and Applied Immunology 82: 440-443.

Björheden, R 1989. Integrerad avverkning (Integrated harvesting). Swedish University of Agricultural Sciences, Skogsfakta konferens # 16.

Björheden, R. 1993. Coordinating forest fuel production with other activities of a divisionalized forest concern. Proceedings of the IUFRO Symposium on Managerial Economics in Lithuania, August 29-September 4, 1993.

Björheden, R. 1997. Studies of large scale forest fuel supply systems. Dissertation. Acta Universitatis Agriculturae Sueciae/Silvestria 31.

Björheden, R. 2000. Integrating production of timber and energy – a comprehensive view. New Zealand Journal of Forestry Science 30(1/2): 67-78.

Björheden, R and Axelsson, J. 1991. Truck systems for transportation of small trees and forest residues. Proceedings of the IEA/BA Workshop in Oregon and California, 1991.

Björheden, R. and Eriksson, L.O. 1989. Optimal storing, transport and processing for a forest-fuel supplier. European Journal of Operational Research. No. 43.

Björheden, R. and Eriksson, L.O. 1990. The effects on operational planning of changes in energy content of stored wood fuels. Scandinavian Journal of Forest Research 43: 26-33.

Björklund, L. 1983. Lagring av helträdsflis av olika trädslag samt i olika fraktioner (Storage of whole-tree chips of different species and in different fractions). Sveriges Lantbruksuniversitet, Institutionen för virkeslära. Rapport 143. 50 p.

Brunberg, B., Frohm, S., Nordén, B., Persson, J. and Wigren, C. 1994. Forest-fuel technology project, Final report. SkogForsk. Redogörelse Nr. 5.

Brunberg, B., Andersson, G., Nordén, B. and Thor, M. 1998. Forest bioenergy fuel—final report of commissioned project. SkogForsk. Redogörelse Nr. 6.

Bulmer, M. (ed.). 1977. Sociological Research Methods. The MacMillan Press Ltd.

Burdon, R.D. and Miller, J.T. 1992. Introduced forest trees in New Zealand: recognition, role and seed source. 12. Radiata pine (*Pinus radiata* D. Don). Bulletin 24, New Zealand Forest Research Institute Ltd. 59 p.

Caldwell, B.J. (ed.). 1992. The Philosophy and Methodology of Economics. Edward Elgar Publishing Ltd.

Christersson L., Sennerby-Forsse, L. and Zsuffa, L. 1993. The role and significance of woody biomass plantations in Swedish agriculture. Forestry Chronicle 69: 687-693.

Dahlin, B. 1989. The influence of tree dimension on production capacity for different delimbing principles. Scandinavian Journal of For. Research 4: 267-272.

Dahlin, B. 1991. Cradle type multi-stem delimber. Stud For Suec. 185.

Eickhoff, K. 1988. Flisning eller gallring av lövbestånd? (Chipping or shortwood thinning of broadleaf stands?). Results # 1/88, Sw. Logging Res. Found., Skogsarbeten.

Eickhoff, K. 1989. Träddelssystem, en analys från stubbe till industri. SkogForsk, Redogörelse Nr. 5.

Emory, C.W. 1985. Business Research Methods. Irwin Publications.

Eriksson, L.O. and Björheden, R. 1989. Optimal storing, transports and processing for a forest fuel supplier. European Journal of Operational Research 43: 26-33.

Evans, J. 1992. Plantation forestry in the tropics. Second edition. Clarendon Press, Oxford, England. 403 p.

Evans, J. 1998. The sustainability of wood production in plantation forestry. Unasylva 192 (49): 47-52.

FAO. 2001. Future production from forest plantations. Report based on the work of C Brown. Forest Plantation Thematic Papers, Working Paper 15. Forest Resources Development Service, Forest Resources Division. FAO, Rome (*unpublished*).

Farnden, C. 1996. Stand density management diagrams for lodgepole pine, white spruce and interior Douglas-fir. Information Report BC-X-360. Canadian Forest Service, Victoria, BC.

Farrar, J.L. 1995. Trees in Canada. Fitzhenry & Whiteside Limited. Markham, Ontario, Canada and the Canadian Forest Service, Natural Resources Canada. 502 p.

Feist, W.C., Springer, E.L. and Hajny, G.J. 1971. Encasing wood chip piles in plastic membranes to control chip deterioration. Tappi Journal 54(7): 1140-1124.

Fredholm, R. and Jirjis, R. 1988. Säsongslagring av bark från våtlagrade stackar. (Seasonal storage of bark from wet stored logs). Swedish University of Agricultural Sciences, Uppsala, Sweden. Department of Forest Products, Report No. 200. 31 p.

Fredrikson, H. and Rutegård, G. 1985. Lagring av småved och bränsleflis i binge. (Storage of chunk wood and fuel chips in bins). Sveriges Lantbruksuniversitet. Inst. för Virkeslära. Uppsatser No. 151. 32 p.

Gardner, D.N.A. 1996. Management of Logging Systems in South Africa. Sappi Forests, South Africa.

Gislerud, O. 1974. Heltreutnyttelse-V. Lagring av heltreflis. (Whole tree utilization - storage of whole tree chips). Norsk Institutt för Skogforskning. Skogteknologisk avdeling Rapport 4/74. 29 p.

Gjølsjø, S. 1995. Storage of comminuted birch in piles in Norway. *In* Mattsson, J.E., Mitchell, C.P. and Tordmar, K. (eds.). Preparation and supply of high quality woodfuels. IEA/BA Task IX Proceedings of a workshop held in Garpenberg, Sweden 13-16 June, 1994. Sveriges Lantbruksuniversitet. Inst. för skogsteknik. Uppsatser och resultater No. 278: 76-86.

Glöde, D. 2000. Grot & gagnvirkesskördaren- analys av ett framtida koncept för bättre lönsamhet vid grotskörd. SkogForsk Arbetsrapport Nr. 449.

Hakkila, P. 1962. Polttohakepuun kuivuminen metsässä. (Forest seasoning of wood intended for fuel chips). Commun. Inst. Forestalis Fenniae 54(4). 82 p.

Hakkila, P. 1989. Utilization of residual forest biomass. Springer Series in Wood Science, Springer-Verlag, Berlin.

Hakkila, P., Rieppo, K. and Kalaja, H. 1998. Ensiharvennuspuun erilliskäsittely tehdasvarastolla. Metsäntutkimuslaitoksen tiedonantoja #700.

Hammond, D. (ed.). 1995. Forestry handbook. New Zealand Institute of Forestry, Christchurch, New Zealand. 240p.

Harstela, P. 1993. Forest work science and technology, Part 1. University of Joensuu, Faculty of Forestry, Silva Carelica No. 25.

Hatton, J.V. 1970. Precise studies on the effect of outside storage on fiber yield: white spruce and lodgepole pine. Tappi Journal 53(4): 627-638.

Hektor, B. 1998. Assessment of wood fuel prices in integrated operation. Proceedings of the IEA Conference on Wood Fuels in Integrated Forestry, Nokia, Finland.

Hillebrand, K., Marttila, M. and Nurmi, J. 2000. Puupolttoaineiden laadunhallinta (Quality control of woodfuels). Väliraportti 1999. VTT Energia, Tutkimusselostus ENE32/ T0016/ 2000. 32 p.

Hillring, B. 1995 Evaluation of forest fuel systems utilizing tree sections. Swedish University of Agricultural Sciences, Department of Operational Efficiency, Research Notes No. 275, Garpenberg.

Jaakko Pöyry 1998. Executive summary report on the future development prospects for British grown softwood. Jaakko Poyry Consulting.

James, N.D.G. 1981. A history of English forestry. Basil Blackwell Publisher, Oxford, UK. 339 p.

Jenkins, M.B. and Smith, E.T. 1999. The business of sustainable forestry: strategies for an industry in transition. Island Press, Washington, U.S.A.

Jirjis, R. and Lehtikangas, P. 1993. Bränslekvalitet och substansförluster vid vältlagring av hyggesrester. (Fuel quality and dry matter loss during storage of logging residue in a windrow). Sveriges Lantbruksuniversitet. Inst. för Virkeslära. Rapport No. 236. 26 p.

Jirjis, R. and Theander, O. 1990. The effect of seasonal storage on the chemical compositions of forest residue chips. Scandinavian Journal of For. Research 5: 437-448.

Johansson, J. 1994. Excavators and backhoe loaders in forest harvesting operations. Dissertation, Swedish University of Agricultural Sciences, Department of Operational Efficiency, Garpenberg.

Kofman, P. 1994. Storage trial of chips, chunkwood and firewood. *In* Jirjis, R. (ed.). Storage and drying of woody biomass. IEA Task IV Activity 5 workshop, New Brunswick, Canada 1993. Sveriges lantbruksuniversitet, Institution för virkeslära. Rapport No. 241.

Kofman, P. 1998. Review of fuelwood from early thinnings in Denmark. *In* Wood fuel from early thinning and plantation cleaning, an international review. Finnish Forest Research Institute, Research Papers 667.

Kubler, H. 1982. Air convection in self-heating piles of wood chips. Tappi Journal 65(8): 79-63.

Kubler, H. 1987. Heat generation processes as cause of spontaneous ignition in forest products. Forest Products Abstracts 10(11): 299-322.

Læstadius, L. 1990. A comparative analysis of wood-supply systems from a cross-cultural perspective. Doctoral thesis, Report # 180, Swedish University of Agricultural Sciences, Department of Operational Efficiency, Garpenberg.

Lehtikangas, P. 1991. Avverkningsrester i hyggeshögar - avbarrning och bränslekvalitet. Del 1. (Logging residue in piles - needle loss and fuel quality). Sveriges Lantbruksuniversitet. Inst. för Virkeslära. Rapport No. 223. 33 p.

Lehtikangas, P. and Jirjis, R. 1993a. Vältlagring av avverkningsrester från barrträd under varierande omständigheter. (Storage of softwood logging residues in windrows under variable conditions). Sveriges Lantbruksuniversitet. Inst. för Virkeslära. Rapport No. 235. 45 p.

Lehtikangas, P. and Jirjis, R. 1993b. Lagring av buntade hyggesrester. (The storage of softwood residues in bundles). Projekt Skogkraft Rapport Nr. 17, Vattenfall Research-Bioenergi.

Lehtikangas, P. and Jirjis, R. 1998. Storage of logging residues in bales. Pp. 1013-1016 *in* Kopetz, H. *et al.* (eds.). Biomass for energy and industry, Proceedings of 10th European Conference, Würzburg, Germany, 8-11 June 1998.

Ling, E. 1999. Bioenergy, its present and future competitiveness. Doctoral thesis, Acta Univ .Agric. Suec. Silvestria 114, Swedish University of Agricultural Sciences, Department of Forest Management and Products, Uppsala, Sweden.

Maclaren, J.P. 1993. Radiata pine grower's manual. New Zealand Forest Research Institute Bulletin No. 184. 140 p.

Makeschin, F. 1999. Short rotation forestry in Central and Northern Europe – introduction and conclusions. Forest Ecology and Management 121: 1-7.

Malmberg, P., Palmgren, U. and Rask-Andersen, A. 1986. Relationship between symptoms and exposure to mold dust in Swedish farmers. American Journal of Industrial Medicine 10: 316-317

Malmberg, P., Palmgren, U. and Rask-Andersen, A. 1988. Natural and adaptive immune reactions to inhaled microorganisms in the lungs of farmers. Scandinavian Journal of Work Environment and Health 14(Supplement 1): 58-71.

Mathews, J.D. 1989. Silvicultural systems. Clarendon Press, Oxford, England.

Mattsson, J.E. 1988. Handling characteristics of wood fuels – a literature review of the state-of-the-art and suitable measurement methods. Swedish University if Agricultural Sciences, Department of Operational Efficiency, Report No. 174, Garpenberg.

McCallum, B. 1997. Handbook for small scale fuelwood chipping operations. Activity 1.2 (Harvesting) Task XII/IEA Bioenergy. Vantaa, Finland. ISBN 951-49-1588-6. 49 p.

Mead, D.J. and Gadgil, R.L. 1978. Fertiliser use in established radiata pine stands in New Zealand. New Zealand Journal of Forestry Science 8: 105-134.

Mitchell, C.P., Ford-Robertson, J.P., Hinckley, T. and Sennerby-Forsse, L. (eds.). 1992. Ecology of short rotation forest crops. Elsevier Science Publishers, Barking, England.

Mitchell, C.P., Stevens, E.A. and Watters M.P. 1999. Short rotation forestry – operations, productivity and costs based on experience gained in the UK. Forest Ecology and Management 121: 123-126.

Nair, P.K.R. 1989. Agroforestry for biomass energy / fuelwood production. Pp. 59-97 *in* Nair, P.K.R. (ed.). Agroforestry systems in the tropics. Kluwer Academic Publishers, Dordrecht, Netherlands.

Nair, P.K.R. 1993. An introduction to agroforestry. Kluwer Academic Publishers, Dordrecht, Netherlands.

National Academy of Sciences. 1980. Firewood crops. National Academy of Sciences, Washington, USA. 231 p.

National Academy of Sciences. 1983. Firewood crops II. National Academy of Sciences, Washington, USA.

Nilsson, P.-O. 1983. Energy from the forest. Research Results NE 1983:9, Swedish Board of Energy Production.

Nilsson, T. 1987. Jämförande lagringsstudie av småved och bränsleflis. (Comparison of storages of chunk wood and fuel chips). Sveriges Lantbruksuniversitet. Inst. för Virkeslära. Rapport No. 192. 76 p.

Nurmi, J. 1999a. The storage of logging residue for fuel. Biomass and Bioenergy 17: 41-47.

Nurmi, J. 1999b. Hakkuutähteen ominaisuuksista. Metsäntutkimuslaitoksen tiedonantoja 722. 32 p.

Nyland, R.D. 1996. Silviculture: concepts and applications. McGraw Hill, New York, U.S.A.

Nylinder, M. and Thörnqvist, T. 1980. Lagring av grenar och toppar i olika fraktioner (Storing of branches and tops of different fractions). Sveriges Lantbruksuniversitet. Inst. för Virkeslära. Rapport No. 113. 45 p.

Pandey, D. and Ball, J. 1998. The role of industrial plantations in future global fibre supplies. Unasylva 193: 37-43.

Parikka, M. and Vikinge, B. 1994. Quantities and economy at removal of woody biomass for fuel and industrial purposes from early thinning. Report No. 37, Swedish University of Agricultural Sciences, Department of Forest-Industry-Market-Studies, Garpenberg.

Petersen, E.B. and Petersen, M.M. 1992. Ecology, management, and use of aspen and balsam poplar in the prairie provinces. Canadian Forestry Service, Special Report 1, Edmonton, AB.

Petersen, E.B. and Petersen, M.M. 1995. Aspen managers handbook for British Columbia. Canadian Forestry Service, FRDA Report 1SSN 0835-0752:230, Victoria, BC.

Rask-Anderson, A. 1988. Pulmonary reactions to inhalation of mould dust in farmers with special reference to fever and allergic alveolitis. Doctoral thesis, Uppsala University.

Richardson, R. 1986. Evaluation of Bruks off-road chippers. FERIC TR-71.

Rothbaum, H.P. 1963. Spontaneous combustion of hay. Journal of Applied Chemistry 13(7): 291-302.

Samoil, J.K. (ed.). 1988. Management and utilization of northern mixedwoods. Canadian Forestry Service Information Report NOR-X-296. 163 p.

Sanchez, P.A. 1995. Science in agroforestry. Agroforestry Systems 30: 5-55.

Shafritz, J.M. and Ott, J.S. 1987. Classics of organization theory. Second edition, Dorsey Press, Chicago.

Smith, D.W. 1994. The Southern Appalachian hardwood region. *In* Barrett, J.W. Regional Silviculture of the United States. Third edition. John Wiley & Sons Inc.

Springer, E.L., Hajny, G.J. and Feist, W.C. 1971. Spontaneous heating value in piled wood chips. II. Effect of temperature. Tappi Journal 54(4): 589-591.

Sundberg, U. and Silversides, C.R. 1988. Operational Efficiency in Forestry, Volume 1: Analysis, Kluwer Academic Publishers, Dordrecht, The Netherlands.

Swedish Forest Statistics Yearbook 2000. Official Statistics of Sweden, National Board of Forestry, Jönköping, Sweden.

Terho, E.O., Husman, K., Kotimaa, M. and Sjöblom, T. 1980. Extrinsic allergic alveolitis in a sawmill worker: a case report. Scandinavian Journal of Work Environment and Health 6: 153-157.

Thörnqvist, T. 1983. Lagring av sönderdelade hyggesrester (Storing of disintegrated logging residuals). Sveriges Lantbruksuniversitet, Institutionen för virkeslära. Rapport 137. 78 p.

Thörnqvist, T. 1984a. Hyggesresternas förändringar på hygget under två vegetationsperioder. (The change of logging residues during storing at the clear cutting area in two growing seasons). Sveriges Lantbruksuniversitet. Institutionen för virkeslära., Rapport 150. 69 p.

Thörnqvist, T. 1984b. Hyggesrester som råvara för energiproduktion - torkning, lagring, hantering och kvalitet. (Logging residues as a feedstock for energy production - drying, storing, handling and grading). Sveriges Lantbruksuniversitet. Institutionen för virkeslära. Rapport 152. 104 p.

Thörnqvist, T. and Jirjis, R. 1990. Bränsleflisens förändring över tiden vid lagring i stora stackar. (Changes in fuel chips during storage in large piles). Swedish University of Agricultural Sciences, Uppsala, Sweden. Department of Forest Products. Report No. 219.

Turnbull, J.W. 1999. Eucalypt plantations. New Forests 17: 37-52.

Tuskan, G.A. 1998. Short-rotation woody crop supply systems in the United States: what do we know and what do we need to know? Biomass and Bioenergy 14(4): 307-315.

Uusvaara, O. 1984. Hakepuun kosteuden alentaminen ennen haketusta korjuuseen ja varastointiin liittyvin toimenpitein (Decreasing the moisture content of chipwood before chipping, harvesting and storage measures). Folia Forestalia 599. 31 p.

Uusvaara, O. and Verkasalo, E. 1987. Metsähakkeen tiiviys ja muita teknisiä ominaisuuksia (Solid content and other technical properties of forest chips). Folia Forestalia 683. 53 p.

Van Assendelft, A.H.W., Raitio, M. and Turkia, V. 1985. Fuel chips-induced hyper sensitivity pneumonitis caused by *Penicillium* species. Chest 87: 394-396.

Vikinge, B. 1999. Removal of forest fuel in early thinnings. Doctoral thesis, Swedish University of Agricultural Sciences, Silvestria 124.

Watson, W.F. and Twaddle, A.A. 1990. An international review of chain flail delimbing-debarking. IEA/BA Task VI Activity 2 Report/ Aberdeen University Forestry Research Paper 1990:3.

West, G.G. 1995. Oversowing in forests. Pp. 78-79 *in* Hammond, D. (ed.). Forestry Handbook. New Zealand Institute of Forestry, Christchurch, New Zealand.

Wharton, E.H. and Griffith, D.M. 1998. Estimating total forest biomass in Maine 1995. U.S. Forest Service, Resource Bulletin NE-142. 50 p.

White, M.S., Curtis, M.L., Sarles, R.L. and Green, D.W. 1983. Effects of outside storage on the energy potential of hardwood particulate fuels: Part 1. Moisture content and temperature. Forest Products Journal 33(6): 31-38.

White. M.S., Argent, R.M. and Sarles, R.L. 1986. Effect of outside storage on the energy potential of hardwood particulate fuels: Part III. Specific gravity, ash content, and pH of water solubles. Forest Products Journal 36(4): 69-73.

Will, G.M. 1984. Monocultures and site productivity. Proceedings of IUFRO Symposium on site productivity of fast growing plantations. Pretoria and Pietermaritzburg, South Africa, Volume 1: 473-87.

Willebrand, E and T. Verwijst. 1993. Population dynamics of willow coppice systems and their implications for management of short rotation forests. Forestry Chronicle 69: 699-704.

CHAPTER 4

COST OF WOOD ENERGY

A. Asikainen, R. Björheden and I. Nousiainen

Wood is considered to be a local fuel. Nevertheless, on global energy markets it has to compete with other sources of energy. Therefore, since forest biomass is scattered over a large geographic area, procurement of forest fuels calls for cost-effective harvesting and transport practices. The economically-acceptable transport distance for forest fuel, due to its low energy density, is a fraction of that for oil. As a result, the raw material must be gathered from an area with a radius that is typically less than 100 km.

Knowledge of the cost factors associated with harvesting and transport of forest fuels is essential to the design and operation of procurement systems. Mechanization of forest work has stimulated the development of cost-effective harvesting machinery and methods for bioenergy production. Consequently, forest fuels are becoming competitive, especially in the Nordic countries where fossil fuels are subject to high environmental taxes. In North America, however, relative procurement costs are still too high to justify large-scale forest biomass recovery.

Combustion of solid biofuels requires higher investment than gas, oil and coal, which can be handled effectively with simple technology. Variable quality, especially high moisture content, imposes special requirements on design of the combustion chamber. A large amount of vapor will be released from wet chips. In addition, removal of small particles is required to meet standards for flue gas emissions.

New technologies have been developed for co-generation of heat and power using solid forest fuels. These facilitate combined production of electricity, district heat and high-pressure steam for industrial purposes. The proportion of electricity in total energy output is usually higher when natural gas or oil are combusted. If the heat can be used directly, competitiveness is greatly improved.

This Chapter deals with costs associated with the procurement and use of wood-based fuel. Section 4.1 covers the purchase, harvesting and transport of wood for energy. The importance of the procurement organization and the

Richardson, J., Björheden, R., Hakkila, P., Lowe, A.T. and Smith, C.T. (eds.). 2002. Bioenergy from Sustainable Forestry: Guiding Principles and Practice. Kluwer Academic Publishers, The Netherlands.

effect of scale on procurement costs are highlighted. Section 4.2 describes the technologies, cost factors and profitability of combustion of woody biomass.

4.1   COST FACTORS IN FUELWOOD PROCUREMENT

4.1.1   Cost factors in decision-making

Cost factors in forestry can be defined as variables associated with machinery, work conditions, operators, organizations, products and silviculture that affect the costs of productivity and operation. Knowledge of the factors involved is needed for assessment of the applicability of technologies and organizational settings in different operating environments. Cost factors can be derived from conditions that define the environment in which the operations are performed (Harstela 1993):

* Global and regional conditions (e.g. climate, forest characteristics, population density).
* National and local conditions (e.g. landscape features, silviculture, roads and communication networks, taxes).
* Work-site conditions (e.g. timber volume per area and per site, tree size, assortments produced, harvesting season, terrain, transport distances, available methods and machines).

Assessments of harvesting costs are needed at three levels. A decision about the location of a chip-fired plant and the nature of harvesting systems must be made at the *strategic level*. For instance, the system may set limits for the degree of integration of harvesting of industrial roundwood and forest residues. At the *tactical level*, harvesting managers must decide, for example, how much wood will be harvested annually from each district and where it will be processed. In order to optimize harvesting and transport-ation, average costs per district must be known. At the *operational level*, stands to be harvested are identified and a schedule of operations is drawn up. This calls for cost estimates of each system available for fuelwood recovery.

4.1.2   Machine characteristics as cost factors

The characteristics of machines define the limits of their performance in different environments and organizations. Using off-road harvesting machines as an example, characteristics can be grouped under two main headings: vehicle mobility (balance, driving speed, maneuverability, hauling capacity) and material-handling capacity (load capacity, loading, unloading and processing capacity).

Balance or stability is perhaps the most important factor affecting the performance of an off-road machine (Terrängmaskinen 1981). Stability affects driving speed, maneuverability, drawbar pull, ability to move uphill and downhill, and load handling.

Driving speed defines the transport performance of the vehicle together with load capacity. Driving speeds of off-road vehicles are typically only a few kilometers per hour, due to the uneven ground surface and slopes. The stability of the machine (together with ergonomic features) sets the limits for maximum speed on a given type of terrain. On rough ground, high speed places stress on the machine body, power transmission and especially on the wheel mounting.

Maneuverability depends on the dimensions of the machine, and the nature of the steering system (front-wheel steering/frame steering), power transmission system, wheels and controls. A precise steering response allows a higher driving speed, as the optimal path can be chosen precisely.

Hauling capacity describes the ability to overcome forces that resist movement of the machine. Friction, slope, and rut formation reduce hauling capacity. Engine performance and load-volume also affect hauling capacity.

Material handling capacity defines the time taken to load and unload. Loading and unloading speeds are functions of movement and controllability. In addition, lifting capacity and grapple dimensions define the amount of material that can be moved at different lifting distances.

Load capacity has a major effect on productivity, since the output of each work cycle equals the load volume (Figure 4.1-1). Limits to load capacity are set by the gross mass of the machine, which affects vehicle mobility and rut formation. Large load volume causes problems when the machine operates on narrow skid trails or on forest roads in uneven terrain.

In productivity studies, the effects of specific machine characteristics are not usually examined in detail. The machine is treated as a single unit. In *work studies*, however, the work is divided into elements such as loading and unloading, driving loaded and unloaded, and driving during loading. Observation of the distribution of work elements makes it possible to identify those that are most critical for productivity. Different work elements employ machine functions differently. Thus, observations related to a specific work element reflect the capacity of the necessary machine parts to perform that element.

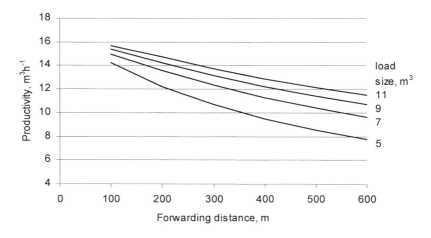

Figure 4.1-1. Effect of logging residue load size on forwarding capacity.

4.1.3   Work-site conditions as cost factors

The productivity of fuelwood harvesting is affected by work-site factors such as the nature of the terrain, the method used for prior or simultaneous harvesting of industrial roundwood, distance to landing, and the amount of biomass left at the logging site. The total amount of fuelwood required will affect the need to move machinery from site to site.

In work studies, the work is broken into elements in order to obtain a detailed view of the work cycle. Different elements require different inputs from an operator or a machine. Time consumption for individual elements is also affected in different ways by specific work-site factors. Using off-road chipping (chipping at stump) as an example, the work cycle of the forwarder can be divided into elements and the factors affecting each element can be defined (Table 4.1-1).

Table 4.1-1. Factors affecting the productivity of a forwarder operation during off-road chipping.

Element	Work-site factors	Machine factors
Driving unloaded	Forwarding distance	Forwarding speed
Driving during loading	Density of material along trails	Forwarding speed
Loading	Stacking of material	Grapple load volume Crane movement speed
Driving with load	Forwarding distance	Forwarding speed
Unloading		Grapple load volume Crane movement speed

Each work element consists of a constant preparation component (fixed time) and a variable component, which is dependent on work-site factors. For instance, when loading, the machine stops beside a pile before the actual loading begins. Loading time depends on the volume and quality of the material. Forwarding speed depends on site conditions, especially the nature and slope of the ground surface. Productivity rates shown in Figure 4.1-2 are based on the relationships presented below. Values presented for each variable are derived from time studies conducted in the Nordic countries.

Driving unloaded:

$$T_{du} = T_{idu} + \frac{l_{du}}{60 v_{du}} c_{du}$$

Where:

T_{du} = *Driving time unloaded (min load^{-1})*;

T_{idu} = *Indirect time for driving unloaded (0.10 min load^{-1})*;

l_{du} = *Driving distance unloaded (m; l_{du} = 2 x average forwarding distance x 0.52)*;

v_{du} = *Driving speed unloaded (1.0 m s^{-1})*;

c_{du} = *Work difficulty factor (in normal conditions c_{du} = 1)*.

Driving during loading:

$$T_d = T_{id} + \frac{100 \dfrac{V}{V_d}}{60 v_d} c_d$$

Where:

T_d= *Driving time during loading (min load^{-1})*;

T_{id} = *Indirect driving time during loading (0.2 min load)*;

V = *Load volume (6 m^3 of solid chips)*;

V_d = *Density of recovered material (12 m^3 100 m^{-1} of skid trail)*;

$100V / V_d$ = *Driving distance during loading (m)*;

v_d = *Driving speed during loading (0.4 m s^{-1})*;

c_d = *Work difficulty factor (in normal conditions c_d=1)*.

Loading:

$$T_l = T_{il} + T_{gl} \frac{V}{V_{gl}} c_l$$

Where:

T_l = *Loading time (min load^{-1})*;

T_{il} = *Indirect loading time (2.5 min load^{-1})*;

T_{gl} = *Direct loading time (0.17 min grapple load^{-1})*;

V = *Load volume (6 m^3 of solid chips)*;

V_{gl} = *Volume of one grapple load (0.06 m^3)*;

c_l = *Work difficulty factor (in normal conditions c_l = 1)*.

Driving with load:

$$T_{dl} = T_{idl} + \frac{l_{dl}}{60v_{dl}} c_{dl}$$

Where:

T_{dl} = *Driving time with load (min load⁻¹);*

T_{idl} = *Indirect driving time with load (0.20 min load⁻¹);*

l_{dl} = *Distance driving with load (m; l_{dl} = 2 x average forwarding distance x 0.48);*

v_{dl} = *Driving speed loaded (0.9 m s⁻¹);*

c_{dl} = *Work difficulty factor (in normal conditions c_{dl} =1).*

Unloading:

$$T_u = T_{iu} + T_{ud}c_u$$

Where:

T_u = *Unloading time (min load⁻¹);*

T_{iu} = *Indirect unloading time (1.3 min load⁻¹);*

T_{ud} = *Time for tipping of container (1.6 min load⁻¹);*

c_u = *Work difficulty factor (in normal conditions c_u = 1).*

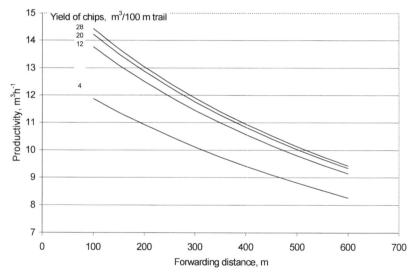

Figure 4.1-2. Effect of logging residue yield and forwarding distance on the productivity of off-road chipping.

Harvesting methods are affected by work-site factors in different ways. For example, off-road forwarder transport of unchipped residues is less sensitive to terrain conditions than chipping and chip transport with a chip harvester. The mobility of vehicles must be considered when machinery is selected for a specific site.

4.1.4   Human factors

The relationship between labor and machine costs affects the optimum degree of mechanization of wood harvesting. In industrialized countires the cost of labor is so high that mechanization is essential. The quality of education in a country affects the degree of mechanization and complexity of technology to be used. Usually, a high level of education is correlated with overall labor cost. In industrialized countries, the cost of labor is typically less than 30% of the total hourly cost of a forest machine unit. In countries where labor costs are low, a lower degree of mechanization is feasible. In Russian Karelia, manual cutting is less expensive than use of a single grip harvester, even for final fellings. In Finnish Karelia, however, manual cutting is competitive only for first thinnings. The difference is largely due to the difference in wage levels in the two countries (Sikanen *et al.* 1996).

The machine operator has an important effect on the productivity of the machine. Factors affecting operator performance are degree of effort, skill, physical and psychological capacity and motivation to perform the work (Harstela 1993). Culture and wage system both affect the motivation of the operator (Gullberg 1995). Independent entrepreneurs seem to be better motivated towards effective use of machines than hired operators.

The development of comparative time study techniques in the Nordic countries has been based on the great variation in forest conditions and differences among operators (Harstela 1993, Gullberg 1995). Harstela (1988) showed that when the same operator was used for comparison of methods or machines, variation in relative time consumption was smaller than the absolute variation in time consumption within each method. Workers respond to different working conditions in different ways. A follow-up study of a chipper operation at road-side landings showed marked differences in productivity between operators using the same machine (Lahti and Vesisenaho 1997, Lahti 1998). With easily-fed material (delimbed small-diameter wood) only small differences were observed, but when logging residues and whole-trees were chipped, productivity differences between operators increased to 20-40% (Lahti 1998). To ensure that the results of time (productivity) studies can be generalized, several operators with similar experience are needed.

4.1.5   Machine interactions

Several studies have emphasized the importance of observing the whole system when harvesting costs are defined (Asikainen 1995, 1998, Asikainen and Nuuja 1999). A supply chain includes use of several machines and vehicles for extraction, processing and transporting of materials. Between most operations, fuelwood is stored in different forms for varying periods. The segments of the supply chain interact either directly or indirectly.

Direct interaction occurs where material is transferred directly from one machine or vehicle to another without intervening storage, e.g. when comminuted material is blown from a chipper directly into a truck. Direct machine interaction is an important factor even in well-balanced systems. Time and simulation studies both indicate that if chipping is done at the landing, interactions can reduce the productivity of a chipper by 10-20% (Kuitto and Rajala 1982, Asikainen 1995). Moreover, when several trucks use the same loading and unloading facilities, they interact with each other during queuing. In a simulation study it was found that when an off-road chipper and one truck worked on a continuous basis, at least 10% of the work time of the chipper was wasted in waiting for the truck (Asikainen and Nuuja 1999) (Figure 4.1-3). In practice, the balance of a system is always affected by changes in work conditions. Hauling distance changes all the time, especially under Nordic conditions where material is collected from small private holdings.

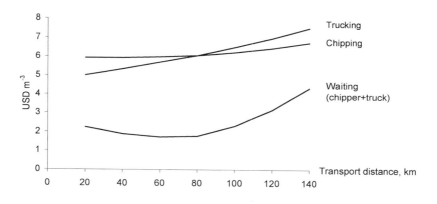

Figure 4.1-3. Effect of machine interaction on off-road chipping and trucking cost.

Indirect interaction occurs where material is stored in piles. Here, pile formation affects productivity of subsequent phases in the system. Productivity of one phase can decrease if careful formation is required in order to accelerate material flow in the next phase. Forwarding of logging residues, tree sections or whole-trees requires the piling of material along the skid-road trails, even though this operation reduces the productivity of the harvester or felling machine. Furthermore, forwarding of industrial roundwood maybe slowed down by the increased lifting distance associated with residue piles (Oijala *et al.* 1999).

4.1.6   Scale of operation

The scale of the operation affects procurement costs in several ways. An increase in annual production diminishes the organizational costs per unit produced. In an optimal situation a single supply chain operates close to its annual capacity.

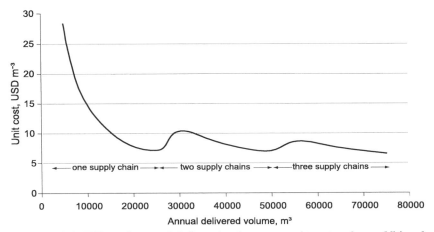

Figure 4.1-4. Effect of annual delivered volume on unit costs when additional supply chains are used.

A procurement organization must hire additional harvesting equipment or an entire supply chain if the maximum capacity of the existing supply chain is exceeded. In theory the resulting direct harvesting costs do not diminish continuously as the volume of the operation increases. Figure 4.1-4 illustrates this for an operation in which the supply chain is based on a truck-mounted chipper and a chip truck. Their annual fixed costs are USD 75,000 and USD 60,000, and variable hourly costs are USD 30 and USD 15 respectively. It is assumed that average productivity per gross effective machine hour of the whole chain is 30 m^3. As delivered volume increases, the unit cost of harvesting decreases, but whenever a new chain is hired, it increases rapidly because all chains then work below capacity. Eventually, when all chains are working close to their upper capacity limit, unit cost declines to a minimum. In practice, hiring machines to harvest a small extra amount often results in lower costs.

Expansion of the use of forest biomass for energy generation can markedly increase harvesting costs for a number of reasons. When the amount of fuelwood recovered is very small in comparison to the amount available, only the best stands are harvested. This means that managers can select stands located near to the utilization plant, aiming at high yields of material per hectare and short forwarding distances. As larger amounts of fuelwood are recovered, harvesting must be extended to more remote and less favorable areas. This effect of scale varies with geographical conditions in both farm- and forest-biomass production (Graham *et al.* 1995, Noon *et al.* 1996, Asikainen and Kuitto 2000).

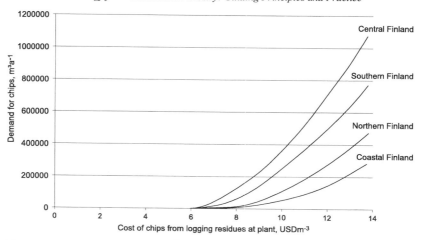

Figure 4.1-5. Effects of cost per m^3 of fuel chips and the demand of a single power plant on the average cost of procurement in different regions of Finland.

Figure 4.1-5 shows the quantity of logging residues from final harvest that can be obtained for a given average cost in various regions of Finland. Power plants located near the coast obtain their wood within a semicircular area, whereas plants in the interior can access stands in all directions. This has a considerable effect on transport costs. Forests located near the coast are dominated by pines and yield less crown mass per hectare than the spruce-dominated stands of central and eastern Finland. The best sources of logging residue are located in the central parts of the country.

4.1.7   Organization of fuelwood procurement

When large forest-industry companies start to recover fuelwood, fairly simple management tools can be used. A computer spreadsheet linked to a logging-site database and map is sufficient. The manager can run this program as long as the number of sites and supply chains is limited, and only a few boiler plants are supplied. As the scale of operation grows, more sophisticated management and control systems are needed. Because recovery of fuelwood should be a part of the operation of every forest industry company, integration with larger management systems becomes a necessity.

The level of integration of forest-fuel recovery with other harvesting operations varies (Björheden 1989). Typically, first-wave systems have a *low level of integration* and are characterized by the use of existing technology, possibly with simple modifications. Organizational relationships with traditional harvesting are limited, fuel recovery being carried out as a separate, subordinate operation. The first extensive Scandinavian fuel-recovery systems, which were based on the collection and forwarding of residues from shortwood logging sites, are good examples of layouts with low-level integration. Systems like this provide management with a good opportunity for testing and designing operations without incurring large economic risks. Only small organizational and technical changes are needed. Two-pass systems with low or moderate levels of integration do not fully utilize the available economic potential.

Higher levels of integration involve extensive development of methods and technology for parallel and coordinated one-pass harvesting operations. Full integration of the harvesting of fuelwood and other industrial assortments provides an opportunity for exploitation of all the positive interacting effects. On the other hand, risk levels and the degree of commitment are higher. For highly integrated systems, the planning of fuelwood recovery is a necessary component of operational planning; decisions on assortment range etc. are based on the net range contribution, and technology and methods are adapted to the task of integrated harvesting. Technology may include multifunction machinery designed to handle several assortments and work tasks, or single-function machinery performing only one work task on a single assortment. Seen as a system, the logging operation includes either or both as purposeful technologies for the complete range of assortments. Whole-tree systems such as feller-skidders/forwarders and whole-tree trucks are examples of highly integrated harvesting systems suitable for short-haul operations to terminals or industrial conglomerates. In remote areas where industries are not located close together, it is often feasible to separate some or all the assortments at the logging site. Some associated examples of highly integrated systems would be tree-section systems yielding sawlogs and undelimbed energy/pulpwood, systems incorporating whole-tree chipping and sorting into fuel and fiber fractions at the mill, and systems where different types of trees are harvested and treated with different machines.

Operations can also be integrated at the organizational level. Considerable savings in organizational costs can be achieved if the same resources are used for bioenergy and industrial roundwood harvesting. When fuelwood harvesting is the only activity, all administrative costs must be assigned to it. Because the total volume and value of forest residues are low, administrative costs become relatively high; 8-10% of the total price of delivered material. At the Nordic price level, this is USD 1-1.5 per m^3.

Marginal-cost pricing can be applied to fuelwood if procurement is integrated into larger industrial roundwood procurement operations. Assuming that a major part of the organization will be maintained regardless of the fuelwood business, and that forest fuelwood harvesting is seen as additional to the main activities, only the extra marginal administrative costs are assigned to energywood. These depend on the degree of integration with the primary operation. If fuelwood harvesting uses existing managerial tools, procurement practices and information systems, the marginal administrative cost can approach zero.

The effects of organizational costs are demonstrated in Figure 4.1-6. In the "Chips" alternative, chipping is done at the roadside. Loose residues and composite residue logs are comminuted at the energy plant. Organizational costs for these systems are relatively high because machinery used for roundwood cannot be used for loose residues or chips, and additional administrative effort is needed to run these operations.

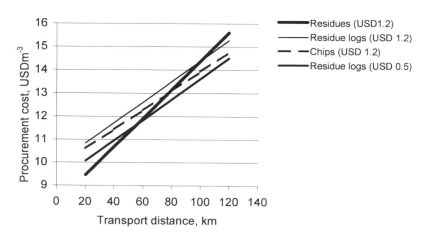

Figure 4.1-6. Effect of organizational costs on the competitiveness of different procurement systems. The figure in parentheses is the organizational cost per m^3: USD 1.2 refers to a situation where forest fuel is harvested by a separate organization, and USD 0.5 to a situation where forest fuel harvesting is integrated with industrial roundwood harvesting.

The FiberPac system (Figure 3.3-7), currently being evaluated in Sweden and Finland, offers an interesting approach to the integration of supply chains for fuelwood and industrial roundwood procurement. Forest residues are fed through a compaction device and bundled into uniform composite residue logs (CRLs) which are easy to load and handle with conventional logging and trucking equipment (see also Section 3.3.2). Although the FiberPac system decreases forwarding and hauling costs, the additional cost of bundling weakens its competitiveness. At the current stage of development FiberPac is not competitive if organizational costs are the same as those for other methods of fuelwood harvesting (USD 1.2 m^{-3}). However, use of conventional forwarders and timber trucks allows the use of information systems and procurement organization designed for the roundwood system. In addition, effective optimization routines designed for log transport can be applied to CRLs. A marginal cost pricing principle can therefore be applied to estimates of organizational cost. When the marginal cost is set at USD 0.5 m^{-3}, harvesting based on bundling becomes competitive.

4.1.8   Chip quality as a cost factor

The quality of fuelwood affects costs in several ways. Contaminants such as stones, metal and sand can damage chipper blades, which then require more frequent replacement or sharpening. Dull blades impair chipper productivity and result in undesirable chip size. Blade costs associated with green logging residues without contaminants are USD 0.3-0.4 per m³. Dry residues increase blade costs by 10-20%, and the presence of contaminants raises the cost even more.

Wood moisture content affects the cost of transport. The more water there is in the material, the lower the volume of fuelwood per load. Furthermore, high or erratic moisture content causes problems during the combustion process.

Reduction of moisture content calls for storage on the logging site, or at the landing, terminal or plant. Storage increases the heating value of the material, especially if it involves protection from rain. In a case study of tree sections, the heating value of woody material increased by 12% during storage. However, storage is also associated with loss of raw material (Brunberg *et al.* 1998). For instance, storage of spruce residues can lead to 20-30% losses due to defoliation. Other causes of dry matter reduction during storage are described in Section 3.4. As a result of these losses, the recovery rate diminishes and the radius of the procurement area increases, raising transport costs. In addition, loading time for forwarders or off-road chippers may be increased due to the lower density of the material (Kärhä 1994).

The size of the energy plant and the combustion principle employed will determine the infrastructures, handling techniques, boiler design, and requirements for volume and quality of raw material. Fuel moisture content, net heating value, energy intensity, and particle size are the most important cost factors. Quality manuals for various solidwood-based fuels have been produced in Finland (Quality Assurance for Solid Wood Fuels 1998, Quality Assurance for Firewood 1998, Quality Assurance for Recovered Fuels 1998).

4.1.9   Cost structure in fuelwood harvesting

Although cost-assessment models differ according to country and type of operation, their basic structure is similar. Variables associated with cost functions depend on cost and wage levels, infrastructure, accounting practices, procurement methods, availability of slack resources, overheads, etc.

Productivity, operational factors and capital costs are usually the most important components of machinery costs. Administrative costs and entrepreneurial risk (profit/loss) are flexible, and depend on the enterprise. For fuelwood harvesting, costs per output unit can be calculated by dividing hourly costs by the productivity of the machine (m³/productive machine hour).

Normally labor costs, including social costs and taxes, can be determined on the basis of the rate paid. Machine costs are divided into fixed and variable costs, and numerical values are determined according to the influence of each factor on the method or equipment used (Harstela 1993).

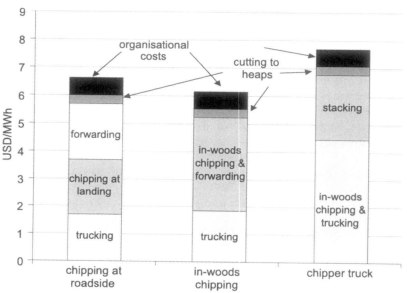

Figure 4.1-7. Cost structure of three logging residue chip procurement systems (Oijala et al. 1999).

The cost structures of different systems vary according to the system design (Figure 4.1-7). Systems based on chipping at the roadside deploy a forwarder, a chipper and trucks. An off-road chipping system (chipping at the stump) includes a terrain chipper and trucks built to take interchangeable containers. A chipper truck undertakes both chipping (on the logging site or at a landing) and road transport. These systems have different cost structures (Figure 4.1-7) which react differently to changes in harvesting conditions. For instance a chipper truck works well on gentle terrain and with short transport distances. A roadside chipping system can be used in difficult terrain, although the landing must be on flat and solid ground. A chipper truck can operate even at poor landings. Terrain chipping is competitive when the forwarding distance is short.

4.1.10 Silvicultural costs and revenues associated with fuelwood harvesting

When performing systems analysis, it is essential to pay attention to all differences between the alternatives (Andersson and Björheden 1990). Apparent changes caused by the addition of a fuelwood harvesting system to roundwood harvesting are:

- A reduced amount of biomass left at the logging site.
- A substantial increase in nutrient removal from the site, especially nutrients in readily-available form. This may be 1.5-3 times the nutrient loss when only roundwood is harvested.
- Increased traffic intensity on the logged areas.
- Increased number of machines and operations on logged areas.

These changes suggest that future tree growth and development may be affected on sites that have been subject to fuelwood harvest. Some effects are likely to be negative from a silvicultural point of view, while others will be favorable. The nature of the effects will depend on whether fuelwood harvesting takes place after final felling or at thinning. From this it can be seen that fuelwood harvesting may incur indirect revenues and costs in the subsequent treatment of the stand or site. These will be referred to here as *silvicultural costs and revenues*. Elsewhere they are known as 'biological', 'ecological' or simply 'extraordinary' costs and revenues.

Silvicultural costs and revenues are mainly indirect and do not arise at time of harvesting. They are the effects of impaired or improved tree development subsequent to logging, and may be defined as the difference in value between a stand that has been subject to fuelwood harvesting and one that has been treated conventionally. Due to different stand development patterns, it is necessary to discount future economic events to a common time, normally the present. For each alternative, present capitalized value may be calculated as follows:

$$\textit{Present capitalized value} = \frac{\sum_{t=0}^{T}\left[(1+i)^{-t}(r-c)_t\right]}{1-(1+i)^{-T}}$$

Where:

T = the time period under consideration (yr);

t = the ordinal number of the year in which a revenue or cost is generated;

i = real rate of interest in decimal form (e. g. for 3 per cent, $i=0.03$);

r = revenues;

c = costs.

The expression above the fraction line denotes the capitalized value for one observation period, while the expression below the line allows for an infinite series of identical repetitions (Mattsson 1999). The capitalized value can be divided by the number of units produced (e.g. m³ chips) to arrive at a number directly comparable to others in an analysis of current cost structures.

Silvicultural costs linked with fuelwood harvesting at final felling

Silvicultural effects of fuelwood harvesting at final felling are:

- Risk of reduced growth rate due to nutrient loss.
- Possibility of earlier stand regeneration work.
- Improved quality (possibly also reduced cost) of soil scarification.
- Improved quality and reduced cost of planting/seeding operations.
- Increased natural regeneration.
- Delay of future thinning and final felling operations.
- Increased soil disturbance and compaction leading to decreased site productivity.
- Altered susceptibility to damage caused by herbivorous animals and insects.

Current studies imply that *decrease in growth rate after whole-tree harvesting* may be equivalent to 1-3 years' prolongation of the regeneration cycle (e.g. Rosén 1991, Mattsson 1999). Experimental treatment in these studies normally includes the extraction of all tree biomass above the stump. If branches are left to dry in order to promote needle-fall, and, as is normally the case in practice, a substantial amount of debris is left on the site, the decline in growth rate can be assumed to be less evident.

The small amount of residue left on the site after fuelwood harvesting provides an opportunity for *acceleration of stand regeneration work*. This indirect operational advantage, rather than direct economic compensation, is a common reason for Swedish forest owners to request slash removal (Björheden 1986). Under Nordic conditions it is often possible to complete regeneration work 1-2 years earlier on slash-free sites. Exploitation of this possibility counteracts the prolongation of the rotation period caused by decreased growth rate.

The *quality of scarification* is improved when it is not impeded by fresh slash (Gyldberg 1993). Although this means that the cost per acceptable planting point is diminished, reduction of total time consumption has not been reported. This means that costs are not reduced at stand level (Mattsson 1999).

The higher quality of scarification *simplifies manual planting*, which is the most common method of stand regeneration in the Nordic countries. Seed sowing results also improve. The number of good planting spots and their distribution over the site is improved (Gyldberg 1993). Plant survival is improved. At the same time the cost of planting decreases. Cost reduction is most substantial for sites with abundant residues. Under Swedish conditions it may be 7-10% on the best sites; 3-5% on normal sites. On the poorest sites with small amounts of residues it is negligible (Prissättning 1985).

Several observations of *improved natural regeneration* after forest fuel harvesting have been reported (e.g. Egnell *et al.* 1998). This means that pre-commercial thinning will not be delayed. On the other hand, this operation is easier in slash-free areas. Brunberg (1990) calculated a cost reduction of 5% for normal stands and 10-20% for the best yield classes. No reduction was evident for the poorest sites.

Reduced growth *prolongs the rotation period.* Prudent planning of early regeneration may decrease this effect but normally the income from thinnings and final felling is delayed. This will result in decreased capitalized value of future generations of trees grown on the site. The magnitude of this effect is very sensitive to rate of interest. As the value of i increases, the imminent costs and revenues become more important and future costs and reductions of revenue become less important. Under Nordic conditions, a real rate of interest between 2 and 4% is commonly used (Wibe 1988). Figure 4.1-8 shows an example of the sensitivity to interest rate. In this study it was assumed that the prolongation of the rotation period is two years for the stand types included; that all biomass from harvested trees is recovered, and that the possibility of earlier regeneration is not realised. The calculation starts at year zero without crediting of revenues from fuel harvesting until a full rotation period is concluded. It can therefore be regarded as a maximum risk calculation.

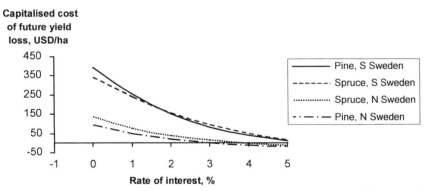

Figure 4.1-8. Sensitivity of capitalized cost of future growth loss to varying rate of interest when losses are due to nutrient depletion associated with fuelwood harvesting at final felling. (Based on Mattsson 1999).

Silvicultural costs associated with fuelwood harvesting at thinning

Concern for stand development and productivity complicate the evaluation of fuelwood harvesting at time of thinning. On the one hand interesting alternatives are offered in terms of the *low cost bulk treatment of small-sized trees* during felling, transportation and processing (Parikka and Vikinge

1994). On the other hand thinning is a tending operation aimed not only at wood procurement but also at improving future stand development. In terms of economics, the goal of the thinning operation is an increase in the capitalized value of the stand.

The risk of reduced growth rate due to nutrient removal seems to be greater with thinning since it takes place in a fully-closed, vigorous tree stand. From literature studies, Mattsson (1999) suggested that the expected prolongation of the rotation period could vary between 1 year under north Swedish conditions and approximately 2 years in the more productive forests of southern Sweden. Yield reduction abates within a 10-15 year period. The losses are considerable in view of the fact that only small amounts are harvested. Normally the volume removed in thinning does not exceed 20 m^3 of 'extra' biomass in the form of branches, tops and foliage.

The effect on the residual stand must also be considered. Whole-tree thinning tends to result in *a different selection of crop trees* when compared with conventional shortwood logging (Fröding 1992a). This is probably due to differences in working pattern caused by the bulkiness of undelimbed crop trees. Impaired selection may result in loss of yield, quality and diameter development. No such effects could be demonstrated in trials designed to isolate yield loss through poor selection at levels close to practical thinning Elfving (1986).

Soil compaction and damage to trees, especially root damage, can be expected to be more severe during whole-tree removal since branch material cannot be used to protect the soil and superficial root systems. Fröding (1992b) attempted an economic evaluation of the effect of damage caused at time of thinning on future revenues. He concluded that there is *an increased risk of damage* in whole-tree thinning. At 'normal' interest rates used for discounting, the economic effect of losses was not substantial, and did not exceed USD 0.25 per harvested m^3 calculated to a present capitalized value.

Mattsson (1999) calculated the effect of interest rate on costs associated with yield loss due to nutrient removal at first thinning (all biomass from thinned trees recovered). Assumptions were that *reduction of volume increment* over a period of 15 years will be 7% in pine trees and 10% in spruce trees. Rotation periods range from 76 to 100 years depending on stand type and are prolonged until loss of production is recovered. Prolongation varies from 1.3 years for pine stands in north Sweden to 2.2 years for spruce stands in southern Sweden. Costs due to damage and soil compaction are not included. Results are shown in Figure 4.1-9.

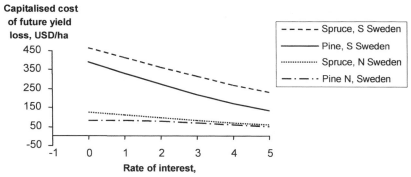

Figure 4.1-9. Sensitivity of capitalized cost of future growth loss to varying rate of interest when losses are due to nutrient removal in whole-tree thinnings. (Based on Mattsson 1999).

4.1.11   Ash recycling

If the ash remaining after combustion of fuelwood is returned to the forest, the nutrient loss and acidification that results from whole-tree harvesting can be counteracted. In addition, problems and costs of ash disposal are reduced and the ash becomes a useful, recyclable resource.

The establishment of an ash-recycling regime encourages a more responsible attitude to maintenance of soil nutrient status in managed forests. Nutrients can be removed when tree demand is low and leaching risk is high, then returned in ash when the trees are more demanding. The balance between different nutrient elements can be adjusted according to localised needs. One example is the practice of adding calcium to ash before it is spread in the forest. Whole-tree harvesting and acid rain both result in soil acidification. Adding calcium (as dolomite or slaked lime) to the ash improves the alkaline buffering capacity of the ash. In Sweden, up to 50% of the weight of dry ash destined for spreading in the forest may consist of added calcium compounds. There is a wide-spread acceptance of the need to counteract acidification through ash recycling. There is also skepticism around the addition of calcium compounds as this may lead to nutrient imbalances , e.g. by increasing potassium leaching.

Recycling of wood ash is a complex operation requiring technical and logistical consideration. The properties of recycled ash can be modified by hardening, granulation or pelletizing into an economic and ecologically-acceptable product.

Ash can be spread *by ground vehicles or from the air*, normally by helicopter. Ground-spreading is cost-efficient and allows precise distribution but a considerable amount of managerial effort is required in planning (Väätäinen *et al.* 2000). It may cause soil damage. The condition of the terrain, trees or soil may make some stands inaccessible. On wet sites, it may only be possible to spread ash on frozen ground. Aerial spreading is expensive but all types of stand can be treated at any time of the year and there is no risk of soil damage. It is easier to plan, although transportation distance must be as short as possible to ensure efficient use of the helicopter. The precision of aerial distribution is lower than that of the ground operation and the devices used are less efficient than ground-based centrifugal spreaders.

The aerial payload varies with the type of helicopter but is normally 900-1500 kg. The number of flight cycles needed per hectare depends on the application rate. In Sweden, 1-3 t of ash ha^{-1} is commonly prescribed. The Finnish recommendation for peatland is 5 t ha^{-1} (Väätäinen *et al.* 2000). A forwarder with disk spreader may carry a payload of 6 t of ash and a smaller four-wheel-drive farm tractor 1-1.5 t. Effective spreading width is 19-25 m for the larger and smaller machine respectively (Ericsson *et al.* 1993).

Systems analysis has been carried out on the *economics of ash recycling*. Andersson and Glöde (1998) presented an overview of existing technologies and estimated costs based where possible on practical trials and operations (Table 4.1-2). In their study, the alternative cost of landfill deposition was also presented and included the current Swedish landfill tax and the variable landfill fee.

Väätäinen *et al.* (2000), included aerial spreading in a study of the logistics of ash recycling. The supply chains investigated are 'hot' meaning that no storage is expected. The cost structure found in this Finnish study is presented in Table 4.1-3. Use of a helicopter with a larger payload would have lowered unit cost. Swedish information is cited which claims that the optimum payload size is between 1200 and 1500 kg. This would be dependent on stand size, road density and the dosage of ash per hectare.

Table 4.1-2. Theoretical cost structures for three possible ash disposal systems: recycling of self-hardened ash; recycling of granulated ash; and landfill deposition. Ranges imply differences due to scale of operation, ash properties or choice of technology.

Cost factor	Self-hardened	Granulated	Landfill
	Cost in USD t^{-1} dry ash		
Ash processing	1–10	30–50	-
Transport to ash store	2.5	2.5	2.5
Storage	0–5	0–5	-
Crushing and screening	5–10	0	-
Loading	0.3	0.3	-
Transport 50 km	5.5	5.5	-
Spreading (3 t dry-matter ha^{-1}, ground-based)	10–20	10–20	-
Landfill tax	-	-	25
Landfill fee (when applicable)	-	-	0–40
Total	24–53	48–83	27.5–67.5

Table 4.1-3. Cost of ash for granulated ash recycling (3 t dry ash ha^{-1}) by helicopter (payload 500 kg) and forwarder-mounted disk spreader.

Cost factor	Ground-based	Aerial
	Cost in USD t^{-1} dry ash	
Ash processing	??	??
Terminal transport from ash processing facility	1	1
Loading	0.4	0.4
Truck transport 50 km	4.1	4.1
Forwarder spreading (0.5 km single distance)	10–20	-
Bobcat + helicopter spreading (0.8 km single distance)	-	48.4
Total	15.5–25.5	53.9

4.2 COMPETITIVENESS OF FOREST BIOMASS IN RELATION TO OTHER FUELS

Energy production costs are generally calculated by dividing the total annual costs of the operation of the plant by the amount of energy generated during the year, thus obtaining the unit cost of the energy produced.

The unit cost is easy to determine when a plant produces only one kind of energy. In a plant producing both electricity and heat, allocation of costs to two different products is a more difficult task because there is no unambiguous criterion for this division. However, cost allocation is not essential for evaluation of the competitiveness of different fuels because the competitiveness is based on total costs.

For comparison and combination of different types of costs, it is necessary to divide total costs into fixed and variable components. This section discusses the cost structure of energy production, especially the cost factors that affect the competitiveness of wood-based fuels.

4.2.1 Fixed costs of energy production

The fixed costs of energy production do not depend on the amount of energy produced by the plant. *Capital costs* account for a major part of the fixed costs. They consist of costs associated with extinction of debt on capital invested in the plant and the interest paid on the capital. The annual capital cost depends on the lifetime of the plant and on the interest rate. The lifetime of the plant can be defined as the technical lifetime, the economic lifetime or the payback period required. If capital costs are assumed to be constant during the lifetime of the plant, then the annual capital costs C_i can be calculated as:

$C_i = aI$

  Where:

  $i = interest;$

  $a = the\ capital\ recovery\ factor,\ i(1+i)^n/((1+i)^n-1);$

  $n = lifetime;$

  $I = investment.$

In cost accounting, the lifetime of a power plant is generally assumed to be 15-25 years. When selecting the interest rate, it is necessary to consider the effect of inflation. A commonly-used real interest rate for energy investments is 5%.

In calculating capital costs, consideration must be given to equipment used for the desulfuration, denitration and limitation of particulate emissions required for environmental protection.

Labor costs of operating personnel, i.e. the number of personnel, depends on the size of the plant, the type of fuel used and the degree of automation.

Fixed service and maintenance costs are a relatively small proportion of the total costs of the plant. It is difficult to draw an exact boundary between fixed and variable service and maintenance costs.

Insurance costs include machinery breakdown insurance, fire insurance and loss of profits insurance on the plant. Normally full fire insurance is taken out but the amount of cover for machinery breakdown and loss of profit varies. In Finland, the total annual insurance premiums generally constitute 0.2-0.5% of the total cost of the power plant.

Inventory costs or *interest on fuel storage* consists of the interest on the delivery price of fuel stored at the plant.

Local and industrial taxes must be paid regardless the volume of the operation. In some countries, environmental taxes are related to annual operation.

4.2.2   Variable costs of energy production

Variable costs are directly proportional to plant output. *Fuel costs* depend on fuel price and plant efficiency. In calculating fuel price, it is necessary to consider the costs of transport and unloading at the plant as well as fuel taxes and storage losses. The cost of conveying fuel to the boiler room is normally included in plant costs.

The efficiency of the plant indicates the proportion of the energy contained in the fuel that can be converted to heat and electricity. Efficiency depends, among other things, on fuel properties, type of boiler and the load factor of the plant.

Other variable costs relate to desulfuration and denitration processes, ash treatment, replacement of fluidised bed sand, supplementary water for the boiler, and variable service and maintenance costs.

4.2.3   Annual costs and energy unit cost

The cost structure of energy production is usually described by expressing the fixed component as annual costs for the design output of the plant, and the variable component as annual costs for the amount of energy produced. In tariff form, total costs are calculated as follows:

$c_{total} = c_{fixed}$ *(USD kW^{-1} yr^{-1})* $+ c_{variable}$ *(USD MWh^{-1})*

> Where:
>
> c_{total} = *total costs;*
>
> c_{fixed} = *fixed costs;*
>
> $c_{variable}$ = *variable costs.*

Annual energy costs will be:

$C_{total,yr} = c_{fixed}$*(USD kW^{-1} yr^{-1})P(kW)* $+ c_{variable}$*(USD MWh^{-1})P(MW)t$_h$(h)*

> Where:
>
> $C_{total, yr}$ = *total annual energy cost;*
>
> P = *design output of the plant;*
>
> t_h = *peak load operating time.*

Peak load operating time is obtained by dividing annual energy production by the design output of the plant.

4.2.4   Technical solutions for wood-fuelled plants

Large-scale conversion of forest biomass to heat or electrical energy is commonly based on grate combustion or fluidised-bed combustion.

Grate combustion represents the traditional technology used for solid fuels. It is seldom used in modern high-capacity boilers, due to higher investment costs, greater emissions levels and limited applicability to multifuel use. For boilers with a capacity less than 5-10 MW$_{th}$, grate combustion is still competitive. A new type of rotating grate boiler has been developed for combustion of wet material such as sawmill refuse (Figure 4.2-1).

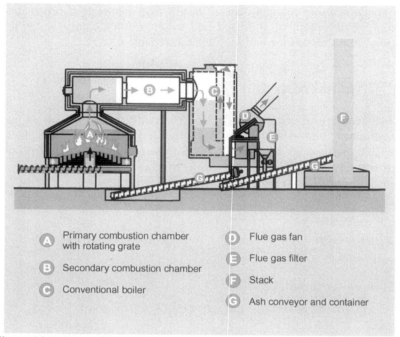

Ⓐ	Primary combustion chamber with rotating grate	Ⓓ	Flue gas fan
Ⓑ	Secondary combustion chamber	Ⓔ	Flue gas filter
Ⓒ	Conventional boiler	Ⓕ	Stack
		Ⓖ	Ash conveyor and container

Figure 4.2-1. Sermet Biograte boiler plant (courtesy Sermet Oy).

New high-capacity solid-fuel boilers are based on *fluidized bed combustion* (FBC) technology, which accomodates a range of fuel types (chips, bark, waste sludge, peat, coal, oil, natural gas) and wide variability in particle size and moisture content. Fuel is fed into a fluidized bed of sand (Figure 4.2-2), where it dries and breaks up into volatile matter, carbon and ash. The bed material is a mixture of sand and fuel ash. As the amount of bed material is large in comparison to the amount of fuel, it effectively stabilizes the combustion process. The sand bed is brought into circulation by blowing air at a high velocity from below. The FBC process can take place either in a bubbling fluidized bed (BFB) or in a circulating fluidized bed (CFB). In a BFB boiler, the fluidizing air is blown at a lower velocity and bed particles behave like a boiling fluid remaining in the bed. In CFB boilers, the air velocity is greater. A large proportion of material leaves the bed and is collected by cyclone separators for recirculation (Alakangas 1997).

In addition to reduced dependence on fuel properties and suitability for multifuel use, FBC technology requires less expensive desulfuration and results in reduced emissions of nitrogen oxides and unburned combustibles. Conversion of grate and pulverized fuel boilers to FBC boilers has increased the potential for utilization of forest residues as fuel.

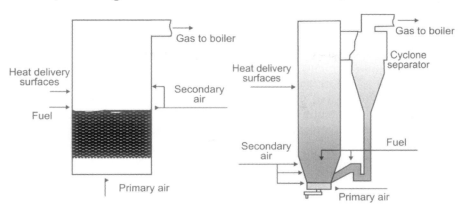

Figure 4.2-2. Fluidized-bed combustion (Alakangas 1997).

When demand for heating is sufficiently high and steady throughout the year, it is often possible to combine power generation with heat generation, thus increasing the efficiency of the plant (Figure 4.2-3). Co-generation of electric power and heat is now possible even in wood-fuelled plants smaller than 1 MW$_e$.

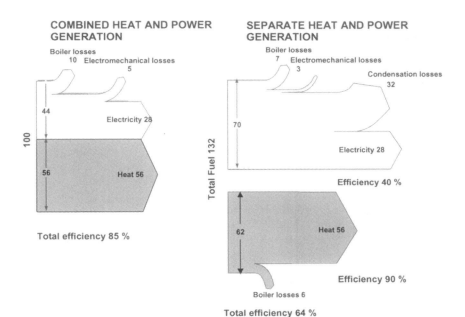

Figure 4.2-3. Energy balance in co-generation of heat and electricity (Alakangas and Flyktman 2001).

In *separate heat production*, 85-88% of the energy input is converted to heat, and the remainder is lost. In a condensing power plant designed for *separate electricity generation*, 40-45% of the energy input is converted to electricity while the remaining heat is lost in cooling water and flue gases (Figure 4.2-3). In a *combined heat and power plant*, power is generated as in a condensing power plant, but heat is recovered and utilised in industrial processes or in a district heat network. The advantage of combined heat and power production is that up to 85-90% of the energy in the fuel can be utilised. With present technology, approximately 20-30% of the energy input is converted to electric power and 55-70% is converted to heat.

4.2.5   Cost structure of wood-fuelled plants

Due to the lower energy density of solid fuel and the need for more complex fuel and ash handling equipment, *investment costs are higher for solid fuel boilers* than for oil and gas boilers. They are particularly high for biomass boilers.

Investment costs depend on the location of the plant and the nature of the technical options used. Two plants of the same type may have different investment costs. In an older plant it is possible to utilize existing equipment, but for a new plant investments have to be made in all the components of the building project. Figure 4.2-4 presents typical investment costs for wood and oil/gas-fuelled heating plants. As plant size increases, the cost of investment in relation to capacity is reduced. For wood-fuelled plants, investment costs are 3-3.5 times higher than those for oil/gas-fuelled plants. For this reason, many countries use investment aid to promote the use of wood energy.

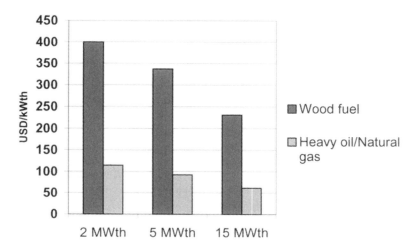

Figure 4.2-4. Typical investment costs for heating plants (Kosunen and Leino 1995, Kosunen and Rajamäki 1999).

Plants using solid fuel require a greater number of operating and maintenance personnel for fuel and ash handling and incur higher labor costs than plants using oil or natural gas. In wood-fuelled heating plants, labor costs are 2.5-3 times as high as those for oil- and natural gas-fuelled plants. Labor costs for operating personnel are approximately 10% higher in wood-fuelled back-pressure power plants than in coal-fuelled plants.

Fixed service, maintenance and insurance costs can be assumed to be directly related to the cost of the power plant. The higher investment costs of wood-fuelled plants therefore result in higher costs in all these categories.

In wood-fuelled plants, circulation of stored fuel is normally so rapid that interest costs associated with storage are negligible.

In plants using solid fuels, the fuel costs are significantly lower than in plants using oil and natural gas, but the boiler efficiency is lower, varying between 80 and 88%. Other variable costs are not dependent on fuel type.

Figure 4.2-5 gives cost structures for energy production from different fuels, using a heating plant with an annual peak load operating time of 5000 h. For fuelwood, fixed costs amount to more than 50% of the total, and fuel costs 40%. With oil or gas, fixed costs contribute 15% and fuel costs more than 80%. The price of oil and gas therefore has a significant effect on the competitiveness of wood-based fuels.

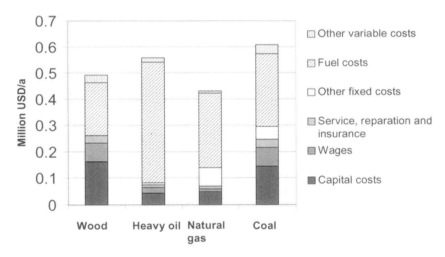

Figure 4.2-5. Cost structure of energy production by a 5 MW heating plant in Finland in 2000.

4.2.6   Production costs of heat and electricity from wood-based fuels

The peak load operating time of the plant has a considerable effect on production costs and on the relative competitiveness of different fuel, particularly in plants requiring high investment costs. Economic efficiency demands a high peak load operating time, especially in plants with high investment costs. The peak load operating time is influenced by the design capacity of the plant in relation to the required heat load and its variability. Figure 4.2-6 shows the heat production costs of a 5 MW heating plant as a function of peak load operating time. Costs were calculated from data presented in Table 4.2-1.

Table 4.2-1. Cost factors for a 5 MW district heating plant in Finland in 2000 (Kosunen and Leino 1995, Kosunen and Rajamaki 1999).

Cost factor	Wood chips	Heavy oil	Natural gas	Coal
Fuel price, USD MWh[-1]	6.9	16.5	10.4	9.6
Investment costs, USD MWh[-1]	337	93	93	301
Fixed costs, USD yr[-1]				
Capital costs	163,000	44,000	50,000	146,000
Other fixed costs	105,000	31,000	93,000	153,000
Total	268,000	76,000	142,000	298,000
Variable costs, USD MWh[-1]				
Fuel costs	8.0	18.3	11.3	11.1
Other variable costs	1.2	0.6	0.3	1.3
Total	9.2	18.9	11.6	12.4

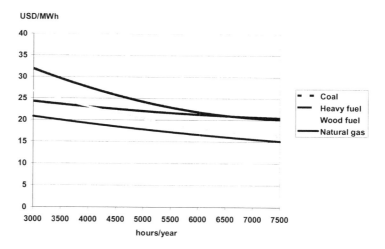

Figure 4.2-6. Heat production costs as a function of peak load operating time at a 5 MW heating plant in Finland.

Under the conditions specified, natural gas is more competitive than wood, oil or coal. However, the availability of natural gas is restricted by its reticulation. If natural gas is not available, wood is the most competitive fuel providing that the peak load operating time exceeds 4000 hours. Figure 4.2-7 shows corresponding prices for oil and gas as a function of the peak load operating time when they are as competitive as wood.

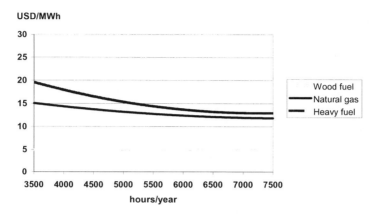

Figure 4.2-7. Competitive price of oil and gas against wood-based fuel at a 5 MW heating plant in Finland.

For heat production in Finland a base load boiler serving a municipal district heating system is generally designed for an annual peak load operating time of about 5000 h. In this case, the price of wood-based fuel should not be more than 30% of the price of heavy fuel oil.

4.3   CONCLUSIONS

The profitability of fuelwood procurement can be improved if harvesting operations are confined to stands where conditions are favorable for recovery. Ecological considerations must also be taken into account. Operational efficiency calls for knowledge of the factors that influence procurement costs. Information about how each system works under different conditions is required when machinery is selected.

The scale of operations has a major impact on procurement costs. Expensive machines incur high capital costs, and to be economic, they must be used close to their full capacity. Large boiler plants must obtain wood from a wide geographical area. Availability of fuelwood varies, depending on land-use pattern, structure and management practice of the forests, and the shape of the procurement area.

The cost of forest fuel is strongly dependent on the operating environment in which the material is recovered. Careful investigation of the availability of fuelwood in a specific geographic procurement area should precede any investment in energy plants fuelled with forest biomass. This is because fuel availability, expressed in terms of volume or mass of recoverable fuelwood in a certain area, has a direct effect on costs of harvesting and long-distance transport. Availability of forest fuel is dependent on fluctuations in the harvesting of industrial roundwood. To minimize raw-material costs, plants should be located at sites surrounded with abundant fuelwood resources and forest industries. From the standpoint of co-generation, consumers of heat should also be available.

Integration of fuelwood supply with flows of other wood materials offers opportunities for cost saving in administration, harvesting and transportation. If a fuelwood harvesting organization can use existing planning and control systems, marked savings can be gained in operational costs. If machinery in current use is appropriate for recovery of fuelwood, the startup phase will involve smaller risks and offer additional job opportunities for entrepreneurs already associated with the organization.

Selection of the most competitive harvesting technology requires information about the operating environment and particularly about the availability of biomass. If the density of the raw material is relatively low, systems with high load capacity, e.g. those based on composite residue logs or wood chips, will be more competitive than systems with a low load capacity (transport of loose residues, chipper trucks with small load volume).

The indirect costs and revenues of forest fuel activities are closely linked to sustainability issues. Sustainability is a major consideration in forest-fuel-based energy supply. Methods and technology can be aimed at more efficient use of resources, e.g. recycling of 'waste products', rather than consignment to sinks such as landfills. All harvesting regimes result in removal of tree nutrients from the forest ecosystem. The increased intensity of harvesting associated with forest fuel recovery does not therefore introduce a new problem but elevates an existing one to a new level (see Sections 5.2.4 and 5.5).

From a silvicultural perspective, forest fuel harvesting is likely to affect timing and need for different treatments, and also future tree stand development. Although the same basic principles apply for different types of forests in different areas, decisions will vary according to regional conditions. Other knowledge is still shallow and patchy. In the young boreal forest ecosystem it seems very likely that harvesting can be intensified without risk of serious decline in yield and vitality, especially if recovery of biomass is confined to the final felling phase. In most cases, the effects of temporary yield loss can be countered by earlier and more efficient stand regeneration.

The availability of fuelwood is not a decisive factor when the size of an energy plant is under consideration. Plant design and size are determined by the demand for energy. The majority of power plants designed for solid fuel combustion actually co-fire several fuel types. This means that fuel mixtures can be varied according to price and availability. The proportion of forest biomass chips can be optimized within constraints set by the combustion process and emission limits set by the public authority.

In wood-fuelled plants, fixed costs make up a higher proportion of the total costs of oil- or gas-fuelled plants. For this reason, the prices of oil and gas have a large effect on the competitiveness of wood-based fuels. The peak load operating time of the plant also has a considerable effect on production costs and on the relative competitiveness of different fuels. Particularly in plants with a high investment cost, economic efficiency requires a high peak load operating time. This can be influenced by appropriate design capacity of the plant in relation to the required heat load and its variability.

4.4   REFERENCES

Alakangas, E. 1997. Bioenergy in Finland II. Final report of Finland-Phase II. AFB-NETT. VTT Energy.

Alakangas, E. and Flyktman, M. 2001. Biomass CHP Technologies. SAVE programme. VTT Energy Reports 7/2001.

Andersson, G. and Glöde, D. 1998. Beskrivning av askproduktion vid biobränsleeldade värmeverk och återföring av askan till skogen (Ash production in bio-fuelled heating plants and recirculation of the ash to the forest). SkogForsk, Arbetsrapport Nr. 402.

Andersson, S. and Björheden, R. 1990. Analys av skogliga driftssystem. (Analysis of productions systems in Forestry, in Swedish). Swedish University of Agricultural Sciences, Department of Operational Efficiency Textbooks 1990. 94 p.

Asikainen, A. 1995. Discrete-event simulation of mechanized wood harvesting systems. Research Notes 38. University of Joensuu. Faculty of Forestry. Joensuu. Finland. 86(8).

Asikainen, A. 1998. Chipping terminal logistics. Scandinavian Journal of Forest Research 3(13): 385-391.

Asikainen, A. and Kuitto, P-J. 2000. Cost factors of energy wood harvesting. New Zealand Journal of Forestry Science. Vol. 30 No. 1/2(2000):79-87.

Asikainen, A. and Nuuja, J. 1999. Palstahaketuksen ja hakkeen kaukokuljetuksen simulointi. Metsätieteen aikakauskirja 3/1999:491-504.

Björheden, R. 1986. Jämtlandsbränslen - en utvärdering av tillgängliga produktionssystem för trädbränslen i Östersundsregionen. (An evaluation of accessible production systems for forest fuels in the county of Jämtland) Swedish University of Agricultural Sciences, Department of Operational Efficiency Results # 68. 63 p.

Björheden, R 1989. Integrerad avverkning. Swedish University of Agric. Sci., Skogsfakta konferens nr 16, 1989.

Björheden, R. 2000. Integrating production of timber and energy – A comprehensive view. New Zealand Journal of Forestry Science 30(1/2): 67-78.

Brunberg, B., Andersson, G., Nordén, B. and Thor, M. 1998. Uppdragsprojekt Skogsbränsle – slutrapport (Forest bioenergy fuel – final report of commissioned project). Skogforsk redogörelse Nr. 6. 61 p.

Brunberg, T. 1990. Underlag för prestationsmål för motormanuell röjning (Basic data for piece rates for motormanual cleaning, in Swedish). Stencil 1990-06-23, SkogForsk, Uppsala.

Egnell, G., Nohrstedt, H-Ö., Weslien, J. Westling, O. and Örlander, G. 1998. Miljökonsekvensbeskrivning av skogsbränsleuttag. (Environmental evaluation of forest fuel harvesting) Proceedings of the Royal Swedish Academy of Agricultural Sciences Conference June 5 1996, KSLA tidskrift 13: 73-82.

Elfving, B. 1986. Urval, stickvägar och tillväxteffekter i gallring (Selectivity, striproads and yield effects in thinning). Swedish University of Agricultural Sciences, Department of Operational Efficiency Results # 52.

Ericsson, S.O. Fornling, C., Jonsson, T., Lundborg, A. and Oskarsson, R. 1993. Skogsbränsle för miljövänlig energiproduktion (Forest fuel for environmentally friendly production of energy). Vattenfall: Projekt Skogskraft Slutrapport.

Fröding, A. 1992a. Gallringsskador - nuläge och förändringar sedan början av 1980-talet (Thinning damage - current state and changes since the beginning of the 1980's). Swedish University of Agricultural Sciences, Department of Operational Efficiency.

Fröding, A., 1992b. Beståndsskador vid gallring (Thinning damage to coniferous stands in Sweden). Swedish University of Agricultural Sciences, Department of Operational Efficiency Dissertation.

Graham, R.L., Liu, W., Downing, M., Noon, C., Daly, M. and Moore, A. 1995. The effect of location and facility demand on the marginal cost of delivered wood chips from energy crops: A case study of the state of Tennessee. Pp. 1324-1333 *in* Proceedings, Second Biomass Conference of the Americas: Energy, Environment, Agriculture and Industry.

Gullberg, T. 1995. Evaluating operator-machine interactions in comparative time studies. Journal of Forest Engineering 7(1): 51-61.

Gyldberg, B. 1993. Studies of the workings of a powered disc trencher. Swedish University of Agricultural Sciences, Department of Operational Efficiency Dissertation.

Harstela, P. 1988. Principle of comparative time studies in mechanized forest work. Scandinavian Journal of Forest Research 3: 253-257.

Harstela, P. 1993. Forest Work Science and Technology. Part I. University of Joensuu, Faculty of Forestry, Joensuu. 113 p.

Kärhä, K. 1994. Hakkuutähteen talteenotto osana puunkorjuun kokonaisurakointia. University of Joensuu, Faculty of Forestry, Joensuu. 46 p.

Kosunen, P. and Leino, P. 1995. The competitiveness of biofuels in heat and power production. Ministry of Trade and Industry, Studies and Reports 99/1995.

Kosunen, P. and Rajamäki, J. 1999. Biopolttaineiden kilpailykyky sähkön ja lämmön tuotannossa (Bioenergy research program). Publications 26. Finland.

Kuitto, P-J. and Rajala, P. 1982. Kokopuiden välivarastohaketus ja metsähakkeen autokuljetus. Metsätehon tiedotus 372. 14 p.

Lahti, P. 1998. Evolution energiapuuhakkurin käyttöselvitys. (Follow up study of Evolution drum chipper). Bioenergia tutkimusohjelma julkaisuja 21: 493-502.

Lahti, P. and Vesisenaho, T. 1997. Evolution energiapuuhakkurin käyttöselvitys. Energiaselvityksen loppuraportti KTM:n energiaosastolle Nr. 21/854/95. 21 p.

Mattsson, S. 1999. Ekonomiska konsekvenser av tillväxtförluster och billigare beståndsanläggning vid skogsbränsleuttag – exempel på beståndsnivå (The economical consequences of growth loss and less expensive regeneration after forest fuel harvesting - stand level examples). SkogForsk, Arbetsrapport Nr. 425.

Noon, C.E., Daly, M.J., Graham, R.L. and Zahn, F.B. 1996. Transportation and site location analysis for regional integrated biomass assessment (RIBA). Proceedings. Bioenergy '96 – The Seventh National Bioenergy Conference: Partnership to develop and apply biomass technologies, September 15-20 1996. Nashville, Tennessee. 7 p.

Oijala, T., Saksa, T. and Sauranen, T. 1999. Hakkuutähteen korjuumenetelmien vertailu ja vaikutus metsänuudistamiseen. Bioenergian tutkimusohjelma. Julkaisuja 27. Jyväskylän Teknologiakeskus Oy. Finland.

Parikka, M. and Vikinge, B. 1994. Quantities and economy at removal of woody biomass for fuel and industrial purposes from early thinning. Report 37, Swedish University of Agricultural Sciences, Department of Forest-Industry-Market-Studies.

Prissättning. 1985. Prissättning av plantering med rotade plantor och rör samt barrotsplantor och hacka. (Pricing of planting with containerized and bare-root stock). Stencil 1985-04-03. Stora Skog AB, Falun.

Quality Assurance for Solid Wood Fuels. 1998. FINBIO Report 7. Jyväskylä. Finland.

Quality Assurance for Firewood. 1998. FINBIO Report 8. Jyväskylä. Finland.

Quality Assurance for Recovered Fuels. 1998. FINBIO Report 9. Jyväskylä. Finland.

Rosén, K. 1991. Skörd av skogsbränslen i slutavverkning och gallring – ekologiska effekter. (Harvest of forest fuels in final felling and thinning – ecological effects). Swedish Board of Forestry, Skogsstyrelsen, Meddelande # 5.

Sikanen, L., Gerasimov, Yu.Yu. and Sivonen, S. 1996. Cost comparison between alternative thinning technologies in Russian Karelia. Pp. 133-138 *in* Research Reports of Forest Engineering Faculty of Petrozavodsk State University 1996.

Terrängmaskinen. 1981 (The terrain vehicle, part 2, in Swedish) Skogsarbeten. Stockholm. 461 p.

Väätäinen, K., Sikanen, L. and Asikainen, A. 2000. Rakeistetun puutuhkan metsäänpalautuksen logistiikka (Logistics of forest spreading of granulated ash). University of Joensuu, Faculty of Forestry Research Notes 116. 99 p.

Wibe, S. 1988. Hur hög är den "skogliga räntan"? (What is the level of "forestry interest rate?"). Swedish University of Agricultural Sciences, Department of For. Econ., Skogsfakta konferens # 11.

CHAPTER 5

ENVIRONMENTAL SUSTAINABILITY OF FOREST ENERGY PRODUCTION

5.1   ENVIRONMENTAL SUSTAINABILITY

R. J. Raison

Bioenergy systems must be sustainable if they are to produce 'renewable' energy. The sustainability of forests that provide fuel is a critical element of overall sustainability and is the focus of this chapter.

Sustainable forest management (SFM) is an evolving concept, but is widely recognized to comprise social, economic and environmental components. The relative weighting given to these will vary from place to place, and will reflect the goals and outcomes negotiated between those having legitimate interests in or concerns about forest management. The goals and outcomes effectively become a working definition of SFM that takes into account local values and issues.

The environmental component of SFM covers the properties and processes occurring in the various parts of the forest ecosystem. These have been described in a series of sustainability criteria as part of internationally accepted sets of Criteria and Indicators (e.g. those developed under the Helsinki and Montreal Processes). The environmental Criteria cover:

- Forest health.
- Productive capacity.
- Biodiversity.
- Soil and water.
- Carbon budgets.

All of these may be affected by the utilization of forests for bioenergy, especially where more intensive harvesting practices are applied. Potential impacts and ways of minimizing harmful effects are outlined in this chapter.

Richardson, J., Björheden, R., Hakkila, P., Lowe, A.T. and Smith, C.T. (eds.). 2002. Bioenergy from Sustainable Forestry: Guiding Principles and Practice. Kluwer Academic Publishers, The Netherlands.

Sets of indicators are being developed for each of the Criteria. The indicators will be surrogates for properties and processes important to SFM. They should be sensitive and inexpensive to measure and will be used to track temporal changes in forest condition and output (Raison *et al.* 1997).

5.1.1   Achievement of sustainable forest management

Criteria and Indicators can be used to help forest managers to meet growing community expectation that they should demonstrate SFM by quantifying progress against agreed goals and outcomes (targets). It is essential that there should be a shared view and agreement about the validity of indicators, appropriateness of targets, and methods for monitoring and review. Effective stakeholder engagement is needed to achieve this (Raison, Brown and Flinn 2001). It is only through continuous interaction between stakeholders and managers that a shared understanding of the benefits and risks of different forestry strategies and practices can be gained. Monitoring of outcomes provides the foundation for continuing review and improvement, which are essential if there is to be progress towards SFM. While Criteria and Indicators have considerable potential for improvement of forest policy and management, the science underpinning them and their application to forestry is still immature.

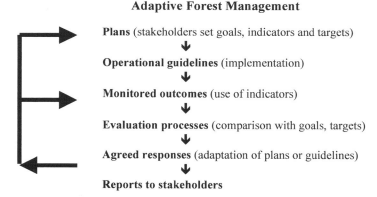

Adaptive Forest Management

Plans (stakeholders set goals, indicators and targets)
⬇
Operational guidelines (implementation)
⬇
Monitored outcomes (use of indicators)
⬇
Evaluation processes (comparison with goals, targets)
⬇
Agreed responses (adaptation of plans or guidelines)
⬇
Reports to stakeholders

Figure 5.1-1. Procedures used in an Adaptive Forest Management system (Raison, Flinn and Brown 2001).

The elements of *Adaptive Forest Management* (Figure 5.1-1) are similar to those of an Environmental Management System (EMS), such as that developed by the International Standards Organisation (ISO). These systems help to ensure structured, consistent and auditable practices. The essential components are summarized in section 5.1.2. They are based on the principles of transparency (stakeholder clarity about decision-making processes), openness (ability of all interested parties to participate in decision-making) and accountability (identification of those responsible for implementing agreed decisions).

5.1.2 Components of environmental management systems relating to sustainable forest management

Commitment, legislation and policy framework. The legislative and policy framework for forest management should ensure that all forest values are protected and maintained in a balanced manner. Obligations to international agreements, treaties and conventions, environmental legislation, and other regulatory instruments must be met.

Planning. The principles of SFM should be reflected in goals and objectives clearly stated in plans. Transparency and openness in planning processes are essential at local (forest management unit), regional and national levels. Adequacy of an information base for environmental risk assessment and planning is a critical issue.

Implementation. Effective implementation of practices designed to achieve SFM involves provision of the support required for achievement of the objectives and targets specified in policies and plans. Key issues are designation of responsibility, capacity to implement (adequate resourcing), operational control, documentation, and staff knowledge and skill levels.

Information, monitoring and evaluation. Forest management should be supported by information at appropriate levels of detail and scale. This information is important for risk assessment and planning, for assessing compliance with management guidelines, and for quantifying the outcomes of forest management practices. Independent audit is becoming increasingly important, as is the reporting of environmental performance to stakeholders. Evaluation of the importance of any environmental change is essential as a basis for forest management, and for the reporting of environmental performance. Evaluation steps are discussed in more detail below.

Review and Improvement. Commitment to and capacity for review and improvement of forest management plans and practices are important elements of SFM. Appropriate data must be collected. Processes facilitating the utilization of recent advances in management methods are needed. Ability to respond to changing community expectations, to new legislation, and to past environmental incidents is also very important.

In practical terms, forest management plans are often implemented through *codes of forest practice* that contain a set of goals and guidelines for environmental care. These give broad directions for the planning and conduct of management practices, often focusing on roading, harvesting, regeneration, and the protection of soil and water values. Important support for codes of forest practice is provided by local management prescriptions (guidelines). These specify actions needed for protection of the environment

– e.g. alteration of the width of buffers along the sides of streams in order to protect aquatic habitats or to maintain water quality where there is a risk of soil erosion.

5.1.3   Monitoring and evaluation steps

A key question in the Adaptive Forest Management cycle is: *'Have objectives and targets for environmental care been met?'* Answers to this question rely on the monitoring and evaluation steps shown in Figure 5.1-1. Monitoring must include both spatial (i.e. the range of ecosystem types and environments) and temporal considerations. From a spatial viewpoint, monitoring should focus on areas where the risk to environmental values from forest operations is greatest. Risk analysis based on available information and expert input should be included in the development of forest management plans. Monitoring should be varied temporally so that rates of change in environmental values, the effects of changing forest practice, and the effectiveness of so-called Best Management Practice can be recorded. A strategic approach to monitoring is required in order to maintain focus on representative higher-risk situations, and to assess the effectiveness of practices used to mitigate the risks.

Evaluation of the importance of any measured trend in environmental values provides a basis for implementation of Adaptive Forest Management. Ideally, it should be guided by the performance measures and targets agreed upon during the forest management planning process. Definition of performance measures is a neglected area, but generally performance measures or targets take the following forms (Raison and Rab 2001):

- Scientifically-based standards. These are currently rare in forestry.
- Interim standards, based on currently-available scientific opinion and adopted until rigorous, scientifically-based standards have been developed. These need to be calibrated to establish 'threshold' values for specific forest ecosystems.
- Compliance with plans, guidelines and prescriptions. This commonly-used approach concentrates on management inputs rather than on environmental outcomes. There is a clear need to move to the outcome-oriented approach shown in Figure 5.1-1.

The evaluation step must take account of the size of the area affected by important environmental change, the temporal pattern of change, and the integration of consequences of change in a number of values. These issues have been discussed from the perspective of the soil resource by Burger and Kelting (1999) and Kelting *et al.* (1999).

The Adaptive Forest Management approach is used in the remainder of Chapter 5 as the framework for review of environmental implications of forest fuel harvesting. Conclusions are synthesized into a set of interim guidelines for forest managers.

5.1.4   Summary

Sustaining environmental values in forests that provide biofuels is central to the production of *renewable* energy. The relative weight given to social, economic and environmental values in forest management will vary from place to place and must reflect goals and outcomes negotiated between interested parties.

Forest environmental values are described by a set of Criteria and related Indicators covering health, productive capacity, biodiversity, soil and water, and carbon budgets. Indicators can be used to track temporal change in forest condition and output.

Effective stakeholder input is needed to clearly define environmental goals and outputs (targets) at the local scale. These can then be incorporated into an EMS that involves the steps of planning, implementation, monitoring, review and improvement, and reporting. The EMS provides a framework for Adaptive Forest Management, by ensuring structured, consistent and auditable processes. Adaptive approaches to forest management are essential because our knowledge of the long-term effects of forest utilization on environmental values is incomplete.

5.1.5   References

Burger, J.A. and Kelting, D.L. 1999. Using soil quality indicators to assess forest stand management. Forest Ecology and Management 122: 155-166.

Kelting, D.L., Burger, J.A., Patterson, S.C., Aust, W.M., Miwa, M. and Trettin, C.C. 1999. Soil quality assessment in domesticated forests – a southern pine example. Forest Ecology and Management 122: 167-185.

Raison, R.J., McCormack, R.J., Cork, S.J., Ryan, P.J. and McKenzie, N.J. 1997. Scientific issues in the assessment of ecologically sustainable forest management, with special reference to the use of indicators. Pp. 229-234 *in* Bachelard, E.P. and Brown, A.G. (eds.). Proceedings of the 4th Joint Conference of the Institute of Foresters of Australia and the New Zealand Institute of Forestry, Canberra.

Raison, R.J. and Rab, M.A. 2001. Guiding concepts for the application of indicators to interpret change in soil properties and processes in forests. Pp. 231-258 *in* Raison, R.J., Brown, A.G. and Flinn, D.W. (eds.). Criteria and Indicators for Sustainable Forest Management. IUFRO Research Series, No. 7. CABI, Wallingford, UK.

Raison, R.J., Brown, A.G. and Flinn, D.W. (eds.). 2001. Criteria and Indicators for Sustainable Forest Management. IUFRO Research Series, No. 7. CABI, Wallingford, UK. 443 p.

Raison, R.J., Flinn, D.W. and Brown, A.G. 2001. Application of Criteria and Indicators to support sustainable forest management: some key issues. Pp. 5-18 *in* Raison, R.J., Brown, A.G. and Flinn, D.W. (eds.). Criteria and Indicators for Sustainable Forest Management. IUFRO Research Series, No. 7. CABI, Wallingford, UK.

5.2   SOIL AND LONG-TERM SITE PRODUCTIVITY VALUES

J. A. Burger

Soil is the earthen foundation upon which forest biomass is produced. As part of the forest bioenergy production system, it supports root anchorage, supplies water and mineral nutrients for tree growth, provides a surface for operating harvesting machinery, and creates favorable conditions for the decomposition and recycling of forest residues and wood ash. Bioenergy production systems are usually components of forest management systems that provide multiple products and services, e.g. wood, water, biodiversity, water quality control, and carbon capture. Soil plays a role in each of these. In the process of forest management for bioenergy production, soils must be managed to sustain the forest values that are important to human communities. The purpose of this section is to define the role of soils in forest systems managed for bioenergy production, and to present approaches for managing them sustainably.

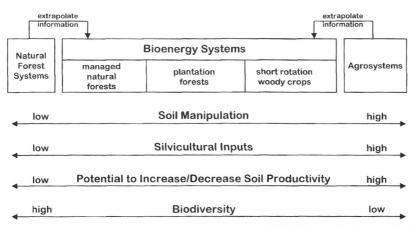

Figure 5.2-1. Spectrum of managed systems from which biomass is harvested for energy.

A forester's concept of soil usually depends on his or her forest land-use objectives and intensity of management. Gradients of use and management intensity have been recognized and illustrated by several researchers (Stone 1975, Nambiar 1996, Burger 1997). Figure 5.2-1 shows the spectrum of forest management systems that may include biomass harvesting for energy. In long-lived native forests that are managed extensively and regenerate naturally, higher emphasis is usually placed on biodiversity and non-biomass values. Silvicultural inputs are focused on the composition and quality of the tree stand. Drainage, fertility and physical structure of soils are not usually manipulated in native forests. Soil is perceived as a secondary ecosystem resource and may simply be defined in the context of the broader forest site, land type, or ecotype (Carmean 1975). At the other end of the spectrum where short rotation woody crops are managed intensively for bioenergy and fiber, soil fertility, water availability, organic matter content and structure are often modified to maximize productivity. Here, soil is perceived as the primary ecosystem resource, and its inherent composition and function is recognized separately from other site factors such as topography and climate. A forester's connotation of soil, the degree to which it is managed, and the role it plays in sustainable forestry, will depend on where he or she works in the spectrum of management intensity.

Soils should be defined and understood in terms of their formation, composition, function, and fragility. In their introduction to the Handbook of Soil Science, Sumner and Wilding (2000) define soil as:

"...an evolving, living organic/inorganic layer at the Earth's surface in dynamic equilibrium with the atmosphere and biosphere above, and the geology below. Soil acts as an anchor and purveyor of water and nutrients for roots, as a home for a vast and still largely unidentified community of micro-organisms and animals, as a sanitizer of the environment, and as a source of raw materials for construction and manufacturing. Soil is the long-term capital on which a nation builds and grows. It is the basic component of ecosystems and ecosystem management. Soil serves as the foundation essential for continued human welfare, the well-spring of other renewable natural resources."

This broad concept is appropriate to the consideration of expectations of forest soil function which include plant productivity, hydrologic control, carbon sequestration, biodiversity, and water quality. In the context of intensive forest management for energy, wood, and fiber, the productive function is the primary concern of most foresters. However, the expectation of most industrial societies is that forest land management will have a positive influence on the amount and quality of surface- and ground-water, and promote carbon capture and biodiversity. Soils must be managed simultaneously with tree stands in order to meet these expectations.

5.2.1   Soil composition

Soils consist of several layers called horizons. These layers are composed of aggregates called peds that are made up of mineral and organic particles of various sizes. Soil particles and larger aggregates are often coated with colloidal-sized particles of crystalline clays, amorphous iron and aluminum hydrous oxides, precipitated salts, and organic matter polymers. Approximately 50% of soil volume consists of pore space of various shapes and sizes. As soils become wet or dry, different proportions of the pore space are filled with water and air. Soils are occupied by a myriad of plants and animals ranging in size from bacteria and other one-celled organisms to large tree roots, small mammals and reptiles. The composition and structure of soil determine the extent and rate of its dynamic properties, including water and solute movement and retention, gas exchange, mineral weathering, organic matter decomposition and mineralization, shear strength resistance and failure, nutrient and energy transfers, erosion, oxidation/reduction, and root growth and proliferation.

The individual unit of soil known as a pedon is usually defined horizontally by an area of $1m^2$ at the surface and vertically by soil depth. A group of similar adjoining pedons is a polypedon which represents a body of soil in the landscape. There are thousands of unique soil types which differ in terms of physical, chemical and biological characteristics determined by age, parent material, surface relief, climate, and the organisms that live in them. In the USA over 17,000 different soil types have been recognized; the number defined throughout the world has not been determined (Boul 1995).

Because of the large number of soil types, most countries use a hierarchal classification system based on similarities and differences. In an attempt to provide a universal system for recognition of soil types, FAO and UNESCO produced a soil map of the world based on a common taxonomy. A biocategorical system was developed from criteria used in several countries (Dudal 1974). Soil names in the highest-order category are listed in Table 5.2-1.

Table 5.2-1. Major soil groups of the world (FAO/UNESCO classification system)

1. Arenosols - formed from sand	10. Phaeozems - dark surface, leached
2. Andosols - dark, from volcanic ash	
3. Combisols - light color	11. Planosols - abrupt A-B contact
4. Chernozems - black, high humus under prairie	12. Rankers - thin soil over sand
	13. Regosols - thin soil over unconsolidated material
5. Fluvisols - water deposited	
6. Gleysols - mottled or reduced horizons	14. Rendzinas - shallow soil over limestone
7. Greyzems - dark surface, bleached E, fine textured B	15. Solonchaks - salt accumulation
	16. Solonetz - high sodium
8. Kastanozems - brown color, steppe vegetation	17. Vertisols - inverting soils rich in 2:1 clays
9. Luvisols - high base status, high clay in B	18. Xerosols - dry, semiarid
	19. Yermosols - desert soils

Individual soil types are identified on the landscape, mapped, and described in terms of various uses or functions. For soils which support forests used for wood, fiber, and bioenergy production, interest centres on productivity, trafficability, response to silviculture, and limitations or fragility. Understanding the silvicultural potential of a soil from its character, extent, productivity, and limitations is essential if soil management is to be included in harvesting and silvicultural prescriptions. Synchronization of management options with soil capability is fundamental to soil sustainability.

5.2.2   Soil function

From a human perspective, soil function refers to the role soils play in the provision of products and services. Soil has three major roles: (1) A medium for plant growth. (2) A water-transmitting mantle. (3) An ecosystem component (Stone 1975). As a medium for plant growth, soil supports the physical structure of plants, controls root and rhizosphere temperatures, receives, accepts, retains and releases water and provides a source of essential nutrient elements (Boul 1995). As a water-transmitting mantle, it absorbs rainfall and determines the rate and quality of water flow from a watershed. As an ecosystem component, it has a profound influence on energy flows, the cycling of chemical elements, the rate of organic matter decomposition, the amount of carbon stored, and the number and diversity of organisms. Soil as a medium for tree growth, or *forest soil productivity*, will be examined further in Section 5.2.3. The influence of soil on forest hydrology and biodiversity is covered in Sections 5.3 and 5.4.

5.2.3   Forest soil productivity: foundations and principles

Forest soil productivity is the capacity of a soil to contribute to forest biomass production. This capacity depends on its physical, chemical and biological qualities. It is one component of overall forest productivity, which is a function of many other site factors, as well as tree genetic potential, and silviculture. Soil productivity can be estimated by measuring net primary production per unit area per unit time. Foresters usually estimate it from measurements of standing above-ground biomass per hectare at a given tree age. Above-ground biomass at one point in time is not an absolute measure of net primary production because it does not include below-ground production or biomass lost to herbivory or mortality.

Overall forest productivity is a function of genetic potential, silvicultural inputs, and site factors that include climate, topography and soil, and can be described in terms of biomass accumulation as a function of time (Burger 1996). The relationship in the following equation is illustrated in Figure 5.2-2.

dP/dt=bP (K-P)/K

   Where:

   P = production, or tree growth;

   b = a constant;

   K = carrying capacity, or soil productivity.

Production increases exponentially with time during the early part of the forest cycle when tree growth is not severely resource-limited (*dP/dt=bP* below the point of inflection). Later there is a logistic decrease (*1-(K-P)/K*) as the supply of light, water and nutrient resources is exploited in an increasingly-occupied space. A maximum production level (*P_max*), the limit to biomass accumulation, is equal to the carrying capacity of the site (*K*), which is largely dependent on soil potential or productivity.

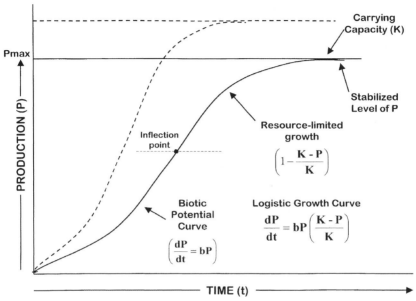

Figure 5.2-2. Forest biomass production curve. Broken lines show the potential for increasing productivity by increasing soil carrying capacity and by shortening the time required to reach carrying capacity.

Silviculturists increase biomass productivity in three ways: (1) Increasing the biotic potential of the trees *bP* (e.g. by use of genetically-improved tree stock). (2) Increasing the growth rate of the trees (e.g. by fertilizer application, weed control). This increases the slope of the curve above the point of inflection. (3) Increasing site carrying capacity *K* (e.g. by soil tillage, drainage). Genetic improvement and silvicultural manipulation of light, water, and nutrient resources effectively shorten the time required to reach site carrying capacity. Improvement of soil quality increases the site carrying capacity. The dashed line in Figure 5.2-2 shows how the combination of these practices can improve forest productivity. To take a hypothetical example: the increasing part of the growth curve, *bP*, below the point of inflection, will rise more rapidly if an improved genotype is used; in the dashed curve the growth rate will decrease less rapidly if weed control shifts growth-limiting water, nutrient, and light resources from competing plants to crop trees; maximum production or site carrying capacity (*P_max*) will be raised by modification of soil physical and chemical properties that otherwise limit biomass production.

Management of forest sites for plantations and short rotation woody crops usually includes some combination of site clearing, burning, soil tillage, bedding, drainage, fertilizer application, weed control, and thinning to shorten the growth cycle and increase site carrying capacity. A wealth of research and published management guidelines demonstrates and describes techniques and short-term benefits of these forest management practices (Nambiar and Brown 1997, Evans 1992, Duryea and Dougherty 1991, Bowen and Nambiar 1984). Practices that modify soil productivity or site carrying capacity are reviewed briefly below.

Use of forestry practices to enhance soil productivity

A natural spectrum of soil productivity results from different combinations of physical and chemical soil properties. Some of these properties can be modified, and some are fixed. For example, soil depth, texture, position on slope, and parent material all influence productivity, but cannot be changed by forest management. On the other hand, soil structure, density, organic matter content, water table, drainage, aeration porosity, nutrient content, and temperature can be modified to improve soil productivity and site carrying capacity.

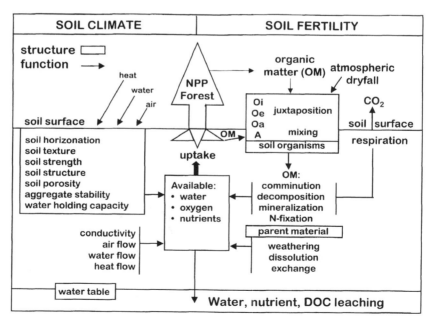

Figure 5.2-3. Conceptual model of soil properties and processes influencing soil productivity.

Two important soil attributes (soil productivity master variables) that confer productive potential are *soil climate* and *soil fertility*. Soil climate includes the amounts and balance of soil water, air, and heat and is largely controlled by soil architecture. Soil fertility includes all factors that influence the total amount, availability, and concentration in solution of essential plants nutrients. Figure 5.2-3 depicts soil structural and functional components that control the flow of water, heat, gases, nutrients, and energy. Forestry practices influence the majority of these properties and processes.

Soil climate

Land drainage, soil tillage, soil scalping and fire are forestry practices that affect soil climate. Large plantations of *Pinus radiata* in Australia and New Zealand, *P. patula* in South Africa, *Eucalyptus* spp. in Brazil, *P. rubra* and *P. banksiana* in Canada, *Picea abies* in Europe, and *Pinus taeda* and *P. resinosa* in the USA are examples of forests that were established in conjunction with some combination of land clearing, soil tillage, scalping or soil drainage. Soil productivity is modified by changes in soil architecture and physical properties that influence soil climate (water, air, and heat balance).

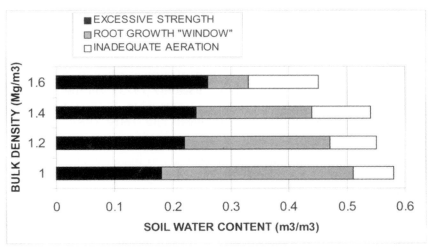

Figure 5.2-4. Soil strength and aeration limits for root growth as a function of soil bulk density and water content (Reicosky et al. 1981; Childs et al. 1989).

Several of the complex interactions involving soil physical properties, soil climate, and root growth potential are depicted in Figure 5.2-4. The extent of the lowest, longest bar represents total pore space in a soil with a bulk density *(D$_b$)* of 1.0 Mg m^{-3}. If totally saturated, this soil would have a volumetric water content *(θ)* of 58%. The extent of the black left hand portion of the bars depicts the θ below which root growth is limited due to excessive soil strength; the shaded portion depicts the range of θ where root growth is possible; and the open portion depicts the range of θ where root growth is limited by inadequate aeration. As soil bulk density increases, perhaps as a result of some management activity, the favorable range of θ, or the 'root growth window', decreases. Therefore, as the soil wets and dries during the growing season, the time over which θ resides within the growth window is shorter at high D_b (narrow window of root-growth opportunity), than at low D_b (wide window of root-growth opportunity). Some soils have a naturally high D_b which can be lowered by subsoiling, bedding, or plowing. These practices decrease soil strength, increase water holding capacity and aeration, and increase root growth potential (Terry and Campbell 1984, Kelting *et al.* 2000).

Soil temperature is an important component of soil climate and is greatly influenced by forest practices (Nambiar *et al.* 1980). Land clearing, tillage, and scalping practices that remove organic layers and expose mineral soil result in higher soil surface temperatures. Increased soil temperature due to scalping can stimulate root and shoot growth at high latitudes and elevations where organic layers insulate the soil (Bassman 1989). On the other hand, retention of organic layers or use of mulches is needed to reduce high diurnal soil temperature if these cause damage to tree seedlings (Lal *et al.* 1980).

The balance between water and air affects root respiration and water and nutrient uptake (McKee and Shoulders 1970). Water/air balance can be manipulated by a number of practices which improve the soil climate and thus tree growth. Clearcut harvesting in marginal wetlands or low-elevation sites often raises the water table which may remain at the soil surface during the dormant season. Bedding (mounding of soil in rows 30-50 cm high) aerates soil in which to plant seedlings (Schultz 1973). As the trees grow, water uptake and transpiration lower the water table. Soil temperature and soil fertility are higher in the beds (Duncan and Terry 1983, Burger and Pritchett 1984).

One of the most effective single practices used to increase soil productivity through soil climate change is drainage by ditching. Today, strict regulations limit land drainage in many countries. In the USA, prior to 1980, approximately 3 million hectares along the Atlantic and Gulf coasts were drained and planted with pine trees. These plantations are now some of the most productive forests in the USA (Allen and Campbell 1988). Drainage was accomplished with a system of open ditches that lowered the water table and improved the water/air balance of the soils (Terry and Hughes 1975, Duncan and Terry 1983).

Soil fertility

Soil fertility is determined by the nature of the soil parent material (e.g. acid shale, calcareous limestone, marine sand dunes, river alluvium, etc.), organic matter content and decomposition rate, reaction (acidity or alkalinity), colloid type and amount, and interactions with vegetation and hydrologic systems. Broad spectra of soil fertility may exist within a geographic region due to different combinations of these factors. When fertility limits forest growth, it is usually due to factors that limit the availability of plant nutrients, particularly nitrogen (N), phosphorus (P), the base cations including calcium (Ca), magnesium (Mg) and potassium (K), and one or more micronutrients such as boron (B), manganese (Mn), or zinc (Zn).

The two nutrient elements that most often limit growth rate in forests are N and P. Because decomposing organic matter is a major source of soil N, deficiencies occur when decomposition processes are limited by soil climate, or when soil organic matter content is low (Keenan *et al.* 1993, Vitousek and Matson 1985). Phosphorus deficiency is usually related to the inherently low P content of some parent materials or to high levels of iron and aluminum oxides that render P unavailable to plants (Pritchett and Smith 1974). Base cations (Ca, Mg, and K) are deficient in some soils, especially those that weather slowly, or receive large amounts of acid precipitation, or are intensively harvested (Adams 1999). Micronutrients limit the growth of forests on soils associated with certain geologic formations and parent materials.

Forest soil fertility is commonly enhanced by the addition of fertilizers containing specific plant nutrient elements. Researchers in many countries have developed site-specific diagnostic and prescription methods for correction of nutrient deficiencies. When combined with weed control, fertilizer application in plantation forests is a silvicultural practice that can greatly increase forest productivity. The addition of fertilizer N, especially during or after canopy closure, is a practice that reduces the time required to reach site carrying capacity (Figure 5.2-2). Plant-available nitrogen ions are mobile element in plant/soil systems; when added in fertilizer, only a portion is captured by the plant/soil system, and the effect is short-lived (3 to 8 years) (Ballard 1984). Repeated applications are needed to synchronize supply and demand during the growth cycle. Phosphorus compounds are less mobile; even relatively small additions of fertilizer P combine with soil minerals to alleviate plant P deficiency for decades. Due to its long-term effect, P fertilizer application is a practice that increases site carrying capacity (Figure 5.2-2). Potassium is applied less frequently than N and P, but with increases in harvesting intensity it will probably be used more regularly on some soils. The reaction, availability and longevity of K in soils depend on the amount and nature of soil colloids present. There are few reported incidences of Ca deficiency. An imbalance of the K:Mg ratio is considered to be the cause of a condition in radiata pine called 'upper mid-crown yellowing' (Beets *et al.* 1993). Intensive harvesting is likely to exacerbate this condition. Calcium and Mg are usually applied to forest soils as crushed dolomite, especially where the soil reaction is acidic. Micronutrients can be added in small amounts using balanced or specialized fertilizers. For example, B deficiency causing stem and leader malformation occurs in conifer plantations in several parts of the world. Plantations at risk are treated prophylactically with slowly-soluble B sources (Hunter *et al.* 1990). In the past 20 years, diagnosis of tree nutrient deficiencies and prescription of remedies have been greatly improved. The main constraint to maintenance of forest soil fertility is the cost of materials and their application.

5.2.4   Negative side-effects of forestry practices on soil productivity

A great deal is known about practices that improve soil productivity in the short-term (Ballard and Gessel 1983), but less is known about their long-term effects on site carrying capacity or soil productivity (Squire *et al.* 1985, Morris and Miller 1994). Intensive forest management may have unintended side effects on soil quality that decrease soil productivity. Low-quality sites, when intensively managed for high productivity, are usually most susceptible to productivity decline. Good sites are also susceptible if management does not fully compensate for negative effects of harvesting and site preparation (Dyck and Skinner 1990). Reduction of soil productivity is usually due to soil erosion; organic matter and nutrient depletion caused by tillage, fire, and biomass removal; deterioration of soil structure and increased soil strength associated with compaction by heavy machinery; inadequate soil drainage, soil puddling and loss of macroporosity again often caused by heavy machinery; and increased acidity and salinity associated with base depletion and water table modification after biomass removal. Loss of soil organic matter and nutrient depletion are of special concern in management systems associated with whole-tree harvesting for bioenergy.

Soil organic matter depletion

Soil organic matter (SOM) is a vital component of productive soils (Chen and Aviad 1990). It serves as a food and energy source for beneficial soil organisms, moderates soil temperatures, and increases water infiltration and soil water-holding capacity. It is an important source of plant nutrients, especially N and P, and it has a major influence on soil structure, aggregate stability, macroporosity, and the water/air balance, all of which affect the growth of roots and soil organisms. It is replenished in soils by continuous inputs of plant and animal detritus, and it is maintained through complex biotic decomposition processes regulated by substrate quality, soil moisture and temperature. In natural systems, it reaches a quasi-equilibrium level commensurate with characteristics of the vegetation and climate that control the rates of input and decomposition (Paul and Clark 1989).

As a result of agricultural systems practised in North America, the inherent fertility of forest and grassland soils that were converted to grain crop production has declined due to inadequate return of organic matter in the form of crop residues or animal manure (Doran and Smith 1987). Soil organic matter levels declined by 40-60%, erosion losses increased, and net mineralization of soil organic nitrogen fell below the rates needed to support sustained production (Mann 1986, Rasmussen and Collins 1991). Practices such as minimum tillage, residue management, agroforestry, use of manures and inorganic fertilizers, crop rotation and intercropping have stabilized and reversed SOM losses (Edwards *et al.* 1990). Lessons learned from agricultural experience and their implications for intensive forest management have been reviewed by Vance (2000).

The extent to which forestry operations might result in a decrease of SOM depends on the quantity of biomass removed, the amount of displacement and removal of the forest floor, and the degree to which substrate availability, soil moisture and soil temperature are modified. It also depends on the ratio of the amount of biomass removed to the amount recycled during the rotation. For example, a plantation grown on a 25-year cycle and reaching maximum leaf and root production by age 10 may return much more organic matter to the soil through litterfall, root turnover, and mortality than is removed in harvesting. The whole-tree component removed at time of harvest represents a relatively small proportion of the total biomass produced by the trees.

In a review of the literature on tree roots and root development, Bowen (1984) showed that a large proportion of net primary production is directed to roots that are continually developing and decomposing. Loblolly pine root production ($9 \text{ t ha}^{-1} \text{ yr}^{-1}$) can be 2.8 times greater than net normal wood production. More than 50% of the CO_2 assimilated by *Pinus sylvestris* can be utilized in fine root production, and this amount is twice as great as that found in fine roots at any one time. These examples illustrate the underappreciated fact that large amounts of organic matter enter the soil through root decomposition as well as in litterfall. In managed natural and plantation forests grown for solidwood products and pulp, the amount of branch and leaf biomass removed for energy production is relatively small when compared with the amount cycled from tree to soil during the life of the stand. Data from Heding and Loyche (1984) show that only 20% of the total above-ground biomass is contained in branches and leaves of 50-year-old Norway spruce (Figure 5.2-5). The proportion is usually less in pine species. Removal of biomass for energy in short rotation wood crops grown on 3 to 10-year rotations has greater potential for SOM depletion. However, short rotation crops are usually grown on deep, fertile soils, or soils treated with manures or biosolids. From an extensive review of the literature on harvesting effects, Johnson (1992) concluded that harvesting causes no overall loss of SOM unless it is followed by intense burning, mechanical disturbance, or tillage.

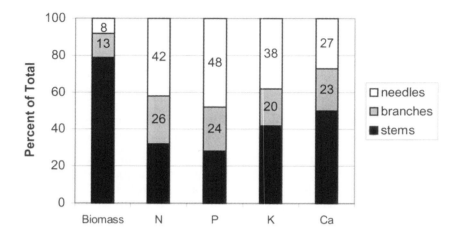

Figure 5.2-5. Distribution of biomass and nutrient elements among needles, branches, and stems in 50-year-old Norway spruce (Heding and Loyche 1984).

Fire, raking, piling, windrowing, bedding and disking are site preparation practices that directly remove or accelerate the oxidation or decomposition of soil organic matter (Burger and Pritchett 1984, Morris *et al.* 1983). Site preparation for reforestation may have a greater effect on SOM than harvesting (McColl and Powers 1984). The extent and spatial distribution of this effect depends on the intensity of the operation and the degree of soil disturbance (Balneaves *et al.* 1991).

This was demonstrated in a recent study by Cerchiaro and Burger (unpublished data), who compared SOM in loblolly pine plantations 18 years after use of different establishment techniques. Sites were prepared using six levels of slash and litter removal. In the surface 20 cm of mineral soil, SOM ranged from 20,600 kg ha^{-1} in the most intensive treatment (shear, windrow, disk harrow) to 24,000 kg ha^{-1} where no material had been removed (Figure 5.2-6). This represented a 14% decrease in SOM due to site preparation; however, there was no correlation between tree volume and SOM content. Using a similar approach, Smith *et al.* (2000) studied the response of radiata pine forest to residue management across a fertility gradient in New Zealand. Tree growth was reduced at one of the three sites where logging residues and the forest floor had been removed. The authors attributed the different growth responses to inherent soil fertility (particularly total N levels) at the three sites, all of which had very different soils and geology. Content and cycling of SOM is fundamental to soil productivity, yet well-defined relationships between SOM, tree growth, other soil properties and forest productivity have not been demonstrated (Morris and Miller 1994).

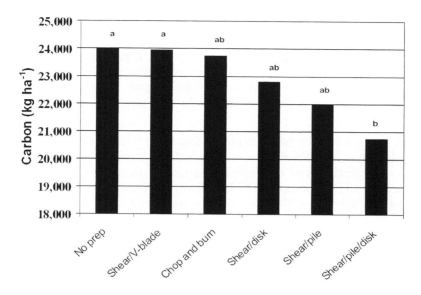

Figure 5.2-6. Influence of site preparation soil carbon content (upper 20 cm). Bars with the same letter are not significantly different at the 0.05 probability level (Duncan's Multiple Range Test).

Treatment of harvest slash and litter influences many tree growth-controlling factors including soil temperature, water and nutrient availability, soil tilth, competition from weeds, and risk of disease and pest infestation. Some responses are positive, while others are negative. Where responses are confounded with each other it is difficult to determine cause and effect (Burger 1996). Furthermore, as in the study reported by Smith *et al.* (2000), tree response to management practice is site-specific. Systematic study of a variety of site and soil types is still needed to assist the understanding of relationships between forest productivity and organic matter addition/depletion.

Nutrient depletion

The effects of intensive harvesting and reforestation practices on nutrient loss from sites and soils is site-, forest-, and treatment-specific. Nutrient loss is a function of species, stand age, season of harvest, nature of the biomass components removed, and original soil nutrient reserves. Although much is known about the influence of these factors on nutrient removal at a variety of sites (Leaf 1979, Johnson 1983), there is little information about the effects of nutrient removal on long-term soil and forest productivity (Morris and Miller 1994). Some broad generalizations can be made: (1) deciduous tree species contain more nutrients per unit of biomass than conifers, but can be harvested without their nutrient-rich leaves. (2) Whole-tree harvesting in shorter rotations removes greater amounts of nutrients (Ca may be an exception) because younger trees contain a greater proportion of nutrient-rich biomass components. (3) Whole-tree harvesting removes more nutrients than stem-only extraction due to the high nutrient content of leaves, bark, and branches (Figure 5.2-5). (4) Site preparation that removes harvesting slash, litter, and topsoil reduces the size of active nutrient pools. (5) Soil tillage increases rates of SOM decomposition and nutrient mineralization; and (6) harvesting and site preparation disturbances desynchronize nutrient supply and demand.

Use of nutrient budgets and computer simulation has led several researchers to predict that long-term intensive harvesting practices will result in soil productivity decline and non-sustainable management (Boyle and Ek 1972, Aber *et al.* 1978, Kimmins 1977). However, these methods often fail to account for nutrient replenishment processes and therefore underestimate system resilience. Soil weathering rates, N fixation rates, and the amounts of nutrients derived from lower soil layers are usually underestimated. Improved genotypes and mixed stands may utilize nutrients more efficiently. Non-target vegetation may sequester and cycle greater amounts of nutrients than previously thought. The potential for nutrient depletion is still uncertain, but it will certainly be a function of many complex and interacting soil and forest processes.

Across the spectrum of intensively-managed forests from which bioenergy may be removed, several nutrient elements may be limiting, and deficiencies can be aggravated by harvesting. Provided that physical, chemical and biological soil quality attributes are maintained, harvesting and management-induced nutrient deficiencies can be economically remedied by application of fertilizers (Jacobson *et al.* 2000, Smith *et al.* 2000, Fox 2000) or wood ash (Korphilahti *et al.* 1999). Of all the macro- and micronutrients removed from forest sites in harvested wood, N and Ca appear to be the most difficult to manage. Despite high anthropogenic inputs, N is still the major growth-limiting nutrient in most coniferous forests.

A large proportion of the N taken up by trees has been mineralized during decomposition of organic matter. Forest management operations influence the amount of total N in the soil, the rate at which it becomes available, and the amounts retained in tree biomass or lost from the site (Fox *et al.* 1986). Synchronizing N supply and demand through management of net N mineralization is difficult. Harvesting disturbances accelerate net mineralization and amounts of available N exceed those required by young stands. This means that demands after canopy closure may not be met (Burger and Pritchett 1984, Dyck and Skinner 1990).

After canopy closure, thinning is used to increase the timber value of crop trees. The amount of biomass removed in pre-commercial thinnings can be fairly large and it can be an important source of bioenergy. However, thinning may reduce site productivity due to nutrient removal (Tamm 1969).

Pre-commercial thinning coincides with the time of maximum leaf area development and maximum stand N content. Thinning releases residual trees for rapid growth at a stage when net N mineralization from the logging residues of the previous rotation has declined (Staaf 1984). Pre-commercial thinning for bioenergy therefore coincides with high N demand by residual trees and an inadequate soil N supply. In Denmark, thinning is done in winter to reduce disease problems, increase the value of the thinned trees, and to increase the amount of N left on the site. Cut trees are left on-site to dry for 4-6 months. Reduction of moisture content increases the energy value and price (see Section 4.1.8). In the process, nutrient-rich needles and small branches fall off, leaving one-half to one-third of total tree nutrients on-site (Figure 5.2-5).

Supplementing of N through fertilizer application is a difficult and costly process. Usually less than half of the total N applied is retained and growth responses are short-lived. Repeated applications are needed to meet tree demand. Ash remaining after combustion of forest biomass does not contain N. Fertilizer N requires high energy input during the manufacturing process.

There are few, if any reports of Ca deficiency associated with tree biomass removal. However, nutrient budget analysis suggests that Ca is the nutrient most likely to be depleted as a result of intensive biomass harvesting.

The fundamental question is whether or not nutrient inputs are greater than harvesting removals. Harvesting and leaching outputs, both of which are relatively easy to measure, are commonly compared with atmospheric inputs. Mann *et al.* (1988) summarized data on nutrient removal in conventional and whole-tree harvesting from eight forest stands located throughout the United States. The sites were in managed natural forests and plantations, and species were oak-hickory at Oak Ridge, Tennessee; Douglas-fir at the Pack Forest, University of Washington; spruce-fir at Weymouth Point Watersheds, Maine; oak-maple-birch at Cockaponset State Forest, Connecticut; maple-birch-beech at Mt. Success, New Hampshire; oak-poplar at Coweeta Hydrologic Lab, Otto, North Carolina; loblolly pine at Clemson University, South Carolina; and slash pine at the University of Florida.

Across all eight sites, annual amounts of N, P, K, and Ca removed in conventional harvesting were less than 3, 0.5, 2, and 5 kg ha^{-1}, respectively. Whole-tree annual harvest removals were 2 to 3 times as great as conventional harvesting removals. Annual leaching and stream losses were lower than harvesting removals and returned to pre-harvest levels within 3 years. The effects of harvesting on leaching and stream losses are therefore almost negligible on a long-term basis. Nutrient inputs were calculated by subtracting leaching outputs from precipitation inputs. No account was taken of dryfall or weathering inputs. About three-quarters of the sites showed net losses of P, K, and Ca, indicating that hydrologic inputs did not compensate for harvesting and leaching removals. There were no known nutrient deficiencies in any of the stands. The authors concluded that Ca showed the greatest potential for depletion under intensive management.

Follow-up studies in the Tennessee and New Hampshire stands showed that the amount of available Ca remained nearly constant under whole-tree harvesting (Johnson and Todd 1998). This conclusion was based on nutrient budget analyses that included estimates of weathering inputs. Nutrient budget analysis based on hydrologic fluxes alone overestimates the rate at which soil Ca will be depleted. Rock and soil weathering inputs and nutrient uptake from lower soil layers are not fully accounted for in most analyses because they are difficult to measure.

Other studies of nutrient removal in the eastern United States also found that Ca was the element most likely to be depleted by harvesting (Weetman and Webber 1972; Boyle *et al.* 1973; Silkworth and Grigal 1982; Federer *et al.* 1989). However, Ca deficiency has never been shown to limit growth in these forests. It is possible that Ca deficiency will develop in time, or it may be that all Ca inputs were not accounted for.

New warnings of base cation depletion have recently been reported. In Sweden, computer models show that removal of logging residues results in a negative balance for base cations at many forest sites (Jacobson *et al.* 2000). For whole-tree harvesting without base cation return, calculations indicate losses averaging 0.62 kg ha^{-1} yr^{-1}. There is a risk of soil cation depletion in less than one to two rotation periods throughout Sweden (Sverdrup and Rosén 1998). In the USA, Huntington *et al.* (2000) argue that southeastern forest soils are especially prone to Ca depletion due to their base-poor geologic origin, their age (several hundred thousand to 2 million years), history of abusive land-use practices, and harvest removals. They maintain that the weathering rate of non-exchangeable Ca in primary minerals within the rooting zone is insufficient to compensate for losses due to tree uptake and soil leaching. Adams *et al.* (2000) reviewed evidence showing effects of intensive harvesting and acidic deposition on budgets for base cations. They concluded that ambient levels of N and sulfur (S) deposition are leading to N and S saturation and increases in leaching of cations in some eastern US forests. It is clear that continuing research and monitoring are required to clarify the impacts of harvesting on nutrient budgets.

In summary, forest management practices are designed, in part, to improve soil productivity by mitigating specific growth-limiting factors. Unintended side effects may decrease soil productivity. Soil quality in managed forests will be determined by the balance between positive and negative influences on soil climate and fertility imparted by management during the course of the tree growth cycle.

Soil erosion and displacement

Soil erosion and displacement by machinery decrease site productivity through removal of nutrient-rich topsoil (Morris *et al.* 1983). Both can cause sedimentation in waterways and reduced water quality. Erosion of forest soils is usually caused by water flow on sloping terrain that has been cleared of vegetation and the organic layers above the mineral soil. Prior to the development of herbicides for forestry use, weeds were commonly controlled by tillage. Control of unwanted vegetation with chemicals has reduced the need for tillage and minimized the potential for soil erosion. Maintenance of an intact forest floor layer across the site helps to prevent soil erosion, and orientation of skid trails and site preparation operations along land contours minimizes or eliminates this soil degrading process (Dissmeyer and Foster 1981).

Soil compaction

Soil productivity is strongly affected by the degree to which roots can grow unimpeded to exploit soil resources. Machinery used in logging and site preparation operations can compact soils, increasing their bulk density and strength, and reducing soil porosity, aeration, root growth and productivity (Morris and Lowery 1988, Sands 1982). Soils are especially vulnerable to compaction when θ is near or at field capacity (the θ after a soil has drained). When soils are nearly saturated they approach their liquid limit (the θ at which soils flow under traffic pressure). In this case they are especially vulnerable to puddling. Puddling occurs when soil structure collapses, and clay particles become oriented. Puddled soils drain slowly and are poorly aerated (Miwa *et al.* 2000).

5.2.5   Soil-specific application of forest practices

An understanding of the nature of soils and how they respond to various practices is crucial to sustainable management. Soil productivity varies greatly across landscapes according to parent material, landscape relief, climate, vegetation, and past land-use history. Detailed soil classification systems have been developed in most industrialized countries. Each soil type has a specific combination of physical, chemical, and biological properties that create different levels of soil climate and fertility and thus influence productivity. The extent to which soil productivity is enhanced by forest management practices, and the degree to which a soil is susceptible to the negative side effects of these practices is a function of the unique properties of each soil type.

Responses of specific soil orders (groups of soil types) to forestry practices are illustrated in Figure 5.2-7. The diagram shows sustainable soil productivity as a function of change in two master variables, *soil fertility* and *soil climate*. Soil fertility on the y-axis extends from depleted to conserved, and soil climate on the x-axis from imbalanced to balanced. A soil productivity gradient increases from lower left to upper right, the highest level of being at the point where soil fertility is conserved and soil climate balanced. The response of some soil types is largely influenced by fertility (Entisols), in others it is governed by soil climate (Histosols). Responses may be influenced equally by fertility and soil climate (Ultisols, Inceptisols). This is indicated by the area occupied by each soil order. Within a soil order, the productivity of some soil types is at high risk from management-induced change (e.g. sand hill Entisols). For others (e.g. river bottom Entisols receiving an annual sediment load) this risk is low. Within the response surface for a given soil order, e.g. Entisols, high-risk soils are less resistant and resilient and thus have more potential for change (long vector), while productivity of low-risk soils is difficult to change and the effects of management practice are small (short vector).

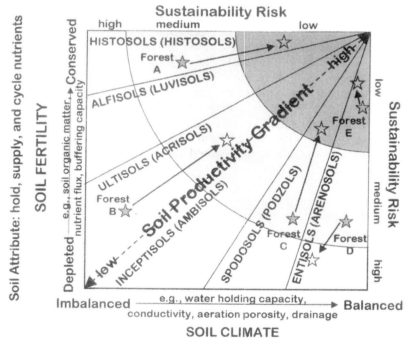

Figure 5.2-7. Forest management effects on sustainability of soil productivity in different soil groups (after Burger and Kelting 1999).

Several examples are shown; in each case, management-induced change is indicated by the direction and length of the vector. Forest A is located on deep ditched Histosols on the US Atlantic coastal plain. Productivity was increased through management (Terry and Hughes 1975). Forest B was planted on abused agricultural land in the Appalachian Piedmont that was made more productive as a result of reforestation (Cerchiaro and Burger unpublished data). Forest C is on a naturally P-deficient Spodosol made productive by P fertilizer application (Pritchett and Smith 1974). Forest D was sand pine (*Pinus clausa*) but conversion to slash pine (*P. elliottii*) using mechanical techniques, resulted in depletion of organic matter and fertility (Brendemuehl 1967). Forest E is a bottomland hardwood forest made slightly more productive as a result of harvesting which increased sediment trapping, despite serious physical soil disturbance (Aust and Lea 1992).

Soil productivity is largely determined by the balance between soil fertility and climate. Because of their unique properties, some soils are more sensitive to forest management practice than others, and some are at greater risk from damaging impacts. Productive soils are usually resistant to change, while less productive soils are more easily improved and more easily damaged.

5.2.6   Summary

The research record and years of practice suggest that it is possible to harvest biomass for energy from forests world-wide without compromising soil quality and forest productivity. Natural resources, forest practice, energy needs, and socio-economic conditions differ so widely that it is difficult to recommend specific guidelines for the sustainable management of soil and productivity. The following general points, based on the above review, are applicable to most forest environments and should be considered by forestry practitioners.

1. Forest productivity and bioenergy production depend on the type, amount, and quality of the soil on a forest site. Soils can vary greatly, even within a forest management unit. Some soils are more sensitive to management inputs than others, and specific tree growth-limiting properties can be improved by soil management. There is a fundamental need for identification and mapping of soils on any land used for intensive forestry. This will assist the prediction of productivity, responses to silvicultural practices and responses to heavy machinery use.

2. In some cases, inherent soil productivity can be increased through soil management. Regulation of water content within the surface meter, correction of nutrient deficiencies, and afforestation of abused agricultural land are examples of practices that can also increase productivity.

3. Soil organic matter content can be maintained by increasing forest productivity and by minimizing disturbance during site preparation. Higher productivity increases organic matter inputs to soils, particularly through root turnover. Herbicides, when safe and acceptable, should be used to replace fire and mechanical weed control. This will reduce

residue removal and physical disturbance, both of which accelerate organic matter decomposition.

4. Harvesting procedures should be aimed at minimizing nutrient removal from forest sites. Nutrient removals in biomass can be minimized by leaving nutrient-rich leaves and needles behind. On-site drying of forest fuel allow most of the leaves and needles to fall off. This reduces nutrient losses, increases the value of the fuel, and can reduce the incidence of diseases and pest attacks. On-site drying is especially beneficial in precommercial thinning, which is carried out at a time when tree nutrient demand is high.

5. Wood ash from the combustion of forest biomass should be returned to forest sites to counteract long-term nutrient depletion. Further work is needed to define and describe positive and potentially-negative responses to wood ash application. In the meantime, the principles and practices of wood ash application to forest sites appear to be sound.

6. After all possible nutrient conservation measures have been taken, inorganic fertilizers should be used to replace nutrients removed during harvesting in order to maintain soil fertility. Many diagnostic and prescriptive practices have been developed for major commercial tree species. Careful benefit/cost analysis should be performed before wood is harvested for energy if fertilizer application is likely to be required.

7. Nitrogen supply often limits growth in intensively managed forest systems. Practices that conserve organic matter also conserve nitrogen and should be encouraged. Herbaceous legumes and other non-competitive vegetation should be managed to add and cycle N in ways that help to synchronize N supply and tree demand.

8. Base cation depletion, particularly of Ca could become a problem in intensively-managed forests, especially those receiving high levels of acid deposition from air pollution. The combination of harvesting removal and leaching on shallow, siliceous soils with low levels of weatherable minerals could be a problem even in the short term. Conserving Ca by de-barking wood and leaving the needles at the site is a preventative measure. Additions of wood ash, lime, or gypsum may be necessary on some sites, but will be expensive.

9. Machine traffic compacts, puddles, and ruts forest soils, especially when they are wet. These physical changes have a negative influence on soil water, air, and heat balance and reduce the extent and duration of root growth. Many other processes controlled by soil fauna and flora are also affected. Traffic impacts can be minimized or eliminated if operations are confined to periods when soils are dry. Information on susceptibility of the soils involved should be considered and used in harvest planning. Specialized operations such as shovel logging and use of high-flotation equipment should be carried out when soils are moist or wet. Ripping, subsoiling, and bedding should be used to repair damaged soils.

10. Soil erosion causes severe site degradation and should be prevented. Soils on sloping sites should never be laid bare by harvesting operations. Tillage should be done along the contour, and skid trails should be repaired as soon as possible after harvest.

11. Site-specific management is essential for the maintenance of soil productivity. Each soil has unique characteristics, and different management approaches are needed if soil fertility and a favorable air-water balance are to be maintained in intensively-harvested forests.

5.2.7   References

Aber, J.D., Botkin, D.B. and Melillo, J.M. 1978. Predicting the effects of different harvesting regimes on forest floor dynamics in northern hardwoods. Canadian Journal of Forest Research 8: 308-316.

Adams, M.B. 1999. Acidic deposition and sustainable forest management in the central Appalachians, USA. Forest Ecology and Management 122: 17-28.

Adams, M.B., Burger, J.A., Jenkins, A.B. and Zelazny, L. 2000. Impact of harvesting and atmospheric pollution on nutrient depletion of eastern U.S. hardwood forests. Forest Ecology and Management 138: 301-319.

Allen, H.L. and Campbell, R.G. 1988. Wet site pine management in the southeastern United States. Pp. 173-184 *in* Hook, D.D. *et al.* (eds.). The Ecology and Management of Wetlands. Volume 2: Management, Use and Value of Wetlands. Timber Press, Portland, Oregon.

Aust, W.M. and Lea, R. 1992. Comparative effects of aerial and ground logging on soil properties in a tupelo-cypress wetland. Forest Ecology and Management 50: 57-73.

Ballard, R. 1984. Fertilization of plantations. Pp. 327-360 *in* Bowen, G.D. and Nambiar, E.K.S. Nutrition of Plantation Forests. Academic Press, London.

Ballard, R. and Gessel, S.P. 1983. IUFRO Symposium on Forest Site and Continuous Productivity. USDA Forest Service General Technical Report PNW-163.

Balneaves, J.M., Skinner, M.F. and Lowe, A.T. 1991. Improving the re-establishment of radiata pine on impoverished soils. Pp. 137-150 *in* Dyck, W.J. and Mees, C.A. (eds.). Long-Term Field Trials to Assess Environmental Impacts of Harvesting. IEA/BE T6/A6 Report No. 5. Forest Research Institute Bulletin No. 161. Rotorua, New Zealand.

Bassman, J.H. 1989. Influence of two site preparation treatments on ecophysiology of planted *Picea engelmannii* x *glauca* seedlings. Canadian Journal of Forest Research 19: 1359-1370.

Beets, P.N., Payn, T.W. and Jokela, E.J. 1993. Upper mid-crown yellowing (UMCY) in *Pinus radiata* forests. New Zealand Forestry 38: 24-28.

Boul, S.W. 1995. Sustainability of soil use. Annual Review of Ecological Systems 26: 25-44.

Bowen, G.D. 1984. Tree roots and the use of soil nutrients. Pp. 147-179 *in* Bowen, G.D. and Nambiar, E.K.S. (eds.). Nutrition of Plantation Forests. Academic Press, New York.

Bowen, G.D. and Nambiar, E.K.S. (eds.). 1984. Nutrition of Plantation Forests. Academic Press, London.

Boyle, J., Phillips, J. and Ek, A.R. 1973. Whole-tree harvesting: nutrient budget evaluation. Journal of Forestry 71: 760-762.

Boyle, J.R. and Ek, A.R. 1972. An evaluation of some effects of bole and branch pulpwood harvesting on site measurements. Canadian Journal of Forest Research 2: 407-412.

Brendemuehl, R.H. 1967. Options for management of sandhill forest land. Southern Journal of Applied Forestry 5: 216-222.

Burger, J.A. 1996. Limitations of bioassays for monitoring forest soil productivity: Rationale and example. Soil Science Society of America Journal 60: 1674-1678.

Burger, J.A. 1997. Conceptual framework for monitoring the impacts of intensive forest management on sustainable forestry. Pp. 147-156 *in* Hakkila, P., Heiro, M. and Puranen, E. (eds.). Forest Management for Bioenergy. The Finnish Forest Research Institute Research Paper 640.

Burger, J.A. and Kelting, D.L. 1999. Using soil quality indicators to assess forest stand management. Forest Ecology and Management 122: 167-185.

Burger, J.A. and Pritchett, W.L. 1984. Effects of clearfelling and site preparation on nitrogen mineralization in a southern pine stand. Soil Science Society of America Journal 48: 1432-1437.

Carmean, W.H. 1975. Forest site quality evaluation in the United States. Advances in Agronomy 27: 209-269.

Chen, Y. and Aviad, T. 1990. Effects of humic substances on plant growth. Pp. 161-186 *in* McCarthy, C.E., Malcolm, R.L. and Bloom, P.R. (eds.). Humic Substances in Soil and Crop Sciences: Selected Readings. American Society of Agronomy, Madison, Wisconsin.

Childs, S.W., Shade, S.P., Miles, W.R., Shepard, E. and Froelich, H.A. 1989. Soil physical properties: importance to long-term forest productivity. Pp. 53-66 *in* Perry, D.A. *et al.* (eds.). Maintaining the Long-Term Productivity of Pacific Northwest Forest Ecosystems. Timber Press, Portland, Oregon.

Dissmeyer, G.E. and Foster, G.R. 1981. Estimating the cover-management factor (C) in the universal soil loss equation for forest conditions. Journal of Soil Water Conservation 36: 235-240.

Doran, J.W. and Smith, M.S. 1987. Organic matter management and utilization of soil and fertilizer nutrients. Pp. 53-72 *in* Mortredt, J.J. and Buxton, D.R. (eds.). Soil Fertility and Organic Matter as Critical Components of Production Systems. Soil Science Society of America Special Publication No. 19. American Society of Agronomy Inc. Publishers, Madison, WI.

Dudal, R. 1974. Key to soil units for the soil map of the world. Volume 1. Legend. UNESCO. FAO, Rome.

Duncan, D.V. and Terry, T.A. 1983. Water management. Pp. 91-111 *in* Stone, E.L. (ed.). Managed Slash Pine Ecosystem Symposium Proceedings. School of Forest Resources and Conservation, University of Florida, Gainesville.

Duryea, M.L. and Dougherty, P.M. 1991. Forest Regeneration Manual. Kluwer Academic Publishers, Boston, Massachusetts. 433 p.

Dyck, W.J. and Skinner, M.F. 1990. Potential for productivity decline in New Zealand radiata pine forests. Pp. 318-332 *in* Gessel, S.P., Lacate, D.W., Weetman, G.F. and Powers, R.F. Sustained Productivity of Forest Soils. University of British Columbia, Faculty of Forestry Publication, Vancouver, British Columbia, Canada.

Edwards, C.A., Lal, R., Madden, P., Miller, R.H. and Horst, G. 1990. Sustainable Agricultural Systems. Soil and Water Conservation Society, Ankeny, IA.

Evans, J. 1992. Plantation Forestry in the Tropics. Second edition, Clarendon Press, Oxford, England. 403 p.

Federer, C.A., Hornbeck, J.W., Tritton, L.M., Martin, C.W., Pierce, R.S. and Smith, C.T. 1989. Long-term depletion of calcium and other nutrients in eastern U.S. forests. Environmental Management 18: 593-601.

Fox, T.R. 2000. Sustained productivity in intensively managed forest plantations. Forest Ecology and Management 138: 187-202.

Fox, T.R., Burger, J.A. and Kreh, R.E. 1986. Effects of site preparation on nitrogen dynamics in the southern Piedmont. Forest Ecology and Management 15: 241-256.

Heding, N. and Loyche, M. 1984. Volume and nutrient content of Norway spruce needles. Dansk Skovforenings Tidsskrift. Zhefte, August. Dansk Skovforening, Kobenhavn.

Hunter, I.R., Will, G.M. and Skinner, M.F. 1990. A strategy for the correction of boron deficiency in radiata pine plantations in New Zealand. Forest Ecology and Management 37: 77-82.

Huntington, T.G., Hooper, R.P., Johnson, C.E., Aulenbach, B.T., Cappellato, R. and Blum, A.E. 2000. Calcium depletion in a southeastern United States forest ecosystem. Soil Science Society of America 64: 1845-1858.

Jacobson, S., Kukkola, M., Malkonen, E. and Tveite, B. 2000. Impact of whole-tree harvesting and compensatory fertilization on growth of coniferous thinning stands. Forest Ecology and Management 129: 41-51.

Johnson, D.W. 1983. The effects of harvesting intensity on nutrient depletion in forests. Pp. 157-166 *in* Ballard, R. and Gessel, S.P. (eds.). IUFRO Symposium on Forest Site and Continuous Productivity. USDA Forest Service General Technical Report PNW-163.

Johnson, D.W. 1992. The effects of forest management on soil carbon storage. Water, Air, and Soil Pollution 64: 83-120.

Johnson, D.W. and Todd, D.E. Jr. 1998. Harvesting effects on long-term changes in nutrient pools of mixed oak forest. Soil Science Society of America Journal 62: 1725-1735.

Keenan, R.J., Prescott, C.E. and Kimmins, J.P. 1993. Mass and nutrient content of the forest floor and woody debris in western redcedar and western hemlock forests on northern Vancouver Island. Canadian Journal of Forest Research 23: 1052-1059.

Kelting, D.L., Burger, J.A. and Patterson, S.C. 2000. Early loblolly pine growth response to changes in the soil environment. New Zealand Journal of Forestry Science 30(1/2): 206-224.

Kimmins, J.P. 1977. Evaluation of the consequences for future tree productivity of the loss of nutrients in whole-tree harvesting. Forest Ecology and Management 1: 169-183.

Korphilahti, A., Moilanen, M. and Finer, L. 1999. Wood ash recycling and environmental impacts: state-of-the-art in Finland. Pp. 82-89 *in* A.T. Lowe and C.T. Smith (eds.). Developing systems for integrating bioenergy into environmentally sustainable forestry. Forest Research Bulletin No. 211. New Zealand Forest Research Institute, Rotorua, New Zealand.

Lal, R., DeVleeschauwer, D. and Malafa, N.R. 1980. Changes in properties of a newly cleared tropical Alfisol as affected by mulching. Soil Science Society of America Journal 44: 829-833.

Leaf, A.L. (ed.). 1979. Impact of intensive harvesting on forest nutrient cycling. State University of New York, Syracuse. 421 p.

Mann, L.K. 1986. Changes in soil carbon storage after cultivation. Soil Science 142: 279-288.

Mann, L.K., Johnson, D.W., West, D.C., Cole, D.W., Hornbeck, J.W., Martin, C.W., Reikerk, H., Smith, C.T., Swank, W.T., Tritton, L.M. and Van Lear, D.H. 1988. Effects of whole-tree and stem-only clearcutting on post-harvest hydrologic losses, nutrient capital, and regrowth. Forest Science 42: 412-428.

McColl, J.G. and Powers, R.F. 1984. Consequences of forest management soil-tree relationships. Pp. 379-412 *in* Bowen, C.D. and Nambiar, E.K.S. (eds.). Nutrition of Plantation Forests. Academic Press, London.

McKee, W.H. and Shoulders, E. 1970. Depth of water table and redox potential of soil affect slash pine growth. Forest Science 16: 399-402.

Miwa, M., Aust, W.M., Burger, J.A. and Patterson, S.C. 2000. Morphological and physical characterization of disturbed forest soils in the lower coastal plain of South Carolina. Soil Science Society of America Journal *(in press).*

Morris, L.A. and Lowery, R.F. 1988. Influence of site preparation on soil conditions affecting seedling establishment and early growth. Southern Journal of Applied Forestry 12: 170-178.

Morris, L.A. and Miller, R.E. 1994. Evidence for long-term productivity change as provided by field trials. Pp. 41-80 *in* Dyck, W.J., Cole, D.W. and Comerford, N.B. Impacts of Forest Harvesting on Long-Term Site Productivity. Chapman & Hall, London.

Morris, L.A., Pritchett, W.L. and Swindel, B.F. 1983. Displacement of nutrients into windrows during site preparation of a flatwoods forest. Soil Science Society of America Journal 47: 591-594.

Nambiar, E.K.S. 1996. Sustained productivity of forests is a continuing challenge to soil science. Soil Science Society of America Journal 60: 1629-1642.

Nambiar, E.K.S. and Brown, A.G. (eds.). 1997. Management of Soil, Nutrients, and Water in Tropical Plantation Forests. ACIAR Monograph No. 43. Canberra, Australia. 571 p.

Nambiar, E.K.S., Bowen, G.D. and Sands, R. 1980. Root regeneration and plant water status of *Pinus radiata* D. Don seedlings transplanted to different soil temperatures. Journal of Experimental Botany 30: 1119-1131.

Paul, E.A. and Clark, F.E. 1989. Soil Microbiology and Biochemistry. Academic Press, Inc., San Diego, California.

Pritchett, W.L. and Smith, W.H. 1974. Management of wet savanna forest soils for pine production. Florida Agricultural Experiment Station Technical Bulletin 762. Florida Agricultural Experiment Station, Gainesville.

Rasmussen, P.E. and Collins, H.P. 1991. Long-term impacts of tillage, fertilizer, and crop residue on soil organic matter in temperate semiarid regions. Advances in Agronomy 45: 93-134.

Reicosky, D.C, Voorhees, W.V. and Radke, J.K. 1981. Unsaturated water flow through a simulated wheel track. Soil Science Society of America Journal 45: 3-8.

Sands, R. 1982. Physical changes to sandy soils planted to radiata pine. Pp. 146-152 *in* Ballard, R. and Gessel, S.P. (eds.). IUFRO Symposium on Forest Site and Continuous Productivity. USDA Forest Service General Technical Report PNW-163.

Schultz, R.D. 1973. Site treatment and planting method alter root development of slash pine. USDA Forest Service Southeastern Forest Experiment Station, Asheville, North Carolina. Research Paper SE-109. 11 p.

Silkworth, D.R. and Grigal, D.F. 1982. Determining and evaluating nutrient losses following whole-tree harvesting of aspen. Soil Science Society of America Journal 46: 626-631.

Smith, C.T, Lowe, A.T., Skinner, M.F., Beets, P.N., Schoenholtz, S.H. and Fang, S. 2000. Response of radiata pine forests to residue management and fertilization across a fertility gradient in New Zealand. Forest Ecology and Management 138: 203-223.

Squire, R.D., Farrell, P.W., Flinn, D.W. and Aeberli, B.C. 1985. Productivity of first and second rotation stands of radiata pine on sandy soils. Australian Forestry 48: 127-137.

Staaf, H. 1984. Harvesting of logging slash on highly productive sites. Buffering of negative effects. Pp. 96-97 *in* Andersson, B. and Falk, S. (eds.). Forest Energy in Sweden. Swedish University of Agricultural Sciences, Uppsala.

Stone, E.A. 1975. Soil and man's use of forest land. Pp. 1-10 *in* Bernier, B. and Winget, C.H. (eds.). Forest Soils and Forest Land Management. Proc., Fourth North American Forest Soils Conference, Laval University, Quebec, Canada.

Sumner, M.E. and Wilding, L.P. (eds.). 2000. Handbook of Soil Science. CRC Press, New York.

Sverdrup, H. and Rosén, K. 1998. Long-term base cation mass balances for Swedish forests and the concept of sustainability. Forest Ecology and Management 110: 221-236.

Tamm, C.D. 1969. Site damages by thinning due to removal of organic matter and plant nutrients. Thinning and mechanization. Pp. 175-177 *in* Proc. IUFRO Meeting, Stockholm.

Terry, T.A. and Campbell, R.G. 1984. Soil management considerations in intensive forest management. Pp. 98-106 *in* Forest Regeneration: Proc. Symposium on Engineering Systems for Forest Regeneration. American Society of Agricultural Engineers, St. Joseph, Michigan.

Terry, T.A. and Hughes, J.H. 1975. The effects of intensive management on planted loblolly pine growth on poorly drained soils of the Atlantic Coastal Plain. Pp. 351-377 *in* Bernier, B. and Winget, C.H. (eds.). Forest Soils and Forest Land Management. Proceedings, Fourth North American Forest Soils Conference, Laval University, Quebec, Canada.

Vance, E.D. 2000. Agricultural site productivity: principles derived from long-term experiments and their implications for intensively-managed forests. Forest Ecology and Management 138: 369-396.

Vitousek, P.M. and Matson, P.A. 1985. Disturbance, nitrogen availability, and nitrogen losses in an intensively managed loblolly pine plantation. Ecology 66: 1361-1376.

Weetman, G.L. and Webber, B. 1972. The influence of wood harvesting on the nutrient status of two spruce stands. Canadian Journal of Forest Research 2: 351-369.

5.3   HYDROLOGIC VALUES

D. G. Neary

The production of energy from forests is a cyclical process that places environmental burdens and impacts on the hydrologic system at various stages (Mälkki *et al.* 2001). Traditionally, most concern has been focused on forest harvesting and site preparation operations. Their effects are transitory, but the road network that supports harvesting operations is largely permanent. Although roads have a significant effect on the routing of water in a watershed and are a major source of sediment, they are included with harvesting-related disturbance in this analysis. Industrial installations for processing conventionally harvested wood as well as their storage facilities, can make significant, localized, and permanent impacts on water resources. Although industrial sites are an important component of the forest bioenergy production cycle, they are not considered here.

5.3.1   Hydrologic processes affected by harvesting

The hydrologic cycle consists of interactions between the atmosphere, geosphere, biosphere, and hydrosphere (Figure 5.3-1). Water is a primary driving force in ecosystem processes and fluxes. It is a necessity of life in all ecosystems. Water quality and quantity are affected by processes and disturbances occurring in watersheds. The quantity and quality of water emanating from watersheds can be used to assess ecosystem condition. The hydrologic cycle can be represented by a simple equation (Brooks *et al.* 1997):

$I = O + dS$

   Where:

   I = *Water input;*

   O = *Water output;*

   dS = *Change in amount of water stored.*

Principal outputs of water from undisturbed watersheds are streamflow (4-58% of the total), evaporation and transpiration (often referred to in combination as evapotranspiration or ET). Storage includes temporary water retention in depressions or bodies of water not connected to stream channels, and losses to soil and geologic formations. Where ET is high due to aridity, latitude, altitude, or wind, streamflow is minimal. Below 480 mm of precipitation, most streamflow is ephemeral, occurring only during large precipitation events or during periods of the year when ET is low (Clary *et al.* 1974). Water can be temporarily stored in the litter layer (litter holds 0.5 mm cm^{-1} depth), in soil macro- and micropores; in surficial, unconfined saprolite and porous rock formations; and in unconsolidated channel sediments (Velbel 1988). It can also be stored for long periods of time in the geosphere in deep, unconsolidated sediments, and as confined groundwater in regional aquifers (amounts vary with local geology).

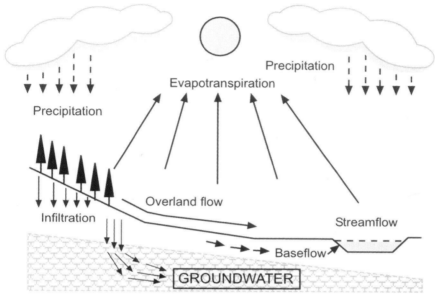

Figure 5.3-1. Components of the hydrologic cycle.

Figure 5.3-2 shows the inputs, fluxes, and outputs of water in undisturbed forested watersheds in humid regions (Hewlett 1982). Partitioning varies in arid shrub, grassland, and woodland ecosystems, and in watersheds disturbed by climate change, harvesting, burning, insect defoliation, windthrow, land-use conversion, mining, and agriculture. Precipitation inputs consist of rain, snow, and sleet. Fluxes (processes influencing water movement) are interception, evaporation, transpiration, stemflow, throughfall, infiltration, surface runoff, interflow, baseflow, and stormflow.

In forests, about 7% of total precipitation is intercepted by leaf, branch, upright stem, and woody debris surfaces. Most of this evaporates and returns to the atmosphere, but initial precipitation from a storm event is retained. This may amount to 2 mm on trees, 1 mm on shrubs, and 3 mm on litter, the amount intercepted being a function of ground cover. More water is retained from small storms than from large ones. The reduction of intercepting surfaces as a result of harvesting increases the amount of rainfall reaching the forest floor or soil surface. Variable amounts of intercepted rainfall flow over foliage, down stems, through logging slash and litter and into the soil.

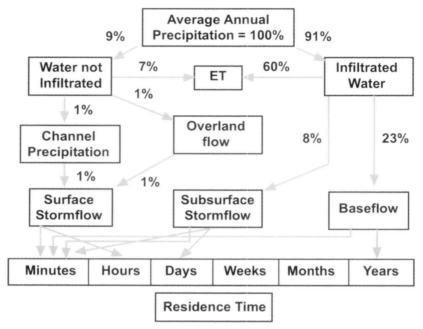

***Figure 5.3-2. Partitioning of precipitation into output components in undisturbed
forest watersheds in humid regions (adapted from Hewlett 1982).***

Throughfall is precipitation that falls through a plant canopy and lands on
the litter or the soil surface. In undisturbed forests, 91% of precipitation
passes directly into the litter and soil, only 1% flowing over the soil as
surface runoff (Figure 5.3-2). Infiltration is the process whereby water passes
through the soil surface into the soil profile. When rainfall intensity exceeds
infiltration capacity, surface runoff occurs. Rates of infiltration are a function of
soil texture, vegetation cover, litter cover, and soil porosity. Infiltration rates
can often exceed 160 mm hr^{-1} where litter is present. Infiltration rates decrease
where soil particle size is smaller or cover is reduced. Rainfall infiltration rates
can be greater than 25 mm hr^{-1} in exposed sandy soils; less than 5 mm hr^{-1} in
clay soils. Water infiltrating the soil moves vertically and laterally as interflow.
This movement can be rapid if there are large pore spaces in the soil
(macropores) but is slow through small pores (micropores). Interflow accounts
for storage of water in the soil mantle and for baseflow that supports the
perennial flow of streams between storm events.

Streamflow consists of baseflow between storm events and stormflow from
precipitation. Baseflow is water that has not been affected by evaporation,
transpiration or storage. Stormflow is greater than baseflow because it results
from processes such as in-channel precipitation and surface runoff that are
minimally affected by evaporation, transpiration, and soil moisture storage
capacity.

Stormflow, which occurs during and immediately after storm events, originates from precipitation which falls directly into stream channels (expanding source areas), from rapidly-moving subsurface flow in macropores, and surface runoff. Once a soil becomes saturated (moisture storage at maximum level), additional rain results in surface runoff. Water repellency, which develops in some soils after site preparation fires, greatly reduces infiltration rates and percolation capacity after rainfall (DeBano *et al.* 1998).

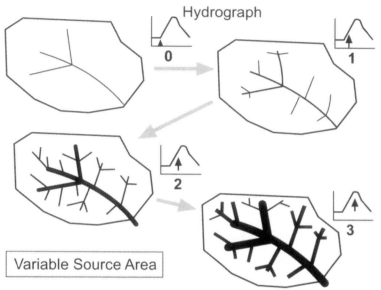

Figure 5.3-3. Variable Source Area concept of watershed and streamflow response to precipitation (after Hewlett 1982).

The Variable Source Area concept (Hewlett and Hibbert 1967) describes the expansion of the perennial channel system during precipitation due to saturation of soil at the head of the watershed and adjacent to perennial channels (Figure 5.3-3). This concept assists the understanding of hydrologic responses in a watershed to disturbances such as harvesting, site preparation, and fire. Important hydrologic and geomorphic processes (sediment transport, channel scour and fill, streambank erosion, fish habitat damage, riparian vegetation damage, nutrient and woody debris transport and deposition) occur during stormflows when water volumes and velocities are at their highest. Although the cutting of trees shifts some of the flow from ET to baseflow, forest harvesting *per se* does not affect the magnitude of the fluxes shown in Figure 5.3-2. However, a high degree of soil disturbance (e.g. physical exposure, removal, tilling or development of chemical water repellency) can decrease the amount of infiltrated water and increase non-infiltrated water. The resulting overland flows and surface stormflows often cause damaging peak flows or floods. Water in surface stormflow does not contribute to long-term soil moisture levels and baseflow. Soil moisture is essential for the regeneration of vegetation, and baseflow is important for the maintenance of aquatic habitats and water supplies.

After harvesting, the magnitude and duration of stormflow are determined by climate characteristics and by watershed condition. The probability of observing stormflow due to extreme climatic events increases with the length of the recording period. The response of a forested watershed to extreme events depends on precipitation (amount, intensity, and duration), season, topography, vegetation cover, litter, soils, and geology. Forest harvesting may have a large effect on stormflow or no effect at all.

Watershed condition is a term that describes the ability of a watershed system to receive, route, store, and transport precipitation without ecosystem degradation. When a watershed is in good condition, rainfall infiltrates the soil, and baseflow is sustained between storms. In this situation, precipitation does not contribute to increased surface runoff and erosion (Figure 5.3-2). Well-vegetated watersheds in good condition do not usually produce damaging peak flows (flash floods). However, in some regions of the world, destructive streamflows are common, irrespective of watershed condition. Severe fires, poor harvesting practices, over-grazing, conversion to agriculture and urban use, and other disturbances can reduce watershed condition to a moderate or poor level. The percentage of infiltrated rainfall is reduced, water runs over the surface of the soil, and there is little or no baseflow between storm events (Figure 5.3-2). Erosion may be considerable during high stormflows since water flow at the soil surface can detach and transport sediments.

Surface factors that contribute to watershed condition include: (1) Presence or absence of an organic litter layer and coarse woody debris. (2) Presence or absence of herbaceous, shrub, or woody vegetation. (3) The nature of the geologic material (soil and rock). Disturbances that destroy, remove, redistribute, or increase plant litter and vegetation, and change soil physical properties, alter the infiltration and percolation capacity of soil. When watershed conditions deteriorate, flood flows and erosion increase.

Plant litter has a major influence on watershed condition. In a forest, the organic 'floor' above the mineral soil consists of layers which are commonly referred to in descending sequence as L, F and H. In other nomenclatures these layers are known as O_i, O_e, and O_a or O1, O2, and O3 (Buol *et al.* 1989). The L layer consists of freshly-fallen tree litter (leaves, branches, cones). The F layer is made up of partially-decomposed litter, and the H layer consists of well-decomposed organic matter. In shrubland ecosystems, distinct L, F, and H layers only form under woody vegetation. In grassland ecosystems the layers are much thinner and difficult to distinguish. Some grasslands have only bare soil between plants. Organic material on the soil surface moderates the impact of rain drops, allowing water to infiltrate rather than run off over the surface. Loss of litter through burning, harvesting, site preparation, or any other means can result to adverse changes to hydrologic condition.

Forest harvesting influences many processes in the water cycle. Specific hydrologic effects are summarized in Table 5.3-1. Changes to baseflow and stormflow affect the quantity of water delivered from forested catchments, and can ultimately alter water quality. Section 5.3-2 summarizes the effects of forest biomass removal.

Table 5.3-1. Changes in hydrologic processes caused by tree harvesting.

Hydrologic process	Type of change	Specific effect
1. Interception	Reduction	Moisture storage smaller Greater runoff in small storms Increased water yield
2. Litter storage of water	Litter reduced Litter not affected Litter increased	Less water stored No change Storage increase
3. Transpiration	Temporary elimination	Baseflow increase Soil moisture increase
4. Infiltration	Reduced	Overland flow increase Stormflow increase
	Increased	Overland flow decrease Baseflow increase
5. Streamflow	Variable	Increase in most ecosystems Decrease in snow systems Decrease in fog-drip systems
6. Baseflow	Variable	Decrease with less infiltration Increase with less transpiration Summer low flows (+ and -)
7. Stormflow	Increased	Volume greater Peak flows larger Time to peak flow shorter
8. Snow accumulation	Variable	Cuts <4 ha, increase snowpack Cuts > 4 ha, decrease snowpack Snowmelt rate increase Evaporation/sublimation greater

5.3.2   Water quantity and quality responses

The occurrence and magnitude of the effects of forest biomass removal depend on climate, precipitation, aspect, latitude, severity of disturbance, and the percentage of the watershed harvested.

Water quantity – total yield

Watershed responses to forest harvesting are ecosystem-specific. Increases in water quantity are normally highest during the first year after harvesting (Brooks *et al.* 1997). Thereafter, water yield declines as vegetation growth recovers and leaf area index returns to pre-harvest levels. The recovery period is very short in forests with high evapotranspiration rates and low precipitation (3-4 years; Brown *et al.* 1974), but may be more than 10 years in ecosystems with high rainfall and low evapotranspiration.

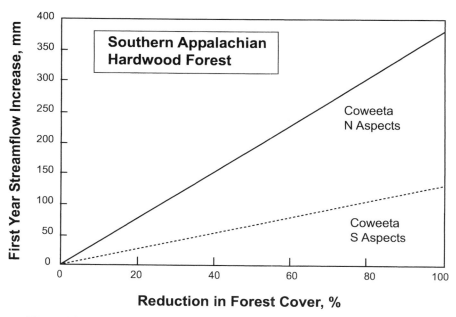

Figure 5.3-4. Effect of aspect on first year streamflow after forest harvesting in the Southern Appalachians of North Carolina.

Water yield increase after forest harvesting depends on the proportion of the watershed from which trees are removed, the amount of precipitation, and site factors such as aspect, soil, and vegetation characteristics. Aspect, a good indicator of potential evapotranspiration, has a strong effect on streamflow response to forest harvesting (Figure 5.3-4). Slopes lying at right angles to solar radiation (south-facing slopes in the Coweeta example) receive higher solar loadings, and evapotranspiration rates are therefore higher. In general, mean annual streamflow increases as the harvested proportion of a forest stand or watershed approaches 100%. Streamflow is usually negligible at the low end of the precipitation range associated with forest ecosystems (Figure 5.3-5). It is substantial in forests growing in high precipitation zones.

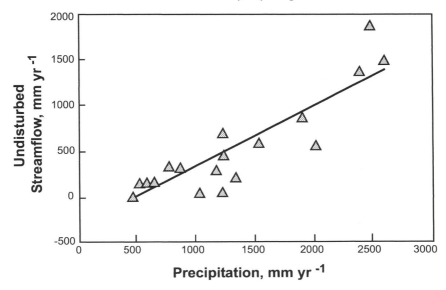

Figure 5.3-5. Effect of precipitation on streamflow in forest ecosystems.

A considerable amount of research has been conducted on the effects of forest harvesting on watershed hydrology (Tables 5.3-2 and 5.3-3). These studies have been very costly and required considerable dedication to the maintenance of record continuity. The earliest watershed experiments were installed in Switzerland, Japan, and the USA in the first decade of the 20th century (Neary 2000). Many have been in existence since the 1930s. The effects of harvesting intensity, configuration and timing have been examined. The following review examines known maximum consequences of clearcut harvesting.

Clearcutting of 100% of the watershed area usually results in a first year water yield increase of 21-80%. An exception to this generalization occurs under conditions of high evapotranspiration and low rainfall (<480 mm); here harvesting does not increase streamflow (Clary *et al.* 1974). Another exception occurs in areas with high evapotranspiration and high rainfall where increase in water yield of 106-280% have been reported (Bren and Papworth 1991, Neary *et al.* 1982). These higher values probably resulted from removal of vegetation with high transpiration rates in both overstorey and understorey. Although absolute water yield during the first year after harvesting increases with total precipitation, the percentage increase is not related to the amount of precipitation (Figure 5.3-6). The absolute amount is strongly related to the average annual rainfall.

Table 5.3-2 First year streamflow responses to harvesting in forest ecosystems receiving 450 to 1200 mm yr^{-1} precipitation.

Forest type	Location	Ppt^n mm	Mean annual streamflow mm	Clear-cut %	Increase mm	Increase %	Reference
Pinyon-juniper	Arizona USA	457	20	100	0	0	Clary *et al.* 1974
Spruce-fir	Alberta Canada	513	147	100	84	57	Swanson & Hillman 197
Aspen-conifer	Colorado USA	536	157	100	34	22	Reinhart *et al.* 1974
Eucalyptus spp.	Victoria Australia	596	86	100	20	23	Burch *et al.* 1987
Ponderosa pine	Arizona USA	570	153	100	96	63	Brown *et al.* 1974
Oak woodland	California USA	635	144	99	33	23	Lewis 1968
Pine-Spruce	Sweden	732	271	100	371	119	Rosén 1984
Spruce-fir-pine	Colorado USA	770	340	40	84	25	Leaf 1975
Aspen - birch	Minnesota USA	775	107	100	45	42	Verry 1972
Spruce-fir	Alberta Canada	840	310	100	79	25	Swanson *et al.* 1986
Slash pine	Florida USA	1020	48	74	134	280	Neary *et al.* 1982
Hardwood	Japan	1153	293	100	209	18	Nakano 1971

In semiarid areas with less than 500 mm annual precipitation, clearcutting does not increase mean annual streamflow (Tables 5.3-2 and 5.3-3; Clary *et al.* 1974). Here evaporation is the dominant factor and watershed vegetation management has little effect on streamflow.

Harvesting of forest stands has been used to augment municipal water supply (Bosch and Hewlett 1982, Brooks *et al.* 1997). The duration of the response is dependent on a number of factors. Generally, the increase in total water yield is not of sufficient magnitude to produce adverse hydrologic or ecosystem effects. Flood peak flows are a matter of greater concern.

Table 5.3-3 First year streamflow responses to harvesting in forest ecosystems receiving 1200 to 2600 mm^{-1} precipitation.

Forest type	Location	Ppt^n mm	Mean annual streamflow mm	Clear %	Increase mm	Increase %	Reference
Coastal redwoods	California USA	1200	686	67	94	15	Keppler & Ziemer 1990
Mixed hardwoods	Georgia USA	1219	467	100	254	54	Hewlett & Doss 1984
Northern hardwoods	New Hampshire USA	1230	710	100	343	48	Hornbeck *et al.* 1987
Loblolly pine	Arkansas USA	1317	214	100	101	47	Miller *et al.* 1988
Slash pine	Florida USA	1450	169	74	134	79	Swindel *et al.* 1982
Dry sclerophyll Eucalpytus	Victoria Australia	1520	330	95	350	106	Bren & Papworth 1991
Mixed hardwoods	W. Virginia USA	1524	584	85	130	22	Patric & Reinhart 1971
Mixed hardwoods	N. Carolina USA	1900	880	100	362	41	Swift & Swank 1980
Montane forest	Kenya Africa	2014	568	100	457	80	Pererira 1962, 1964
Cascade Douglas-fir	Oregon USA	2388	1376	100	462	34	Rothacher 1970
Coastal Douglas-fir	Oregon USA	2483	1885	82	370	20	Harr 1976 Harris 1973
Beech and podocarp forest	New Zealand	2600	1500	100	650	43	Pearce *et al.* 1980

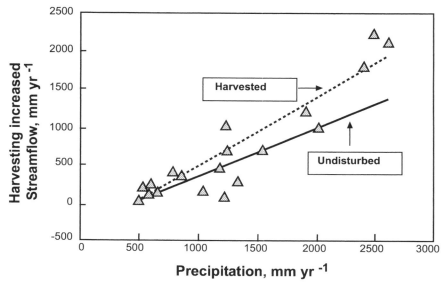

Figure 5.3-6. Effect of forest harvesting on first year streamflow.

Water quantity – peak flow

Forest harvesting produces a mixture of peak flow responses (Table 5.3-4). Where snowmelt runoff is an important component of annual hydrographs, 35% declines in peak flow have been reported (Pierce *et al.* 1970, Verry 1972). Some investigators have reported no peak flow response to harvesting (Bren and Papworth 1991, Kochenderfer *et al.* 1997). In other locations, watershed, vegetation, and climatic characteristics combine to produce peak flow increases of up to 1400%. Such large responses are rare, and are more likely to be associated with disturbances such as wildfire. Some combinations of terrain and geology may create localized hazards (Neary and Hornbeck 1994).

It is important to understand the significance of flood peak flow responses in watershed management, human health, and safety. Major changes to stream channel geomorphology and damage to cultural resources can result from flood peak flows. Peak flow response to forest harvesting may be less than one order of magnitude greater than peak flows in undisturbed and forested areas (Table 5.3-4). They are similar to those at the lower end of responses associated with wildfires (DeBano *et al.* 1998). Floods after severe wildfires often cause channel degradation, sedimentation of reservoirs, damage to transportation systems, personal property damage, and human and livestock death and injury. Although peak flows increase after forest harvesting, they are of less concern than those associated with wildfire.

Table 5.3-4. Effects of forest harvesting on peak flow (Q_{pt}).

Forest type	Location	Ppt[n] mm	Q_{pt} change 1-3 yr post harvest	References
Ponderosa pine	Arizona USA	570	Rainfall: Q_{pt} +167%	Brown *et al.* 1974
Lodgpole pine, Engelmann spruce	Colorado USA	762	Snowmelt: Q_{pt} -23 to + 50%	Goodell 1958
Aspen	Minnesota USA	775	Rainfall: Q_{pt} doubled Snowmelt: Q_{pt} -35 to +1,400%	Verry 1972
Loblolly pine	Georgia USA	1219	Rainfall: Q_{pt} tripled	Hewlett & Doss 1984
Northern hardwoods	New Hampshire USA	1230	Rainfall: Q_{pt} -13 to +170% Rain + snowmelt: Q_{pt} -21 to +23%	Pierce *et al.* 1970
Loblolly and shortleaf pine	Arkansas USA	1317	Rainfall: Q_{pt} +16 to +247%	Miller *et al.* 1988; Beasley & Granillo 1988
Slash pine	Florida USA	1450	Rainfall: Q_{pt} increased six-fold	Swindel *et al.* 1983
Dry sclerophyll eucalyptus	Victoria Australia	1520	Rainfall: no effect on Q_{pt}	Bren & Papworth 1991
Mixed hardwoods	W. Virginia USA	1524	Rainfall: no effect on Q_{pt}	Kochenderfer *et al.* 1997
Mixed hardwoods	North Carolina USA	1900	Rainfall: Q_{pt} + 9%	Hewlett & Helvey 1970
Douglas-fir and hemlock	Oregon USA	2300	Rainfall: Q_{pt} + 1% Snowmelt: Q_{pt} -16%	Harr & McCorison 1979

Water quality

A large number of research studies have examined the effects of forest harvesting on water quality (Neary and Hornbeck 1994). Commonly-measured variables are concentration of nitrogen (NO_3-N) and other nutrients, the amount of sediment, and temperature. These parameters are of particular concern in streams used for water supply, and may affect aquatic biota.

Nitrate-nitrogen concentration is often used as an indicator of watershed health and water quality. It is a good indicator of disturbance (Swank 1988), and of danger to human health. For the most part, large increases in NO_3-N levels have not been observed in streams draining harvested watersheds (Table 5.3-5). There is no general indication that water quality standards (10 mg l^{-1} NO_3-N maximum) are threatened by harvesting. The largest increases were noted where herbicides had been used to suppress vegetation regrowth (Pierce *et al.* 1970), where severe fire had occurred (DeBano *et al.* 1998);

where nitrogenous fertilizers had been used during regeneration (Neary and Hornbeck 1994); or where there was a danger of nitrogen saturation due to atmospheric deposition (Aber *et al.* 1989).

Table 5.3-5. Effects of forest harvesting on NO$_3$-N concentrations in streamwater.

Forest type	Location	NO$_3$-N, mg l^{-1}			Reference
		Mean Uncut	Mean Cut	Max. Cut	
Northern hardwoods	New Hampshire USA	0.3	11.9	17.8	Pierce *et al.* 1970
Oak-maple	Pennsylvania USA	0.1	5.0	8.4	Corbett *et al.* 1975
Mixed hardwoods	West Virginia USA	0.1	0.5	1.4	Aubertin & Patric 1974
Hardwoods	North Carolina USA	<0.1	0.1	0.2	Swank & Douglass 1975
Loblolly pine	Georgia USA	0.1	<0.1	-	Hewlett & Doss 1984
Douglas-fir	Oregon USA	<0.1	0.2	0.6	Fredricksen *et al.* 1975
Douglas-fir	Oregon USA	<0.1	<0.1	<0.1	Fredrickson *et al.*1975
Mixed conifers	Oregon USA	<0.1	<0.2	0.4	Fredrickson *et al.* 1975
Douglas-fir, alder	Oregon USA	1.2	0.4	2.1	Brown *et al.* 1973
Mixed conifers	Montana USA	0.1	0.2	-	Bateridge 1974
Mixed conifers	Idaho USA	0.2	0.2	-	Snyder *et al.* 1975
Aspen, birch, spruce (bog)	Minnesota USA	0.1	0.2	-	Verry 1972
Lodgepole pine, spruce, aspen	Alberta Canada	0.2	0.7	-	Singh & Kalra 1975
Spruce, fir	British Columbia, Canada	0.1	0.2	0.6	Hetherington 1976
Beech, podocarps	New Zealand	<0.1	<0.1	0.2	O'Loughlin *et al.* 1980
Pine, spruce, hardwoods	Sweden	0.1	0.2	-	Rosén 1996
Slash pine	Florida USA	<0.1	0.3	-	Riekerk *et al.* 1980
Mixed hardwoods, conifers	New Brunswick Canada	-	-	1.6	Krause 1982
Western hemlock	British Columbia, Canada	-	-	0.5	Feller & Kimmins 1984

Increases of other nutrients in streamwater after harvesting have been examined by Neary and Hornbeck (1994). Although the tight nutrient cycles characteristic of undisturbed forests are temporarily interrupted by harvesting, the short-term nature of the disturbances and subsequent regrowth of vegetation limit the effect on water quality. Prescribed fire can mineralize nutrients and make them susceptible to leaching or runoff, but amounts involved are low in comparison to losses after wildfire (DeBano *et al.* 1998). Inter-rotation fertilizer application can be an additional source of nutrients in streamwater, but mitigation practices will limit any input (Neary and Leonard 1978). Neary and Hornbeck (1994) concluded that increases in dissolved inorganic ions in streamflow after forest harvesting do not have a significantly adverse effect on water quality.

Sediment yield after forest harvesting is highly variable and depends on such factors as soil type, climate, topography, ground cover, and watershed condition. Increases in sediment yield after harvesting (Table 5.3-6) are related to physical disturbance of soil but are usually transient due to vegetation regrowth. Large increases are associated with post-harvest mechanical site preparation (Beasley 1979); slope instability (O'Loughlin and Pearce 1976); road construction (Swanson *et al.* 1986); and highly erosive soils (Beasley and Granillo 1988). Management practices can reduce sediment yield when properly planned and implemented prior to, during, and after harvesting. Guidelines for these practices relate to design, construction and maintenance of major access roads, logging roads, skid trails, and landings. Most (90%) of the sediment generated by harvesting originates in forest roads and landings (Swift 1988). Principles underlying management practice guidelines involve the minimizing of disturbance in streamside zones; reduction of the erosive power of runoff on bare road surfaces; and maintenance of the normally high infiltration capacity of forest soils.

Although sediment movement to streams is a constant environmental concern in managed forest watersheds, it also occurs where there is no active management. Watersheds vary greatly in their natural sediment load characteristics. After initial deposition, natural and anthropomorphic sediment can relocate into stream channels and move downstream over long time periods and considerable distances. Cumulative effects of erosion and sedimentation that occurred centuries ago can present forest managers with many challenges (Terrene Institute 1993). Sediment content is an important indicator of water quality since large amounts can harm aquatic organisms and habitats, and render water unacceptable for drinking or for recreational purposes.

Forest vegetation protects stream channels from solar radiation, thereby reducing the magnitude and variability of stream temperature (Neary and Hornbeck 1994). Increases in water temperature resulting from forest harvesting affect many physical, chemical, and biological processes. Effects on aquatic biota vary considerably, depending on whether or not individual species are eurythermic and the degree to which stream temperature is controlled by solar heating or stream baseflow. A more complete discussion of the effects of stream temperature has been presented by Brown (1980).

Temperature changes in stream water can be moderated by use of buffer strips. Temperature often increases or decreases as streams merge with those from other harvested or uncut areas.

Table 5.3-6. Effects of forest harvesting and related disturbances on water sediment content in USA and New Zealand (adapted from Neary and Hornbeck 1994; Neary and Michael 1996).

Forest type	Location	Treatment	Sediment increase %	Reference
Northern hardwoods	New Hampshire USA	Clearcut	769	Hornbeck *et al.* 1987
Loblolly pine	S. Carolina USA	Clearcut	655	Van Lear *et al.* 1985
Loblolly pine	Mississippi USA	Clearcut, bed	2,198	Beasley 1979
Slash pine	Florida USA	Clearcut, windrow	1,100	Riekerk *et al.* 1980
Loblolly pine, shortleaf pine	Arkansas USA	Clearcut	65,000	Beasley & Granillo 1988
Beech, podocarps	New Zealand	Clearcut	42	O'Loughlin *et al.* 1980
		Clearcut, skid	700	
Beech, podocarps	New Zealand	Clearcut	2,100	O'Loughlin & Pearce 1976
Lodgepole pine, Engelmann spruce	Colorado USA	Strip Cut	199	Leaf 1966
Douglas-fir	Oregon USA	Clearcut	296	Swanson *et al.* 1986
		Roads	12,400	
Douglas-fir	Oregon USA	Clearcut	267	Swanson & Dyrness 1975
		Roads	4,817	
Mixed conifers	Arizona USA	Clearcut	38	Heede 1987
		Cut, skid	600	
		Cut, road	1,012	

5.3.3   Sustainability Criteria and Indicators: Santiago Declaration

Assessment of sustainability of water resources

The Helsinki Ministerial Conference in 1990 initiated the development of management guidelines and criteria designed to ensure the conservation and sustainable management of forests in Europe and elsewhere (Helsinki 1994). The Montreal Process was a parallel, but independent Canadian initiative, joined by countries with temperate or boreal forests (Anon. 1995). Criterion

5 of the Helsinki Process relates to maintenance and development of the role of forests in water supply and protection against erosion. Criterion 4 of the Montreal Process covers the conservation of soil and water resources and the protective and productive functions of forests. Because the chemical, physical, and biological characteristics of aquatic systems are excellent indicators of the condition and sustainability of the land around them (Breckenridge *et al.* 1995), specific attributes of soil and water resources were selected as indicators of sustainability.

Eight of the 67 indicators selected in the Montreal Process and endorsed by the ten nations that drafted the Santiago Declaration in 1995 pertain to Criterion 4. Those specifically related to hydrologic impacts of harvesting are:

- No. 19. Area and percent of forest managed primarily for protective functions (e.g. watershed, flood protection, avalanche protection, riparian zones).

- No. 20. Percent of stream kilometers in forested catchments in which stream flow and timing has significantly deviated from the historic range of variation.

- No. 23. Percent of water bodies in forest areas with significant variance of biological diversity from the historic range of variability.

- No. 24. Percent of water bodies in forest areas with significant variation from the historic range of variability in pH, dissolved oxygen, levels of chemicals, sedimentation, or temperature change.

Sustainability is the stewardship goal of forestry. Further specific definition of its goals and attributes is complex and open to interpretation (Moir and Mowrer 1995). Many ecologists have attempted to answer the 'what', 'what level', 'for whom', 'biological or economic', and 'how long' questions of sustainability. Allen and Hoekstra (1994) discussed the emergence of the concept of sustainability and the difficulty experienced in defining it. They concluded that there is no absolute definition of sustainability. It must be viewed within the context of human conceptual frameworks and societal decisions about the type of ecosystem to be sustained and the spatial and temporal scales over which attainment of sustainability is to be judged. It is also defined in terms of the needs of society, the experiential frame of reference of ecosystem managers, and the ecological models used to predict the future attributes of natural resources. Our ability to predict these attributes is confounded by encounters with extreme events, poorly understood ecological processes and linkages, surprises associated with the law of unintended consequences, the development of critical thresholds, and chaotic system behavior. Another approach to the definition of sustainability is the identification of conditions that draw attention to ecosystem deterioration (Moir and Mowrer 1995). Criteria and Indicators are, in essence, warning flags designed to attract the attention of land managers before decline into unsustainability occurs.

Soil compaction, erosion, and organic matter loss are the major factors affecting decline of ecosystem productivity (Powers *et al.* 1990). They are discussed in more detail in Section 5.2. These factors can alter carbon allocation and storage, nutrient content and availability, water storage and flux, rhizosphere processes, and insect and disease dynamics. Disturbances which affect them are wildfire, insect and disease outbreaks, climate extremes, vegetation management (wood harvesting, grazing, prescribed fire, chemical weed control, manual removal of plant species), and recreation (foot traffic and vehicles) (Hart and Hart 1993). Management activities that eliminate natural disturbance (e.g. fire suppression, insect control) or alter ecosystem properties can also affect ecosystem sustainability.

Applicability of the Santiago indicators to harvesting of forest residues

Of the indicators listed under Criterion 4 of the Santiago Declaration (Conservation and Maintenance of Soil and Water Resources), Nos. 19–20 and 23–25 deal with water resources. Their applicability to forest harvesting will be discussed in this section. Indicators 18, 21, and 22 relate to soil processes that can ultimately affect water quantity and quality, attributes which are discussed in section 5.2.

INDICATOR 19: Area and percent of forest managed primarily for protective functions. The main reservation about this indicator concerns the assumption that protective management guarantees sustainability. Natural disturbance (drought, fire, insects), and uncontrolled wildlife, feral livestock, and domestic livestock populations (e.g. introduced ungulates) can cause temporary or long-term sustainability decline. Protected areas (water supply watersheds, flood protection areas, avalanche protection zones, riparian ecosystems, and wildlife reserves) are generally in a sustainable ecological state and functioning condition. This indicator is more a measure of society's attitudes to protection, the degree to which Best Management Practices are employed and appropriate management than a guarantor of forest sustainability. Even if harvesting and domestic livestock grazing were prohibited, water resources could be degraded as a result of browsing by wild ungulates or infestation by insects. Watershed areas protected from forest harvesting and prescribed fire could be more vulnerable to wildfire that would have a detrimental effect on sustainability (DeBano *et al.* 1998).

INDICATOR 20: Percent of stream kilometers in forest catchments in which stream flow and timing has significantly deviated from the historic range of variation. Forests play an important role in hydrological cycles and the functioning of riparian and aquatic ecosystems. Properly functioning soils, watersheds, and riparian areas are important for the retention and supply of the water that supports ecosystem sustainability (Medina *et al.* 1996). Abrupt or large fluctuations in stormflow and baseflow can indicate disturbances deleterious to sustainability. They can also be 'natural' variations that occur from time to time but have not been previously observed.

Some ecosystems have natural, long-term oscillations of climate and hydrology that transcend the measured 'historic range of variation' (Grissino-Mayer 1996). Unless these oscillations are understood, it is difficult to discern significant deviations from the historic range of variability, and to determine the consequences. Annual water yields, seasonal peak flows and low flows, flood frequencies, and timing or duration of peak flows and low flows are only useful indicators if they are related to changes in land management or to natural disturbances that could affect ecosystem sustainability. The historic range of variation depends on length of record, and will alter continuously as the length of hydrologic records increases or as climate changes occur. Temporal and spatial variations in biological diversity make this indicator difficult to interpret and apply. Two questions need to be addressed: firstly, 'Has the hydrologic regime been changed from a naturally dynamic state to a permanently disturbed state or a temporarily disturbed state?' (Kauffman *et al.* 1997); secondly, 'Is there any relationship between deviation from the historic range of variation and the sustainability of the forest ecosystem under the current management system?' If there are no linkages and the change is only temporary, then this indicator is not useful.

INDICATOR 23: Percent of water bodies in forest areas with significant variance of biological diversity from the historic range of variability. This indicator is an indirect measure of sustainability since it deals with biological factors that are sensitive to ecosystem change but do not directly predict sustainability. Its usefulness in assessing forest sustainability is limited. Aquatic organisms, particularly benthic macroinvertebrates, are sensitive to changes in sediment level, oxygen concentration, temperature, and pH (Webster *et al.* 1983). Water quality variables can be sensitive to watershed disturbances which may or may not affect long-term sustainability. All ecosystem processes are dynamic and changes in biological diversity may or may not be related to sustainability. Aquatic and riparian ecosystems are highly dynamic and disturbance-prone since they are constantly subjected to variation in the hydrograph. Biological diversity is poorly understood, inadequately documented, and only indirectly correlated with sustainability. Disturbances, both natural and artificial, produce changes in the flora and fauna that temporarily increase or decrease biodiversity. Temporal and spatial variability make this indicator difficult to interpret and apply. Even if this were not so, the number of data sets covering a sufficiently long period of record is too small to allow useful comparisons.

INDICATOR 24: Percent of water bodies in rangeland areas with significant variation from the historic range of variability in pH, dissolved oxygen, levels of chemicals, sedimentation, or temperature change. This indicator is linked directly to Indicator 23 and is similarly subject to reservation. Water quality monitoring over large areas can provide an early indication that disturbances potentially damaging to sustainability are occurring within forest watersheds (Swank 1988). This evidence is only indirect and sources and causes may be difficult to identify. There is an abundance of data, but water quality can be highly variable within and between watersheds and ecosystems. Careful analysis is needed to separate point source trends from general trends that indicate changes in sustainability. Natural processes that do not compromise sustainability can produce changes in water quality. Temporal and spatial variability in water quality between different bioregions and geomorphic regions make this indicator even more difficult to interpret and apply. There is also a problem with definition of the term 'historic range of variablity', and with identification of significant deviations.

INDICATOR 25: *Area and percent of forest land experiencing significant accumulation of persistent toxic substances.* This is not a useful indicator of forest sustainability since few persistent toxic substances are used in modern forest management and problems tend to be localized. It indicates society's attitudes to pollution rather than ecosystem sustainability. Persistent toxic substances may have general or point sources such as industrial operations, mining wastes, agricultural discharges, urban wastes, and air pollution.

Recommendations on monitoring for water resource sustainability

The condition of watersheds and their drainage networks provides an integrated picture of the conditions, health, and sustainability of ecosystems. Discussions about the most appropriate methods for determining whether forestry operations affect the sustainability of water resources have been in progress for decades. The Santiago Declaration Criteria and Indicators represent only one end of a spectrum of approaches for ensuring sustainability, international accounting and accountability. While the indicators for conservation and maintenance of soil and water resources appear to be laudable, a number of economic, scientific, and technical problems make their implementation difficult, if not impossible. Activities designed to ensure sustainability must be accepted locally and applied by everyone involved in the life cycle of wood production and bioenergy (Mälkki *et al.* 2001). These activities, known as Best Management Practices (BMPs) aim to ensure forest and watershed sustainability and also the economic sustainability of forestry enterprises.

Smith *et al.* (1999) described some of the attributes required in BMPs and indicators of sustainable forest management. They should be easy to implement and measure, cost effective, adaptable to changing conditions, scientifically sound, forest ecosystem specific, but amenable to scaling up, integrative of ecosystem function, and related to management goals. Smith *et al.* (1999) recommended the following approach to measurement of the sustainability of forestry operations:

- Monitor all sites for operational BMP compliance (Compliance Monitoring).

- Test BMP effectiveness at a limited number of locations with site-specific indicators (Effectiveness Monitoring).

- Intensively measure a small number of benchmark sites to validate research recommendations and to adapt BMPs (Validation Monitoring).

5.3.4   Summary

Hydrologic responses to forest harvesting are summarized in Table 5.3-7. After 100% clearcutting, reported first year water yield increases range from 21 to 80%; Exceptions to this generalization occur where there is high evapotranspiration and low rainfall (<480 mm), and in areas with high evapotranspiration and high rainfall.

Table 5.3-7 Summary of hydrologic responses to forest harvesting.

Hydrologic response	Increase, %
Streamflow	0 to 279
Flood peak flow	-23 to 600
Nitrate nitrogen	-300 to 5000
Sediment yield	38 to 65,000

Forest harvesting produces a mixture of peak flow responses. Where snowmelt runoff is an important component of annual hydrographs, 35% declines in peak flow have been reported after harvesting. Sometimes there is no peak flow response. In other locations, peak flow increases up to 1400% have been observed, but this level of response is rare.

There is no evidence that additional dissolved inorganic ions in streamflow after forest harvesting have a significant adverse effect on water quality. Nitrate-nitrogen levels in streams draining forested watersheds do not show large increases after harvesting, and there is no indication that water quality standards (10 mg l^{-1} NO_3-N maximum) are exceeded.

5.3.5   References

Aber, J.D., Nadelhoffer, K.J., Steudler, P. and Melillo, J.M. 1989. Nitrogen saturation in northern forest ecosystems. Bioscience 39: 378-386.

Allen, T.F.H. and Hoekstra, T.W. 1994. Toward a definition of sustainability. Pp. 98-107 *in* Covington, W.W. and DeBano, L.F. (technical coordinators). Sustainable Ecological Systems: Implementing an Ecological Approach to Land Management. USDA Forest Service General Technical Report RM-247. 363 p.

Anon. 1995. Criteria and Indicators for the Conservation and Sustainable Management of Temperate and Boreal Forests – The Montreal Process. Fo42-238/1995E. Canadian Forest Service. Natural Resources Canada, Hull, Quebec, Canada K1A 1G5. 27 p.

Aubertin, G.M. and Patric, J.H. 1974. Water quality water after clearcutting a small watershed in West Virginia. Journal of Environmental Quality 3: 243-249.

Bateridge, T. 1974. Effects of clearcutting on water discharge and nutrient loss. Bitterroot National Forest, Montana. M.S. Thesis. Office of Water Resources Research, University of Montana, Missoula, MT.

Beasley, R.S. 1979. Intensive site preparation and sediment losses on steep watersheds in the gulf Coastal Plain. Soil Science Society of America Journal 43: 412-417.

Beasley, R.S. and Granillo, A.B. 1988. Sediment and water yields from managed forests on flat coastal plain sites. Water Resources Bulletin 24: 361-366.

Bosch, J.M. and Hewlett, J.D. 1982. A review of catchment experiments to determine the effect of vegetation changes on water yield and evapotranspiration. Journal of Hydrology 55: 3-23.

Breckenridge, R.P., Kepner, W.G. and Mouat, D.A. 1995. A process for selecting indicators for monitoring conditions of rangeland health. Environmental Monitoring and Assessment 36: 45-60.

Bren, L.J. and Papworth, M. 1991. Water yield effects of conversion of slopes of a eucalypt forest catchment to radiata pine plantation. Water Resources Research 27: 2421-2428.

Brooks, K.N., Ffolliott, P.F., Gregerson, H.M. and DeBano, L.F. 1997. Hydrology and the Management of Watersheds. Iowa State University Press, Ames, IA. 502 p.

Brown, G.W. 1980. Forestry and Water Quality. Oregon State University Book Stores, Inc., Corvallis, OR. 124 p.

Brown, G.W., Gahler, A.R. and Marston, R.B. 1973. Nutrient losses after clear-cut logging and slash burning in the Oregon coast range. Water Resources Research 9: 1450-1453.

Brown, H.E., Baker, M.B. Jr., Rogers, J.J., Clary, W.C., Kovner, J.L., Larson, F.R., Avery, C.C. and Campbell, R.E. 1974. Opportunities for increasing water yields and other multiple use values on ponderosa pine forest lands. USDA Forest Service Research Paper RM-129, Rocky Mountain Forest and Range Experiment Station, Fort Collins, CO.

Buol, S.W., Hole, F.D. and McCracken, R.J. 1989. Soil Genesis and Classification. Iowa State University Press, Ames, IA.

Burch, G.J., Bath, R.K., Moore, I.D. and O'Loughlin, E.M. 1987. Comparative hydrological behavior of forested and cleared catchments in southeastern Australia. Journal of Hydrology 90: 12-42.

Clary, W.P., Baker, M.B. Jr., O'Connell, P.F., Johnsen, T.N. Jr. and Campbell, R.E. 1974. Effects of pinyon-juniper removal on natural resourc products and uses in Arizona. USDA Forest Service Research Paper RM-128, Rocky Mountain Forest and Range Experiment Station, Fort Collins, CO. 28 p.

Corbett, E.S., Lynch, J.A. and Sopper, W.E. 1975. Forest-management practices as related to nutrient leaching and water quality. Pp. 157-173 in Conference on non-point sources of water pollution proceedings. Virginia Water Resources Research Center; Virginia Polytechnical Institute and State University, Blacksburg, VA, May 1-2, 1975.

DeBano, L.F., Neary, D.G. and Folliott, P.F. 1998. Fire's Effects on Ecosystems. John Wiley & Sons, New York, NY. 333 p.

Feller, M.C. and Kimmins, J.P. 1984. Effects of clearcutting and slash burning on streamwater chemistry and watershed nutrient budgets in southwestern British Columbia. Water Resources Research 20: 29-40.

Fredriksen, R.L., Moore, D.G. and Norris, L.A. 1975. Impact of timber harvest, fertilization, and herbicide treatment on stream water quality in the Douglas-fir regions. Pp. 283-313 *in* Bernier, B. and Winget, C.H. (eds.). Forest Soils and Forest Land Management. Proceedings of the Fourth North American Forest Soils Conference, Laval University, Quebec City, Canada.

Goodell, B.C. 1958. A preliminary report on the first year's effects of timber harvesting on water yield from a Colorado watershed. USDA Forest Service Station Paper No. 36, Rocky Mountain Forest and Range Experiment Station, Fort Collins, CO. 12 p.

Grissino-Mayer, H.D. 1996. A 2129-year reconstruction of precipitation for northwestern New Mexico, USA. Pp. 191-204 *in* Dean, J.S., Meko, D.M. and Swetnam, T.W. (eds.). Tree Rings, Environment, and Humanity. Proceedings of the International Congress, University of Arizona, Tucson, AZ.

Harr, R.D. 1976. Forest practices and streamflow in western Oregon. USDA Forest Service General Technical Report PNW-49, Pacific Northwest Forest and Range Experiment Station, Portland, OR. 18 p.

Harr. R.D. and McCorison, F.M. 1979. Initial effects of clearcut logging on size and timing of peak flows in a small watershed in western Oregon. Water Resources Research 15: 90-94.

Harris, D.D. 1973. Hydrologic changes after clearcut logging in a small Oregon coastal watershed. Journal of Research of the U.S. Geological Survey 1: 487-491.

Hart, J.V. and Hart, S.C. 1993. A review of factors affecting soil productivity in rangelands, and woodlands of the southwestern United States. School of Forestry Report 40-8371-2-0733, Northern Arizona University. 75 p.

Heede, B.H. 1987. Overland flow and sediment delivery five years after timber harvest in mixed conifer forest. Journal of Hydrology 91: 205-216.

Helsinki. 1994. Proceedings of the Ministerial Conferences and Expert Meetings. Liaison Office of the Ministerial Conference on the Protection of Forests in Europe. PO Box 232, FIN-00171, Helsinki, Finland.

Hetherington, E. D. 1976. Dennis Creek: A look at water quality following logging in the Okanagan Basin. Environment Canada, Canadian Forestry Service. 28 p.

Hewlett, J.D. 1982. Principles of Forest Hydrology. University of Georgia Press, Athens, GA. 192 p.

Hewlett, J.D. and Doss, R. 1984. Forests, floods, and erosion: A watershed experiment in the Southeastern Piedmont. Forest Science 30: 424-434.

Hewlett, J.D. and Helvey, J.D. 1970. The effects of clear-felling on the storm hydrograph. Water Resources Research 6: 768-782.

Hewlett, J.D. and Hibbert, A.R. 1967. Factors affecting response of small watersheds to precipitation in humid areas. Pp. 275-290 *in* Sopper, W. and Lull, H. (eds.). International Symposium on Forest Hydrology, Pergamon Press, Oxford.

Hornbeck, J.W., Martin, C.W., Pierce, R.S., Bormann, F.H., Likens, G.E. and Eaton, J.S. 1987. The northern hardwood forest ecosystem: 10 years of recovery from clearcuttting. USDA Forest Service Research Paper NE-596. Northeastern Forest Experiment Station, Broomall, PA. 30 p.

Kauffman, J.B., Beschta, R.L., Otting, N. and Lytjen, D. 1997. An ecological perspective of riparian and stream restoration in the western United States. Fisheries 22: 12-24.

Keppler, E.T. and Ziemer, R.R. 1990. Logging effects on streamflow: water yield and summer low flows at Caspar Creek in northwestern California. Water Resources Research 26: 1669-1679.

Kochenderfer, J.N., Edwards, P.J. and Wood, F. 1997. Hydrologic impacts of logging an Appalachian watershed using West Virginia's Best Management Practices. Northern Journal of Applied Forestry 14: 207-218.

Krause, H. H. 1982. Nitrate formation and movement before and after clear-cutting of a monitored watershed in central New Brunswick, Canada. Canadian Journal of Forest Research 12: 922-930.

Leaf, C.F. 1966. Sediment yields from high mountain watersheds, Central Colorado. USDA Forest Service Research Paper RM-23, Rocky Mountain Forest and Range Experiment Station, Fort Collins, CO. 15 p.

Leaf, C.F. 1975. Watershed management in the central and southern Rocky Mountains: A summary of the status of our knowledge by vegetation type. USDA Forest Service Research Paper RM-142, Rocky Mountain Forest and Range Experiment Station, Fort Collins, CO.

Lewis, D.C. 1968. Annual hydrologic response to watershed conversion from oak woodland to annual grassland. Water Resources Research 4: 59-72.

Mälkki, H., Harju, T. and Virtanen, Y. 2001. Life cycle assessment of logging-residue-based energy. Pp. 92-107 in Richardson, J., Bjorheden, R., Hakkila, P., Lowe, A.T. and Smith, C.T. (compilers). Bioenergy from Sustainable Forestry: Principles and Practice. Proceedings of IEA Bioenergy Task 18 Workshop, 16–20 October 2000, Coffs Harbour, New South Wales, Australia. New Zealand Forest Research Institute, Forest Research Bulletin No. 223.

Medina, A.L., Baker, M.B., Jr. and Neary, D.G. 1996. Desirable functional processes: conceptual approach for evaluating ecological condition. Pp. 302-311 in Shaw, D.W. and Finch, D.M. (technical coordinators). Desired Future Conditions for Southwest Riparian Ecosystems: Bringing Interests and Concerns Together. USDA Forest Service General Technical Report RM-272. 359 p.

Miller, E.L., Beasley, R.S. and Lawson, E.R. 1988. Forest harvest and site preparation effects on stormflow and peakflow of ephemeral streams in the Ouachita Mountains. Journal of Environmental Quality 17: 212-218.

Moir, W.H. and Mowrer, H.T. 1995. Unsustainability. Forest Ecology and Management 73: 239-248.

Nakano, H. 1971. Effect on streamflow of forest cutting and change in regrowth on cutover area. Reprint, Bulletin of the U.S. Government Forest Experiment Station, No. 240. 249 p.

Neary, D.G. 2000. Changing perspectives of watershed management from a retrospective viewpoint. Pp. 167-176 in Ffolliott, P.F., Baker, M.B. Jr., Edminster, C.B., Dillion, M.C. and Mora, K.L. (eds.). Land Stewardship in the 21st Century: The Contributions of Watershed Management. USDA Forest Service Proceedings, RMRS-P-13.

Neary, D.G. and Hornbeck, J.W. 1994. Chapter 4: Impacts of harvesting and associated practices on off-site environmental quality. Pp. 81-118 in Dyck, W.J., Cole, D.W. and Comerford, N.B. (eds.). Impacts of Forest Harvesting on Long-term Site Productivity, Chapman & Hall, London.

Neary, D.G. and Leonard, J.H. 1978. Effects of forest fertilization on nutrient losses in streamflow in New Zealand. New Zealand Journal of Forestry Science 8: 189-205.

Neary, D.G. and Michael, J.L. 1996. Herbicides - protecting long-term sustainability and water quality in forest ecosystems. New Zealand Journal of Forestry Science 26: 241-264.

Neary, D.G., Lassiter, C.J. and Swindel, B.F. 1982. Hydrologic responses to silvicultural operations in southern coastal plain flatwoods. Pp. 35-52 *in* Coleman, S.S., Mace, A.C. Jr. and Swindel, B.F. Impacts of intensive Forest Management Practices. School of Forest Resources and Conservation, University of Florida, Gainesville, FL. 110 p.

O'Loughlin, C.L. and Pearce, A.J. 1976. Influence of Cenozoic geology on mass movement and sediment yield response to forest removal. Bulletin of the International Association of Engineering Geology 14: 41-46.

O'Loughlin, C.L., Rowe, L.K. and Pearce, A.J. 1980. Sediment yield and water quality responses to clearfelling of evergreen mixed forests in western New Zealand. Pp. 285-292 *in* Proceedings of the Helsinki Symposium on the Influence of Man on the Hydrological Regime with Special Reference to Representative and Experimental Basins, PIAHS-AISH Publication No. 130.

Patric, J.H. and Reinhart, K.G. 1971. Hydrologic effects of deforesting two mountain watersheds. Water Resources Research 7: 1182-1188.

Pearce, A.J., Rowe, L.K. and O'Loughlin, C.L. 1980. Effects of clearfelling and slash burning on water yield and storm hydrographs in evergreen mixed forests, western New Zealand. Pp. 119-127 *in* Proceedings on the Influence of Man on the Hydrological Regime with Special Reference to Representative and Experimental Basins, Helsinki, Finland, June 1980. I.A.H.S. – A.I.S.H. Publication No. 130.

Pererira, H.C. 1962. Hydrological effects of changes in land use in some South African catchment areas. East African Agriculture and Forestry Journal, Special Issue 27. 131 p.

Pererira, H.C. 1964. Research into the effects of land use on streamflow. Transactions of the Rhodesian Science Association Proceedings 1: 119-124.

Pierce, R.S., Hornbeck, J.W., Likens, G.E. and Bormann, F. H. 1970. Effect of elimination of vegetation on stream water quality and quantity. International Association of Hydrological Science 96: 311-328.

Powers, R.F., Alban, D.H., Miller, R.E., Tiarks, A.E., Wells, C.G., Avers, P.E., Cline, R.G., Loftus, N.S. Jr. and Fitzgerald, R.O. 1990. Sustaining productivity of North American forests: Problems and prospects. Pp. 49-79 *in* Gessel, S.P., Lacate, D.S., Weetman, G.F. and Powers, R.F. (eds.). Sustained Productivity of Forest Soils, Proceedings of the 7th North American Forest Soils Conference, 24-28 July 1988, Vancouver, B.C., University of British Columbia Faculty of Forestry, Vancouver, British Columbia, Canada.

Reinhart, K.G., Eschner, A.R. and Trimble, G.R. Jr. 1974. Effect on streamflow of four forest practices in the mountains of West Virginia. USDA Forest Service Research Paper NE-1, Northeastern Forest Experiment Station, Broomall, PA. 59 p.

Riekerk, H., Swindel, B.F. and Replogle, J.A. 1980. Effect of forestry practices in Florida watersheds. Pp. 706-720 *in* Proceedings of the Symposium on Watershed Management 1980, ASCE, Boise, ID, July 21-23, 1980.

Rosén, K. 1984. Effect of clear-felling on runoff in two small watersheds in central Sweden. Forest Ecology and Management 9: 267-281.

Rosén, K. 1996. Effect of clear-cutting on streamwater quality in forest catchments in central Sweden. Forest Ecology and Management 83: 237-244.

Rothacher, J. 1970. Increases in water yield following clear-cut logging in the Pacific Northwest. Water Resources Research 6: 653-658.

Singh, T. and Kalra, Y.P. 1975. Changes in chemical composition of natural waters resulting from progressive clearcutting of forest catchments in West Central Alberta, Canada. International Association of Scientific Hydrology. Symposium Proceedings, Tokyo, Japan. December 1970. Publication 117. Pp. 435-449.

Smith, C.T., Lowe, A.T. and Proe, M.F. 1999. Preface: Indicators of sustainable forest management. Papers presented at the IEA Bioenergy Task XII Workshop held in Eddleston, Scotland, 20-25 September, 1997. Forest Ecology and Management 122: 1-5.

Snyder, G.G., Haupt, H.F. and Belt, G.H. Jr. 1975. Clearcutting and burning slash alter quality of stream water in Northern Idaho. USDA Forest Service Research Paper INT-1698, Intermountain Forest and Range Experiment Station, Ogden, UT. 34 p.

Swank, W.T. 1988. Chapter 25. Stream chemistry responses to disturbance. Pp. 339-357 *in* Swank, W.T. and Crossley, D.A. (eds.). Forest Hydrology and Ecology at Coweeta. Springer-Verlag, New York, NY. 469 p.

Swank, W. T. and Douglass, J. E. 1975. Nutrient flux in undisturbed and manipulated forest ecosystems in the Southern Applachian Mountains. International Association of Scientific Hydrology. Symposium Proceedings, Tokyo, Japan. December 1970. Publication 117. Pp. 445-456.

Swanson, F.J. and Dyrness, C.T. 1975. Impact of clearcutting and road construction on soil erosion by landslides in the western Cascade range, Oregon. Geology 1: 393-396.

Swanson, R.H. and Hillman, G.R. 1977. Predicted increased water yield after clear-cutting verified in west-central Alberta. Information Report NOR-X-198, Canadian Department of Fisheries and Environment, Canadian Forestry Service, Northern Forestry Centre, Edmonton, Alberta, Canada.

Swanson, R.H., Golding, D.L., Rothwell, R.L. and Bernier, P.Y. 1986. Hydrologic effects of clear-cutting at Marmot Creek and Streeter watersheds, Alberta. Information Report NOR-X-278, Northern Forestry Centre, Canadian Forestry Service, Edmonton, Alberta, Canada. 27 p.

Swift, L.W. Jr. 1988. Chapter 23: Forest access roads: Design, maintenance, and soil loss. Pp. 313-324 *in* Swank, W.T. and Crossley, D.A. (eds..) Forest Hydrology and Ecology at Coweeta. Springer-Verlag, New York, NY. 469 p.

Swift, L.W. Jr. and Swank, W.T. 1980. Long-term responses of streamflow following clearcutting and regrowth. Pp. 245-256 *in* Proceedings on the Influence of Man on the Hydrological Regime with Special Reference to Representative and Experimental Basins, Helsinki, Finland, June 1980. I.A.H.S. – A.I.S.H. Publication No. 130.

Swindel, B.F., Lassiter, C.J. and Riekerk, H. 1982. Effects of clearcutting and site preparation on water yields from slash pine forests. Forest Ecology and Management 4: 101-113.

Swindel, B.F., Lassiter, C.J. and Riekerk, H. 1983. Effects of different harvesting and site preparation operations on the peak flows of streams in *Pinus elliottii* flatwoods forests. Forest Ecology and Management 5: 77-86.

Terrene Institute. 1993. Proceedings of a technical workshop on sediments. Terrene Institute, Washington, D.C. 143 p.

Van Lear, D.H., Douglass, J.E., Fox, S.K. and Augsberger, M.K. 1985. Sediments and nutrient export in runoff from burned and harvested pine watersheds in the South Carolina Piedmont. Journal of Environmental Quality 14: 169-174.

Velbel, M.A. 1988. Chapter 6. Weathering and soil-forming processes. Pp. 93-110 *in* Swank, W.T. and Crossley, D.A. (eds..) Forest Hydrology and Ecology at Coweeta. Springer-Verlag, New York, NY. 469 p.

Verry, E.S. 1972. Effect of aspen clearcutting on water yield and quality in northern Minnesota. Pp. 276-284 *in* Csallany, S.C., McLaughlin, T.G. and Striffler, W.D. (eds.). Watersheds in Transition. Proceedings of a Symposium, Fort Collins, CO, June 19-22, 1972, American Water Resources Association and Colorado State University. 405 p.

Webster, J.R., Gurtz, M.E., Hains, J.J., Meyer, J.L., Swank, W.T., Waide, J.B. and Wallace, J.B. 1983. Stability of stream ecosystems. Pp. 355-395 *in* Barnes, J.R. and Minshall, G.W. (eds.). Stream Ecology, Plenum Press, New York, NY.

5.4   BIODIVERSITY AND FOREST HABITATS

P. Angelstam, G. Mikusinski and M. Breuss

The biodiversity concept was coined in the late 1980s to highlight the need for action to secure the long-term survival of species. Biodiversity is defined as diversity of habitats, species and genes as well as important ecosystem functions. One of the main threats to species is the loss of the habitats in which they live. In Europe, forest cover has decreased over centuries. In most of Western Europe forest cover has been reduced by more than 70% (Mayer 1984). In Costa Rica, the forest cover declined from 60% in 1940 to 17% in 1983 (Sader and Joyce 1988). Loss of natural forest ecosystems can be very rapid and is considered to be a major threat to global biodiversity (Whitmore 1990, Heywood 1995).

At the Earth Summit in Rio de Janeiro in 1992 (UNCED 1992), long-term maintenance of biodiversity was recognized as one of the main issues of sustainable development on a global scale, and was accepted as an official policy by most participating countries. The maintenance of biodiversity is thus an obligation that should influence the course of development in these countries. Several initiatives relating to sustainable use of all forest resources have been recently undertaken in Europe and throughout the world (e.g. Ministerial Conferences on the Protection of Forests in Europe (Liaison Unit in Lisbon 1998), The European Union's Forestry Strategy, Pan-European Biological and Landscape Diversity Strategy). From a biodiversity standpoint, the most significant result of these actions is a statement that sustainable management of forest resources requires that maintenance of forest biodiversity and timber production have equal value.

The aim of this section is to introduce the reader to an ecological view of forest energy production in managed landscapes. Problems related to sustainable forest management and biodiversity are reviewed in the context of potential intensification of bioenergy production. First, we introduce the basic background of conservation biology and landscape ecology. The landscape scale is emphasized to explain how habitat loss and fragmentation along with changing quality of the landscape, affect the survival and present status of populations. We then discuss the evolutionary past of the different biodiversity components and explain the need for formulating benchmarks against which Criteria and Indicators should be compared. Next, we present a series of practical examples of habitats, species and processes in forested landscapes. The problem of communicating new scientific findings about biodiversity maintenance to landscape managers and other stakeholders is discussed. Finally, attention is drawn to pitfalls and opportunities associated with present and future short-term energy production in terms of biodiversity conservation.

5.4.1 Understanding the mechanisms of extinction

Patches and landscapes

Spatial and temporal dynamics of forest ecosystems cause the amount and distribution of different habitat qualities to vary in time and space. Consequently, organisms dependent on these qualities have developed adaptations to cope with fluctuations in habitat characteristics. Landscape characteristics linked to the use of vital resources by a given species include the following:

- *Habitat patch size.* If the patch is too small, certain species will not be able to access their specific habitat requirements and will disappear, or not succeed in breeding. The size of a habitat patch is dependent on the needs of each species and is consequently highly variable, ranging from a single tree to several hundred hectares or more.

- *Patch quality.* Even if the patch is large enough, the quality of the habitat may prevent successful utilization. Patch quality in forested environments may depend on tree species, age class distribution, structural diversity, amount of dead wood, etc.

- *Number of patches and total habitat area.* If there are not enough patches of suitable quality, the total size of the habitat within a given area is insufficient, and species may disappear due to stochastic events and inbreeding depression.

- *Distance between patches.* The ability to move between patches, to colonize new patches or to exchange genes with populations in other patches is crucial. Some species are only able to disperse over short distances to other suitable patches.

- *Degree of hostility of the matrix between patches.* The nature of intervening areas can influence species dispersal ability. Semi-natural vegetation is likely to be easier to traverse than farmland, or urban areas.

In general, adequate spatial and temporal connectivity of stands of a particular forest type is a prerequisite for the occurrence of viable populations of different species associated with that forest type (e.g. Tilman and Kareiva 1997, Jansson and Angelstam 1999).

Habitat loss and fragmentation

The degree of fragmentation of forest cover is a very crude measurement of forest quality. Many species require specific habitat types produced by natural small and large-scale disturbances, or particular forest components (dead wood, hollow trees, old trees, forests with a stable microclimate). Properties such as the forest cover and patch size, factors such as tree species composition, age class distribution, and amount of snags and fallen dead wood are also important. Transformations of forest composition, structure and function associated with human activity take place at the following levels:

1. Fragmentation of forested areas or landscapes within a region.

2. Fragmentation of different habitats (successional stages, different specific site types, etc.) within the forest landscape.

3. Alteration of habitat quality within forest stands, e.g. by reduction of the amount of coarse woody debris, large trees, etc.

Even if forest cover in the landscape is intact, the quality of the forest at stand or tree level may be reduced to the extent that species are lost. The relationship between habitat change or fragmentation and species diversity is therefore complex. Species preferring old growth forest will decline in or disappear from fragmented woodland. If all or most of the old-growth forest in a larger geographical region is fragmented, overall species numbers are likely to decrease. However, if a large old-growth forest area is fragmented by harvesting or browsed by large herbivores (i.e. regeneration of certain tree species is prevented), there may be an increase in species preferring the forest edge ecotone (edges, bushes, open areas), and also an increase in habitat generalists. The pattern of declining old-growth species may therefore be masked by an increase in the number of generalist species.

Because of the long history of forest loss, there is a need for both conservation and the restoration of habitats in order to maintain forest biodiversity. Available analyses of the amount of habitat required for the maintenance of populations of forest species in Europe suggest that the size of the existing area with high conservation value is insufficient (Nilsson and Götmark 1992, Pressey *et al.* 1996, Angelstam and Andersson 2001). There is a need for habitat restoration if biodiversity is to be maintained. Competition for land to be used for conservation on the one hand, and for habitat restoration on the other, is likely. For example, the ancient system of woodland pastures and meadows provided a habitat that supported the flora and fauna of semi-open woodland (Kirby and Watkins 1998). When this land is abandoned for agricultural use, shrubs will encroach, trees will become more common and the canopy cover will increase. Later the amount of dead wood will increase. These processes represent the restoration of a forest with high conservation value. As both semi-open woodland and forests with old-growth characteristics are scarce, competition among different conservation interests for the same piece of land may eventuate.

In general there is no good match between the definitions of commercial, natural or plantation forests on the one hand and habitats with trees which are important for hosting biodiversity on the other. Forest biodiversity can be found outside traditionally-defined forest, and plantation forests may not sustain components of the original biodiversity.

Functional habitat connectivity

The concept of functional connectivity is a useful aid to mitigation of forest fragmentation problems because it includes the spatial configuration of habitats and the life-history traits of species populations (e.g. Forman 1995, Bennett 1998). Species differ according to the nature and the size of the forest environment that they require. Long-term population viability of specialized larger birds and large mammalian predators with extensive area requirements depends on the availability of a network of suitable forest patches at the landscape level (Mikusinski and Angelstam 1998, Storch 1999). Consideration of the needs of several species representing different forest environments will ensure that more landscape variables are included in problem analysis (cf. Lambeck 1997, Angelstam 1998a, b).

Critical thresholds for habitat loss

Ecological models and field data demonstrate the existence of critical thresholds in the amount and distribution of the habitat required for long-term survival of a given species. These thresholds have been derived from observations of patch utilization and are also a function of ability to move between patches. The former relates to areas where individuals live and reproduce, the latter to opportunity for adjustment to changing conditions in the environment.

Knowledge about the adequacy of natural features in a managed forest landscape is limited, and is mainly derived from observation of vertebrates (e.g. Andrén 1994). Theoretical models and empirical data suggest that there are thresholds in population response to habitat loss at the landscape level (e.g. Tilman and Kareiva 1997, With *et al.* 1997, Mönkkönen and Reunanen 1999). As the proportion of a given habitat declines, the chances of successful colonization of remaining fragments decrease in a stepwise manner. The effects of fragmentation are thus not linear. From models of landscape dynamics, Franklin and Forman (1987) found that patches start to become isolated when 70% of the original habitat remains. When 30% of the original habitat is left, fragmentation is complete. Continued habitat loss leads to an exponential increase in the distance between patches (Gustafson and Parker 1992, Andrén 1994).

Andrén (1994), in a review of the effects of habitat isolation on species richness at the landscape level, found that negative effects of reduced patch size and isolation do not occur until less than 20-30% of suitable habitat remains. This level of habitat loss appears to be critical for long-term maintenance of viable populations, at least for vertebrates.

To conclude: when about 30% of the original amount of suitable habitat remains, the rate of population decline becomes greater than would be expected from a relationship with the amount of habitat alone. When 5-10% of the original habitat remains, species have usually become locally extinct. On the basis of these results, a preliminary answer to the question: *'How much habitat loss is possible without loss of the most area-demanding species?'* would be 10-30% of the original habitat coverage.

Extinction debt

Thresholds of extinction indicate that deterioration of the habitat patch network reaches a point where a viable population cannot be supported. Extinction is a stochastic process and does not occur immediately. For example, two consecutive years with exceptionally unfavorable weather conditions might hasten extinction. The time period during which a species persists after habitat destruction is called 'time delay'. Theoretical considerations show that time delay is greatest in species for which the deteriorating environment is near the threshold for persistence (Hanski 1999). On the other hand, a small time delay is predicted if the quality of the current environment is poor. In nature, time delay also depends on species-specific factors such as the average life span of individuals.

Observation suggests that theoretically-determined time delay underestimates the risk of extinction. Following deterioration of a habitat patch network a number of species may persist even though eventual extinction is inevitable. The number of species expected to become extinct as a result of past environmental change is called 'extinction debt'. Before the extinction debt is replaced by actual extinction, the proportion of rare species will increase. It has been estimated that the magnitude of extinction debt in the forests of southern Finland is currently in the order of 1000 species (Hanski 2000). These species are expected to become extinct even if all remaining old-growth forest is protected. Further destruction of old-growth forest will obviously increase the extinction debt. Inertia associated with attempts to restore connectivity also increases the extinction debt (Tilman and Kareiva 1997).

5.4.2   Maintaining biodiversity in managed forest ecosystems

The evolutionary past

Natural authentic forest systems are usually dynamic and variable. By contrast intensively managed forests should be predictable and uniform. Maintenance of the authentic biodiversity therefore requires a new way of thinking based on management for variability rather than for one specified goal. Understanding of the evolutionary past of species is essential. One way to learn more about the conditions under which different organisms have evolved is to concentrate on the factors that affected the composition and structure of the original ecosystems. In many parts of the world the reference point is the natural forest. However, in regions with a long history of slow change and/or low-intensity management, habitats that have a high conservation value may already be present.

Naturalness as a point of reference.
The diversity of forest habitats in any naturally dynamic landscape is determined by differences in soil type, topography, climate, availability of nutrients and water on the one hand, and by disturbances on the other. Disturbance can be caused by non-living (abiotic) factors as well as by animals, fungi and plants. Abiotic disturbances may be large-scale (e.g. fire, hurricanes) (Hunter 1990, Johnson 1992, Shugart *et al.* 1992, Angelstam 1998a, 1998b, 1999), small-scale or localised (e.g. gap regeneration, flooding) (Engelmark *et al.* 1993, Falinski 1986, Frelich and Lorimer 1991, Röhrig 1991, Lugo *et al.* 1990). Examples of biotic disturbances are the effects of large herbivores on trees and plants; the influence of beavers *(Castor canadensis)* on hydrology and shore vegetation; and the consequences of insect and fungus disease. As a result, naturally dynamic forest habitats are usually complex systems consisting of a large number of different components, structures and processes. These are necessary for the maintenance of species populations. The temporal continuity of forest cover at the landscape level, the presence of an adequate and continuous supply of dead wood, and a humid microclimate are all examples of qualities indispensable to long-term species survival.

Ancient forms of land use as a point of reference.
Even in prehistoric times the gradual development of a cultured landscape created new conditions for fauna and flora (Selander 1957, Birks *et al.* 1988, Thirgood 1989, Peterken 1993, Angelstam 1996, 1997). In order to maintain summer and winter food for cows, sheep and other domestic animals, land has been managed through use of fire, mowing, clearing, pollarding and flooding. During periods of inappropriate land use, political instability or war, land has been severely disturbed or left idle. This range of disturbances has resulted in the maintenance, and even an increase in biodiversity. A long period of complex and extensive land use saw the extraction of relatively few resources, little input of energy and nutrients, and multiple use of several components. It resulted in a high level of species richness. As forestry and farming practices were intensified, biodiversity was reduced due to the loss of cultural diversity (e.g. Tucker and Evans 1997).

The village, with zones of different types of land use radiating from its center to the periphery, is a general and basic unit of European landscapes. Loss of authentic village structure is a threat to both cultural identity and biodiversity. The principles governing use of land in traditional rural society before the agricultural revolution were similar over large parts of Europe

(Birks *et al.* 1988, Stahl 1980). Village activities were strictly regulated either by law or by orally-communicated conventions. Fields, meadows and forests were the basic units of traditional land use. The relationships between different forms of land use varied due to physio-geographical factors, and influenced the economy. In mountain areas and remote parts of the landscape animal husbandry and grazing were predominant, while the use of arable land was more common in plains and valleys. As far as short-term maintenance of cultural biotopes and sustainable use of land was concerned, the rural society was in a kind of balance.

During the early- to mid-20th century, European farmland on less fertile soil was abandoned. Forest trees started to invade the boundaries between forest and farmland, and later were also planted. Simultaneously, smaller fields were combined into larger ones without intervening ditches, hedgerows and wetland. This took place at farm, region and country levels (Hart 1998). Historical maps and other sources provide excellent information about the past history of the cultural landscape (Rackham 1976, Peterken 1996) and allow the reconstruction of site history and land-use continuity. This information can be used to determine which sites are likely to be successfully restored.

Standards

In response to international and political concern about biodiversity, hierarchical standards for sustainable forest management have been developed (e.g. Lammerts van Buren and Blom 1997, Salem and Ullsten 1999). A standard is a set of principles, criteria and indicators that serves as a tool for the promotion of a desired development. In countries where timber production is an important economic activity there has been an increase in the rate of development of Criteria and Indicators (e.g. Elliott 1999). However, standards developed for monitoring and reporting at the regional and national level may not be fully compatible with standards used to assess the quality of forest management at the forest management unit or landscape level. Broadly-based standards should stimulate the development of methods for maintaining and measuring biodiversity (Noss 1999). It is essential to define quantitative operational benchmarks against which biodiversity measurements can be compared in order to assess the rate of progress (Noss *et al.* 1997, Bunnell and Johnson 1998, Angelstam and Andersson 2001). These benchmarks should be adjusted as new scientific information and experience become available; also when changes occur in forest ecosystems and related social systems (Lammerts van Bueren and Blom 1997).

Where are we now?

Management of biological resources, including forests, usually involves an increase in the density or growth rate of some naturally-occurring species at the expense of others, or a change in the structure of plant communities through reduction of the number of vegetation layers or removal of certain components. At some level of forest management intensity or duration, a lack of appropriate substrate or habitat will cause naturally-occurring species to decline to a level where there is a risk of extinction. Combination of sustainable wood production with maintenance of viable populations of all naturally-occurring species sets limits to the intensity and extent of management. Mixtures of tree species, dead logs on the forest floor, standing snags and very large trees are not compatible with the production of pulpwood and timber, which requires short rotation times compared with those of the evolutionary past. Today, increased awareness of the importance of biological diversity, green market pressure, and a shift from mono-functional to multi-functional forests have resulted in requests for guidelines for management maintenance of the structural components of naturally dynamic forests.

Forest history has been used during quantification of the loss of components promoting biological diversity (Linder and Östlund 1998). Records are often insufficiently detailed for assessment of changes to all components of biodiversity. For example, amounts of dead and fallen wood and deciduous trees were often not monitored because they were not considered to have substantial value. Information can be complemented with studies of change in important components and structures associated with the development of different types of forest use and management in different parts of the same ecosystem. With the opening up of the Iron Curtain at the beginning of the 1990s, more information from relevant and interesting studies has become available.

The development of forest management has generally followed a pattern of gradual change and distinct stages. The spatial pattern of areas with different histories has corresponding temporal aspect which can be described in the following sequence:

- *Pristine forest.* Natural structures and processes are intact. Man is a part of the system, but does not dominate it. There is no infrastructure in terms of roads or villages and no signs of commercial logging. These landscapes may have been settled during an earlier period but there is now no major modification of the main disturbance patterns that would endanger species or unbalance natural processes.

- *Early human use.* The most accessible pristine forests are used as a source of furbearers, fodder and eventually of large and valuable trees. Depending on the size of the resources, this stage lasts for decades or for hundreds of years. In many regions selective logging of large trees starts several hundred years before the commencement of any intensive utilization.

- *Forest exploitation.* The timber resource is depleted locally or regionally as a consequence of regional or international demand for wood.

- *Sustainable wood production.* Management for desired products is intensive and the objective is continuous productivity. The sustainability concept is gradually broadened to include maintenance of biodiversity.

- *Sustainable ecosystem management.* This is a scenario of the future in which biodiversity, protective functions, multiple use and sustainable productivity are all maintained.

Study of the development of landscapes on the European continent can assist understanding of the relationships between forest biodiversity and anthropogenic habitat change. For a long time human population expansion and economic growth in Europe has been associated with the gradual modification and removal of forests that once covered most of the continent (Delcourt and Delcourt 1987). Use of these forests has a complex and diverse history. The duration and intensity of impacts such as clearing, grazing and pollution have varied between different regions (Angelstam *et al.* 1997). The geopolitical location of different European countries has had profound effects on the amount of deforestation and the intensity of landscape transformation (Chirot 1989, Gunst 1989). Today, European forest types range from artificial plantations of exotic species on old farmland to remnants of ancient forested landscapes and large wilderness areas with little human intervention.

The present challenge concerning European forest environments is the maintenance of both forest resources and biological diversity in the long term. Because of the extended history of use and consequent loss and simplification of many forest ecosystems, future survival of many specialized forest species is uncertain in many European regions. The future also contains uncertainties associated with global change (Sykes *et al.* 1999) and new forms of land use, such as short-rotation crops.

Due to the factors discussed above, landscapes with authentic forest biodiversity are rare and unevenly distributed among regions in Europe (Dudley 1992). Most of the forested landscapes are now intensively managed and biological diversity is being reduced. Intensified management affected the ancient forest landscape in a similar way but over a longer period (Kirby and Watkins 1998). The role of historical development in maintaining forest biodiversity is gradually being recognized in Europe. Forest species were able to survive outside forested areas because the components they require (e.g. old and large trees, pollarded and lopped trees, etc.) were valued and maintained by the human population.

5.4.3   Case studies - species, habitats and processes

Deciduous forest in the European boreal zone

In the boreal forest the natural spatial and temporal distribution patterns of the deciduous forest component result from combinations of local abiotic factors (e.g. soils and nutrients) and disturbances (e.g. fire, water, wind). The distribution and size of deciduous stands in unmanaged landscapes is consequently very variable. In a frequently-occurring pattern, deciduous trees dominate early and middle stages of the succession after disturbance (e.g. fire), forming linear stands along running water (riverine forest), and finally contiguous stands in specific localities.

Agriculture and forestry have dramatically changed the natural patterns. The proportion of deciduous stands of all types has diminished, and present managed landscapes contain only remnants of the deciduous component which was once widely distributed. In a study of the present distribution of the deciduous component in the southern boreal forest, Mikusinski and Angelstam (1999) found that at the landscape level, the majority of deciduous forests were close to farmsteads, i.e. on land presently or historically used for agriculture. The proportion of deciduous stands declined with distance from human settlements. Decline in the proportion of old deciduous stands was particularly apparent. The presence of deciduous trees in the vicinity of farmsteads may be explained by spontaneous succession in abandoned fields, less intensive forest management in privately-owned forests, and by their aesthetic and practical value. Current scarcity of a deciduous component in forest-dominated parts of the landscape may be explained by the absence of natural fire events, by the promotion of coniferous species for intensive forest management, and by the fact that aesthetic values are less important in remote forests.

The amount and distribution of the deciduous component in the landscape has a profound effect on the occurrence of dependent organisms (e.g. specialized birds, lichens and mosses) (Carlson 2000, Enoksson *et al.* 1995, Jansson and Angelstam 1999, Kuusinen 1996). Consequently several species may find a suitable habitat only in the vicinity of human settlement. More remote forests do not meet the requirements of some deciduous specialist species. Certain types of land use clearly sustain the existence of deciduous stands in a forest landscape.

Browsing and moose

The long history of forest use and management in Sweden has promoted conifer-dominated forests at the expense of deciduous trees such as *Populus tremula*, *Salix caprea* and *Sorbus aucuparia*. The moose (*Alces alces*) is an animal species important to the maintenance of biodiversity associated with these tree species and to the production of high quality Scots pine (*Pinus sylvestris*) timber. In order to maintain biodiversity there is a need to restore the deciduous forest component, which is the preferred food of moose. If the ratio between moose density and its preferred food is too high, this restoration will be difficult.

In order to study interactions between the abundance of preferred moose food, moose density and damage to trees, it is necessary to include landscapes with a broader range of combinations of food abundance and moose density than can be found in Sweden. Landscape and management in Sweden tend to be homogeneous, and the same policies on forestry and moose management are implemented throughout. To include a wide range of factor combinations, Angelstam *et al.* (2000) studied eight landscapes in Sweden, Finland and Russian Karelia.

Moose damage to deciduous trees and to Scots pine in pine-dominated stands was correlated with moose density. Damage was most severe in Sweden, intermediate in Finland and least apparent in Russian Karelia. Winter moose densities ranged from 1.7 km^{-2} in Sweden to 0.2 km^{-2} in Russia. The cover of preferred food species (*Populus, Salix, Sorbus*) was 13 times greater in Russia than in Sweden. The proportion of severely-damaged and dead trees of these species was 36 times greater in the most-affected landscape than in the least-affected area. Damage to Scots pine ranged from 57% in Sweden to 7% in Russian Karelia.

Unless moose damage is reduced in Sweden it is doubtful whether deciduous vegetation can be maintained. Biodiversity is affected in either case. Communication with different interest groups is essential if this socio-economic problem is to be resolved. A feasible solution may develop from co-management case studies based on a holistic landscape view and an objective inventory of the problems perceived by all interest groups.

Red-cockaded woodpecker

The red-cockaded woodpecker (*Picoides borealis)* is a habitat specialist confined to the pine forests of the southeastern United States (Jackson 1994). This small, social species, unlike other American woodpeckers, breeds in nesting holes excavated in old living pine trees. The excavation process can take up to 10 years, and forest stands with cavity trees are of great importance to the species. The age of pine trees selected for foraging is usually more than 60 years (Zwicker and Walters 1999). This woodpecker has low dispersal ability and young males often remain in parental territory (Walters *et al.* 1992). The spatial structure of the population is based on clusters of cavity trees occupied and maintained by family groups (Engstrom and Mikusinski 1998).

Once regarded as a common species, the red-cockaded woodpecker has disappeared from most of its range and was officially listed as endangered in

1968. This decline is a well-documented example of the cumulative effects of human impact on forest habitats at both landscape and stand levels and of the consequences for species persistence (Thompson 1971, Wood 1983, Kulhavy *et al.* 1995).

Southern pine once formed the major forest ecosystem in the southeastern USA. An important natural feature of these pine forests is the frequent occurrence of fires which maintain the shrub-free character if the open woodland. Gradual conversion to alternative land use, fire suppression, and the introduction of modern forestry operations, have eliminated most of the suitable habitats of the red-cockaded woodpecker.

Intensive logging of southern pine forests between 1870 and 1930 caused fragmentation of the formerly contiguous geographic range of the species into several smaller, more isolated populations (Conner *et al.* 1997). Populations continued to decline into the 1980s as a result of fire suppression and the introduction of short-rotation forestry. Encroachment of hardwoods associated with fire suppression increases competition with other cavity-nesting species and eventually makes the habitat unsuitable. Short-rotation forestry reduces availability of food as well as the availability of trees suitable for nesting.

Intensive and costly efforts have been made to manage red-cockaded woodpecker populations on federal land. These include the introduction of prescribed burns, use of cavity restrictors to discourage competitors, development of artificial cavity construction techniques, and transfer of individuals between subpopulations (Thompson 1971, Wood 1983, Kulhavy *et al.* 1995). The problem of species maintenance is still far from being solved. Recent developments in species management may result in improvement of habitat availability and quality not only for the red-cockaded woodpecker but also for other species that are adapted and dependent on southern pine ecosystems (Conner and Dickson 1998, Masters *et al.* 1998, Plentovich *et al.* 1998). The adjustment of silvicultural methods to mimic natural disturbance (e.g. frequent fires) seems to be important in this context (Engstrom *et al.* 1995).

5.4.4 Tools for putting forest biodiversity conservation into practice

Adaptive management and the monitoring loop

A vital part of policy implementation is evaluation of the extent to which desired changes are actually made. The aim of monitoring is to apply short-term techniques in order to attain long-term goals. To be effective, the monitoring of biological systems must have a sound scientific basis, be diagnostic to help with understanding of the system, allow assessment of stated policy objectives and provide feedback as a basis for mid course modification of policy or management. Maintaining simultaneous focus at several levels of forest management is difficult in itself. Moreover, changes in the amount of different age classes, patch size and habitat connectivity are gradual and take time. While measurements may indicate trends in the desired direction, it is not always clear whether the rate and the extent of change are sufficient for maintenance of viable populations of all naturally-occurring species.

Continuous monitoring is a vital component of Adaptive Management. Biodiversity assessments (= indicators) should be part of an iterative process with a secured information flow. Adaptive Management can be described as 'learning by doing'. It represents progression from the 'conventional wisdom' and 'best current data' approaches to a sophisticated 'monitoring and modifying' approach (see also Section 5.1). It has been attempted in the management of natural resources, mainly in North America (Walters 1986). The advantage of Adaptive Management is that it involves a diversity of interest groups, including scientists and practitioners. In this way the gap between science and policy can be reduced, and institutions with incentives for combining active management with conservation can be developed. This should remove unnecessary tension between management and conservation. Adaptive management is designed to improve on trial-and-error learning and to integrate uncertainty with management strategies. It endorses practices that confer resilience i.e. improvement of the capacity of an ecosystem to absorb disturbance and maintain opportunity for renewal, reorganization and recovery. By responding to and managing feedback, instead of blocking it out, Adaptive Management seeks to avoid the crossing of ecological thresholds at scales that threaten the existence of social and economic activities. The iterative process includes the following steps:

1. Define a criterion.
2. Derive a guideline.
3. Find an indicator.
4. Monitor, and then go back to 2.

The landscape as an integrative concept

What is a landscape?
In most European languages the word 'landscape' is related to modification of natural scenery for the benefit of human welfare. In the sciences of biology and geography a landscape is defined as a bio-geographic unit in which abiotic and biotic conditions interact with and set limits to the historical and cultural dimensions of human use. The size of a landscape can be several hundred to several thousand hectares, i.e. larger than a forest management unit and smaller than a region.

The ecological biodiversity of a given landscape or community depends on a spectrum of abiotic, biotic, historical and political factors (Brouwer *et al.* 1991, Turner *et al.* 1990, Angelstam 1997, Hart 1998). This is obvious at all levels from global and regional to a few hundred square meters. Development of a scientific basis for the maintenance and restoration of biodiversity requires the inclusion of a broadly-based set of scientific disciplines in planning analysis activities.

The landscape as a link between policy and practice.
Ideally, planning on different scales ranging from trees-in-stands to landscapes-within-regions should be integrated through top-down and bottom-up procedures. A policy can be viewed as the definition of a vision for subsequent successful implementation. To feed visions into the real world there must be a recipient, and the vision must be put into practice. Conservation planning for the establishment of networks of protected representative areas is an example of an important type of top-down planning (Noss *et al.* 1997, Prendergast *et al.* 1999). This should be complemented by bottom-up integration. Forest certification can be used as an example. In the 1990s, forest certification was initiated by non-governmental organizations (NGOs) to promote development of sustainable forest management and to provide a better market for products meeting certain standards. From an analysis of the development of forest certification programs in Indonesia, Canada and Sweden, Elliott (1999) concluded that certification can be best understood as a policy instrument which promotes and facilitates learning among all participants during the development and implementation of standards. Certification provides direct incentives for improved management practice at forest management unit level. It encourages consensus-building among groups that have often been in conflict with each other, such as NGOs, forest owners, indigenous people and governments. The landscape level provides a compromise between top-down and bottom-up approaches.

Maintenance of biodiversity in the landscape.
To maintain forest biodiversity it is essential to incorporate multiple environmental values into landscape analysis. Several arguments exist for the landscape approach: (1) It can encompass a holistic view including both natural and cultural aspects of forest biodiversity. (2) It can recognize that individuals of many species require large areas and that all populations need large areas for long-term maintenance of viable populations. (3) It can accommodate retention of patch dynamics of different successional stages including old-growth forest requiring management and conservation over large areas.

'Umbrella' species as pedagogic tools

In conservation planning, the 'umbrella species' concept appears to be a particularly interesting tool for facilitating communication of information regarding forest biodiversity to stakeholders (Lambeck 1997, Thompson and Angelstam 1998). The selection of umbrella species should be based on detailed knowledge about the quantitative requirements of the most area-demanding species in each forest environment. Research on several specialized species with large area requirements found in different forest environments supports the umbrella species concept. For example, it has been shown for the capercaillie (*Tetrao urogallus*), white-backed woodpecker (*Dendrocopos leucotos*), three-toed woodpecker *(Picoides tridactylus)*, and long-tailed tit (*Aegithalus caudatus*) that a variety of other species co-occur in the particular forest types that they inhabit (e.g. Martikainen *et al.* 1998, Mikusinski *et al.* 2001).

Analysis of the gaps in habitat

An issue of paramount importance in applied ecology is the achievement of a balance between use of renewable resources and maintenance of biological diversity (Hunter 1999). Forests provide a particularly clear example of this problem. During the 1990s there was a strong international trend for forest management to achieve objectives other than wood production (e.g. Liaison Unit in Lisbon 1998). As a result, several approaches to systematic conservation planning have been proposed (e.g. Noss *et al.* 1997, Margules and Pressey 2000). Estimates of the need for conservation areas are made according to the principle of representativeness of different ecosystems (e.g. Pressey *et al.* 1996) and estimates of gap size in areas of protected forest with high conservation value. Hence the term 'gap-analysis' (cf. Scott *et al.* 1987, 1988, 1989, 1993, Iacobelli *et al.* 1995, Jennings 2000). Ideally, networks of conservation areas would be created to secure viable populations of the most demanding species in each forest ecosystem. Restoration and re-creation of forest environments may be needed to achieve the long-term goal of biodiversity conservation if the present area of forest with conservation values is inadequate.

Recently, Angelstam and Andersson (2001) developed a procedure for determining the need for forest biodiversity conservation reserves representative of the main forest types in a region. The goal was to present quantitative estimates of the need for reserves and for restoration in different biogeographical regions. Their starting point was estimation of the occurrence of different environments found in typical forest landscapes and in ancient cultural landscapes existing before intensification of land use during the industrial and agricultural revolutions. The disparity between the long-term goal of forest protection and the area of forest available for protection underlines the urgency for maintenance of existing remnants of natural and cultural forest environments. Furthermore, forest protection alone will not achieve the biodiversity maintenance goal.

5.4.5 Forest management for bioenergy production and practical biodiversity conservation: pitfalls and opportunities

The aim of traditional forest management is the allocation of resources to individual trees that will grow large enough to be harvested. This is most often done by using species that, under local natural conditions, satisfy the demands of the timber and/or pulp market most efficiently. Forest history has shown that even in very different forest ecosystems, the pattern is quite similar, i.e. species-rich and structurally diverse forests have been replaced by stands consisting of only a few tree species often forming a cohort of individuals of the same age and size. In the most extreme but widespread case, the 'new forest' consists of one exotic tree species.

Rotation length in most forest management systems is generally shorter than the time period between naturally-occurring disturbances. This means that several processes essential for forest biodiversity will not be maintained. As discussed earlier, processes such as ageing of individual trees or production of dead wood of different qualities are limited by forestry practice. Also, at the landscape level, the amounts, spatial distribution, and temporal continuity of habitat patches are altered. In general, the extent of the impact of forestry on biodiversity is huge. In other words, forest management and the maintenance of forest biodiversity have so far been largely incompatible.

Today, scientific knowledge and technological development have reached a level where practical biodiversity management can begin. Since critical habitat thresholds have already been transgressed by forestry, active restoration of forest habitats to an acceptable condition from the standpoint of long-term biodiversity maintenance is a matter of urgency. Land abandonment in countries throughout the world provides a unique opportunity for the creation of forest landscapes and stands which are both 'biodiversity friendly' and economically sustainable (Young 2000). This opportunity must be regarded as one of the most important challenges in the coming years.

Increasing interest in forest biomass as a source of renewable energy will certainly influence development of forestry in the future. In conventional forestry, two major sources of bioenergy are identified. One consists of forest residues, defined either as all above-ground biomass left on the ground after timber harvesting operations (logging residues) or as pre-commercial thinnings in young stands (silvicultural residues). The second source includes stands that should be thinned for silvicultural reasons but are left untreated due to lack of demand for small-sized wood. It also includes individual trees that are rejected because of their species, small size, or inferior quality. Overmature, unmanaged forests contain large amounts of unmerchantable biomass in the form of undersized, rough, rotten, and dead trees (Kirby *et al.* 1998, Majewski *et al.* 1995, Nilsson *et al.* in press). After logging, a large proportion of stemwood is left as residue because of its low quality. Although these residues fulfill the quality requirements for fuelwood, unmanaged forests are often remote and recovery is unprofitable (Hakkila and Parikka 2001).

Increasing demand for bioenergy from commercial forestry will probably speed up the afforestation of abandoned land. Although this process has a large potential in terms of restoration and maintenance of forest biodiversity, the use of residues will intensify. In the following subsections we explore these issues, focusing on present and future land-use changes in Europe and also on the expected effects of forest residue removal on biodiversity at the stand level.

Increased forest cover in Europe - a threat or an opportunity?

Forest cover in Europe reached a minimum a century or more ago. In contrast to the past long history of forest decline, the present trend is an increase in both forest cover and total timber volume (Nilsson *et al.* 1992, UN-ECE/FAO 2000). If data on forest cover are reviewed superficially, increase in forest cover may be interpreted as mitigation of problems related to the maintenance of forest biodiversity. However, most of the increase has resulted from conversion of former farmland and grassland into conifer plantations. These forests and other plantations of woody species are often not well-designed for the hosting of species assemblages typical of former natural forest habitats. Although they usually support a very low proportion of the original forest biodiversity, they could, if carefully designed, provide opportunities for restoration of many habitats in a wide range of European landscapes. This applies in particular to regions that have already lost most of the ancient cultural landscapes and natural forest areas.

Most European forests have been influenced by a variety of changes. Present landscapes diverge greatly from the natural patterns and processes of the original forest landscapes and from cultural landscapes with soft transitions between grasslands and ancient woodlands (McNeely 1994). The difference between the forest components of these two landscape types is smaller than might be expected. Several characteristics important to maintenance of forest biodiversity are now absent or rare where there is a long history of modern forest and agricultural management. These characteristics include old-growth forests and old large trees, diversified tree species composition, a diverse range of structural forest components, dead standing and fallen trees, and balanced natural processes (e.g. browsing, predation, nutrient supply).

Restoration of forest cover should aim to create structurally diverse stands of native species, the juxtaposition of stands being arranged managed to preserve individual species as well as species interactions, and connectivity being restored between different habitat types and characteristics. These requirements can be achieved by forest management activities in specific steps:

- Selection of natural forests that should be protected; in most regions with a long forest history this means all natural or near-natural forests.

- Conservation of semi-natural forests.

- Improvement of artificial forests.

- Restoration of forest cover where biotopes and structures are lacking.

Afforestation has great potential for improvement of the status of biodiversity throughout European landscapes, providing that current scientific knowledge will be communicated, understood and finally put into practice. In the context of bioenergy production, exercise of caution when considering use of unmanaged forests is particularly important. From the biodiversity point of view, overmature, unmanaged forests containing large amounts of undersized, rough, rotten, and dead trees are particularly valuable (Samuelsson *et al.* 1994).

Forest quality - maintenance of natural forest components

Dead wood

The influence of forestry practices on the amount of woody debris in a forest is an example of the alteration of habitat quality within stands. Dead wood provides a foraging substrate, habitat or shelter for various forest species (Samuelsson *et al.* 1994). In naturally dynamic temperate and boreal forest ecosystems, the amount and quality of dead wood is related to disturbance and usually makes up a substantial part of the total wood volume (Harmon *et al.* 1986, Spies *et al.* 1988, Peterken 1996). A large number of species are dependent on a continuous supply of dead wood with specific characteristics (Chandler 1987, Berg *et al.* 1994, 1995, Bredesen *et al.* 1997, Ohlson *et al.* 1997, Jonsell *et al.* 1998, Bunnell *et al.* 1999, Ehnström 2001). It has been shown that removal of natural or anthropogenic dead wood from the forest reduces the number of biodiversity components in the soil and above-ground (Du Plessis 1995, Bengtsson *et al.* 1997, Mahmood *et al.* 1999). By limiting tree age, forest management very efficiently reduces the amount of woody debris. Therefore, the difference in dead wood availability between managed and unmanaged forests is usually very large (Andersson and Hytteborn 1991, Guby and Dobbertin 1996, Kirby *et al.* 1998, Siitonen *et al.* 2000).

Intensification of bioenergy production from forest residues will reduce the amount of dead wood in managed forests. It is therefore necessary to identify critical thresholds in habitat quality for representative species that are dependent on dead wood (Ehnström 2001). The role of larger snags as foraging and breeding substrates (providing cavities) has already been recognized and in some countries forest management includes their retention (Angelstam and Pettersson 1997, Ehnström 2001). Cut stumps and logs in different stages of decay are important substrates for some species (Punttila *et al.* 1991, Kaila *et al.* 1997, Mikusinski 1997, Jonsell *et al.* 1998). The amount of small stems, branches and twigs on the ground seems to be higher in managed (silvicultural residues) than in unmanaged forests. For coarse woody debris the position is reversed. It has been shown that even fine woody debris may be important for forest biodiversity (Kruys and Jonsson 1999). Bioenergy production from forest residues aims to remove large quantities of dead wood, the presence of which is necessary for the long-term maintenance of forest biodiversity. The challenge is to find a balance to secure both ecological and economic sustainability of such operations.

Large and old trees
The number and distribution of large trees in natural forests is closely related to stand productivity and disturbance (Peterken 1996). In past forest use, large trees have been the main targets for logging. At present, intensive forest management with relatively short rotations usually produces individuals that are smaller than those in naturally dynamic stands. Large trees are almost completely absent from managed forests (Kirby *et al.* 1998, Linder and Östlund 1998, Nilsson *et al.*, in press). Since the presence of large and old trees has been identified as one of the most important forest qualities for many plants, invertebrates and vertebrates, the degree of threat to species dependent on such trees is very high (Bunnell and Huggard 1999, Bunnell *et al.* 1999). Particularly important features of such trees are the presence of several microhabitats with associated fauna and flora, presence of natural cavities, and prolonged temporal continuity. There is general agreement that an increase in these characteristics in managed forests is essential for sustainable forestry. However, the amount of these qualities required in different forest types to assure the maintenance of associated biodiversity has not been determined.

Since there is a general scarcity or absence of large trees in managed forests, the use of these trees as a biomass source for bioenergy production should not be permitted. A policy of protecting sizable trees during their life and during their decay is advisable. At present, most trees of this type are found in rural landscapes. Special attention should be given to the preparation of afforestation plans for abandoned agricultural areas that include large and old trees.

5.4.6   Summary

Rapid loss of natural forest ecosystems is considered to be a major threat to global biodiversity. Since the maintenance of biodiversity is a major issue in sustainable development, several initiatives have been recently undertaken to assure sustainable use not only of timber, but also of all other forest resources. In this section we reviewed problems of management of forest biodiversity in the context of potential intensification of bioenergy production. We introduced the reader to principles of conservation biology and landscape ecology that give background for further discussion on the maintenance of forest biodiversity in managed landscapes. We placed special emphasis on the following points:

- Forest ecosystems are dynamic entities with qualities that vary in time and space. Species developed adaptations in response to these fluctuations in habitat characteristics. The distribution and size of habitat patches over time influence the occurrence of viable populations of species associated with forest ecosystems.

- Anthoropogenic transformation of forest ecosystems causes habitat loss and fragmentation that, in some cases, leads to local and global extinction of species. On all scales the reduction in the amount of habitat and changes in habitat quality have been so extensive, that the adaptive potential of many species has been inadequate for survival.

- In order to successfully manage forest biodiversity, we need to understand the degree of tolerance of other species tolerance to man-made changes. The concept of functional connectivity, which encompasses both spatial configuration of habitats, and life-history traits, is a useful aid. The identification of critical habitat loss thresholds for organisms linked to specific forest types would allow quantitative goals to be set for forest biodiversity management and restoration.

- Transformation of forest ecosystems in managed landscapes is often so complete that relationships between natural processes, landscape composition and biodiversity are obscured. Studies in areas with a high degree of naturalness may provide the necessary reference points. Ancient cultural landscapes where an equilibrium between biodiversity and human management has been attained at a sustainable level may also provide information. The European continent, with its clear record of forest history, provides forested landscapes ranging from monocultures of exotic tree species in the West to extensive, naturally dynamic forests in the East.

- Hierarchical standards for sustainable forest management including management of biodiversity have been developed. Principles, Criteria and Indicators are tools which help to answer the basic question: 'How much is enough?' To make them operational and successful in maintaining and restoring forest biodiversity, methods for continuous assessment of forest management based on new scientific information applicable to local conditions must be developed.

- The history of forest use generally has the following pattern: pristine forest; moderate human utilization; human exploitation; sustainable wood production. Today, forestry aims at the next stage, which is sustainable ecosystem management.

- Genes and processes are ecosystem components that define biodiversity in terms of species and habitats. Deciduous forest provides habitats in northern Europe which are now rare in the managed landscape. Several species dependent on this habitat have become endangered and there are clear indications that tolerance levels in terms of degree of habitat loss, have been exceeded. Browsing by unnaturally high moose populations in Scandinavian forests is an example of a natural process that, due to intensive forestry practice, has become a problem in the management of biodiversity and timber production. Highly specialised species are usually the least tolerant of intensive forest management and landscape change. In the southeastern USA, the red-cockaded woodpecker is an example of a species which due to a high degree of dependence on forests subject to a natural fire disturbance, cannot survive without specialized forest management.

- Conservation of forest biodiversity requires operational tools. Adaptive Management which includes a monitoring loop is one of these. Ecological planning uses the landscape concept in order to integrate the activity of groups and disciplines needed for the attainment of sustainability. The 'umbrella species' concept is useful for communicating conservation problems to people with an interest in sustainable forest management.

- Analysis of gaps in the maintenance of biodiversity in managed landscapes shows that areas of natural forest in the form of reserves are too small. Conservation of semi-natural forests, improvement of artificial forests, and forest restoration will be needed in order to provide functional habitat connectivity.

- Forest management has been aimed at allocation of resources to trees that will provide harvestable yield. Attainment of this goal involved the structural simplification of forests, through use of shorter rotation lengths, fewer tree species, etc. These practices have limited or eradicated habitats for many species and caused a decline in forest biodiversity.

- Large and old trees, standing and fallen deadwood, and burned areas constitute important habitats that become rare or absent in intensively managed forests. Species dependent on such habitats become endangered or extinct in these forests.

- On the European continent, afforestation can provide an opportunity for quantitative and qualitative restoration of biodiversity if appropriate habitat requirements can be supplied. Special attention should be given to diverse habitats when new forests are planted or trees are allowed to colonize ancient cultural landscapes.

- In order to maintain biodiversity, forest management for bioenergy must take account of the following pitfalls and opportunities:

  a) The demand for forest energy may accelerate the afforestation of agricultural land, thus providing an opportunity for restoration of forest cover and provision of valuable forest habitats. If these new forests are simple monocultures replacing a semi-natural cultural landscape, opportunity will be lost or the opposite effect may result.

b) Forest energy production should not impoverish biodiversity in managed forested landscapes. In particular, use of large and old trees for energy production should be prohibited since they are increasingly rare structures in most landscapes.

c) Use of dead wood for energy production should be considered carefully. Standing and fallen dead wood in different stages of decomposition is a habitat for many species. Managed forests usually contain a very limited amount of dead wood when compared with natural forests. Larger snags and logs are especially valuable in terms of biodiversity.

5.4.7   References

Andersson, L.I. and Hytteborn, H. 1991. Bryophytes and decaying wood – a comparison between managed and natural forests. Holarctic Ecology 14: 121-130.

Andrén, H. 1994. Effects of habitat fragmentation on birds and mammals in landscapes with different proportions of suitable habitat: a review. Oikos 71: 355-366.

Angelstam, P. 1996. Ghost of forest past - natural disturbance regimes as a basis for reconstruction of biologically diverse forests in Europe. Pp. 287-337 *in* DeGraaf, R. and Miller, R.I. (eds.). Conservation of faunal diversity in forested landscapes. Chapman & Hall.

Angelstam, P. 1997. Landscape analysis as a tool for the scientific management of biodiversity. Ecological Bulletins 46: 140-170.

Angelstam, P. 1998a. Maintaining and restoring biodiversity by developing natural disturbance regimes in European boreal forest. Journal of Vegetation Science 9: 593-602.

Angelstam, P. 1998b. Towards a logic for assessing biodiversity in boreal forest. Pp. 301-313 *in* P. Bachmann, M. Köhl and R. Päivinen (eds.). Assessment of Biodiversity for Improved Forest Planning. Kluwer Academic Publishers, Dordrecht, the Netherlands.

Angelstam, P. 1999. Reference areas as a tool for sustaining forest biodiversity in managed landscapes. Naturschutz report 16: 96-121. Landesanstalt für Umweltschutz, Thuringia, Germany.

Angelstam, P. and Andersson, L. 2001. Estimates of the needs for forest reserves in Sweden. Scandinavian Journal of Forest Research. Supplement 3: 38-51.

Angelstam, P., Anufriev, V., Balciauskas, L., Blagovidov, A., Borgegård, S-O., Hodge, S., Majewski, P., Shvarts, E., Tishkov, A., Tomialojc, L. and Wesolowski, L. 1997. Biodiversity and sustainable forestry in European forests - how west and east can learn from each other. Wildlife Society Bulletin 25: 38-48.

Angelstam, P. and Pettersson, B. 1997. Principles of present Swedish forest biodiversity management. Ecological Bulletins 46:191-203.

Angelstam, P., Wikberg, P.E, Danilov, P., Faber, W.E. and Nygrén, K. 2000. Effects of moose density on timber quality and biodiversity restoration in Sweden, Finland and Russian Karelia. Alces 36: 133-145.

Bengtsson, J., Persson, T. and Lundkvist, H. 1997. Long-term effects of logging residue addition and removal on macroarthropods and enchytraeids. Journal of Applied Ecology 34: 1014-1022.

Bennett, A.F. 1998. Linkages in the landscape: the role of corridors and connectivity in wildlife conservation. Switzerland and Cambridge: IUCN, Gland.

Berg, Å., Ehnström, B., Gustafsson, L., Hallingbäck, T., Jonsell, M. and Weslien, J. 1994. Threatened plant, animal and fungus species in Swedish forests - distribution and habitat associations. Conservation Biology 8: 718-731.

Berg, Å., Ehnström, B., Gustafsson, L., Hallingbäck, T., Jonsell, M. and Weslien, J. 1995. Threat levels and threat to red-listed species in Swedish forests. Conservation Biology 9: 1629-1633.

Birks, H.H., Birks, H.J.B., Kaland, P.E. and Moe, D. (eds.). 1988. The cultural landscape - past, present and future. Cambridge University Press. 521 p.

Bredesen, B., Haugan, R., Aanderaa, R., Lindblad, I., Okland, B. and Rosok, O. 1997. Wood-inhabiting fungi as indicators on ecological continuity within spruce forests of southeastern Norway. Blyttia 54: 131-140.

Brouwer, F.M., Thomas, A.J. and Chadwick, M.J. 1991. Land use changes in Europe. Processes of change, environmental transformations and future patterns. Kluwer Academic Publishers. Dordrecht. 528 p.

Bunnell, F.L. and Johnson, J.F. 1998. Policy and practices for biodiversity in managed forests: The living dance. University of British Columbia Press. Vancouver, Canada. 162 p.

Bunnell, F.L. and Huggard, D.J. 1999. Biodiversity across spatial and temporal scales: problems and opportunities. Forest Ecology and Management 115: 113-126.

Bunnell, F.L., Kremsater, L.L. and Wind, E. 1999. Managing to sustain vertebrate richness in forests of the Pacific Northwest: relationships within stands. Environ. Rev. 7: 97-146.

Carlson, A. 2000. The effect of habitat loss on a deciduous forest specialist species: the White-backed Woodpecker (*Dendrocopos leucotos*). Forest Ecology and Management 131: 215-221.

Chandler, D.S. 1987. Species richness and abundance of Pselaphidae (Coleoptera) in old-growth and 40-years old forest in New Hampshire. Canadian Journal of Zoology 65: 608-615.

Chirot, D. 1989. The Origins of Backwardness in Eastern Europe. Economics & Politics from the Middle Ages until the Early Twentieth Century. University of California Press, Berkeley.

Conner, R.N., Craig, R.D., Saenz, D. and Coulson, R.N. 1997. The red-cockaded woodpecker's role in the southern pine ecosystem: population trends and relationships with southern pine beetles. Texas Journal of Science 49 (3 supplement): 139-154.

Conner, R.N. and Dickson, J.G. 1998. Relationships between bird communities and forest age, structure, species composition and fragmentation in the West Gulf Coastal Plain. Texas Journal of Science 49 (3 supplement): 123-138.

Delcourt, P.A. and Delcourt, H.R. 1987. Long-term forest dynamics of the temperate zone. Ecological Studies 63, Chapter 10.

Dudley, N. 1992. Forests in trouble: A review of the status of temperate forests worldwide. World Wildlife Fund. Gland, Switzerland.

Du Plessis, M.A. 1995. The effects of fuelwood removal on the diversity of some cavity-using birds and mammals in South Africa. Biological Conservation 74: 77-82.

Ehnström, B. 2001. Leaving dead wood for insects in boreal forests – suggestions for the future. Scandinavian Journal of Forest Research. Supplement 3: 91-98.

Elliott, C. 1999. Forest certification: analysis from a policy network perspective. Ecole Polytechnique Federale de Lausanne. Switzerland. 464 p.

Engelmark, O., Bradshaw, R. and Bergeron, Y. 1993. Disturbance dynamics in boreal forest. Opulus Press.

Engstrom, R., Brennan, L.A., Neel, W.L., Farrar, R.M., Lindeman, S.T., Moser, W.K. and Hermann, S.M. 1995. Silvicultural practices and Red-cockaded Woodpecker management: A reply to Rudolph and Conner. Wildlife Society Bulletin 24: 334-338.

Engstrom, R.T. and Mikusinski, G. 1998. Ecological neighborhoods in red-cockaded woodpecker populations. Auk 115: 473-478.

Enoksson, B., Angelstam, P. and Larsson, K. 1995. Deciduous forest and resident birds: the problem of fragmentation within a coniferous forest landscape. Landscape Ecology 10: 267-275.

Falinski, J.B. 1986. Vegetation dynamics in temperate lowland primeval forests. Dr. W. Junk Publishers. Dordrecht. 537 p.

Forman, R.T.T. 1995. Land mosaics the ecology of landscapes and regions. Cambridge University Press. 632 p.

Franklin, J. and Forman, R.T.T. 1987. Creating landscape patterns by forest cutting: ecological consequences and principles. Landscape Ecology 1: 5-18.

Frelich, L.E. and Lorimer, C.G. 1991. Natural disturbance regimes in hemlock-hardwood forests of the Upper Great Lakes Region. Ecological Monographs 61: 159-162.

Guby, N.A.B. and Dobbertin, M. 1996. Quantitative estimates of coarse woody debris and standing trees in selected Swiss forests. Global Ecology and Biogeography Letters 5: 327-338.

Gunst, P. 1989. Agrarian Systems of Central and Eastern Europe. Pp. 53-91 *in* Chirot, D. (ed.). The Origins of Backwardness in Eastern Europe. University of California Press.

Gustafson, E.J. and Parker, G.R. 1992. Relationships between landcover proportions and indices of landscape spatial pattern. Landscape Ecology 7: 101-110.

Hakkila, P. and Parikka, M. 2001.Chapter 2 Fuel resources from the forest. *In* Richardson, J., Björheden, R., Hakkila, P., Lowe, A.T. and Smith, C.T. (eds.). 2001. Bioenergy from Sustainable Forestry: Guiding Principles and Practice. Kluwer Academic Publishers, Dordrecht, The Netherlands.

Hanski, I. 1999. Habitat connectivity, habitat continuity, and metapopulations in dynamic landscapes. Oikos 87: 209-219.

Hanski, I. 2000. Extinction debt and species credit in boreal forests: modelling the consequences of different approaches to biodiversity conservation. Annales Zoologici Fennici 37: 271-280.

Harmon, M.E., Franklin, J.F., Swanson, F.J., Sollins, P., Gerhory, S.V., Lattin, J.D., Andersson, N.H., Cline, S.P., Aumen, N.G., Sedell, J.R., Leinkaemper, G.W., Cromack, K. Jr. and Cummins, K.W. 1986. Ecology of coarse woody debris in temperate ecosystems. Advances in Ecological Research 15. 133-302.

Hart, J.F. 1998. The rural landscape. John Hopkins University Press, Baltimore.

Heywood, V.H. 1995. Global biodiversity assessment. Cambridge University Press, Cambridge. 1140 p.

Hunter, M.L. 1990. Wildlife, forests, and forestry: Principles of managing forests for biological diversity. Prentice-Hall, Englewood Cliffs, NJ. 370 p.

Hunter, M.L. 1999. Maintaining biodiversity in forest ecosystems. Cambridge University Press, Cambridge. 695 p.

Iacobelli, T., Kavanagh, K. and Rowe, S. 1995. A protected areas gap analysis methodology. WWF Canada. Toronto. Canada. 68 p.

Jackson, J. 1994. Red-cockaded Woodpecker. The Birds of North America No 85. The American Ornithologists Union and The Academy of Natural Sciences, Philadelphia. 20 p.

Jansson, G. and Angelstam, P. 1999. Thresholds of landscape composition for the presence of the long-tailed tit in a boreal landscape. Landscape Ecology 14: 283-290.

Jennings, M.D. 2000. Gap analysis: concepts, methods, and recent results. Landscape Ecology 15: 5-20.

Johnson, E.A. 1992. Fire and vegetation dynamics. Studies from the North American boreal forest. Cambridge University Press. Cambridge. 129 p.

Jonsell, M., Weslien, J. and Ehnström, B. 1998. Substrate requirements of red-listed saproxylic invertebrates in Sweden. Biodiversity and Conservation 7: 749-764.

Kaila, L., Martikainen, P. and Punttila, P. 1997. Dead trees left in clearcuts benefit saproxylic Coleoptera adapted to natural disturbances in boreal forest. Biodiversity and Conservation 6: 1-18.

Kirby, K.J. and Watkins, C. 1998. The ecological history of European forests. CAB International, Wallingford, UK. 373 p.

Kirby, K.J., Reid, C.M., Thomas, R.C. and Goldsmith, F.B. 1998. Preliminary estimates of fallen dead wood and standing dead trees in managed and unmanaged forests in Britain. Journal of Applied Ecology 35: 148-155.

Kruys, N. and Jonsson, B.G. 1999. Fine woody debris is important for species richness on logs in managed boreal spruce forests of northern Sweden. Canadian Journal of Forest Research 29: 1295-1299.

Kulhavy, D.L., Hooper, R.G. and Costa, R. 1995. Red-cockaded Woodpecker: recovery, ecology and management. Center for Applied Studies in Forestry, College of Forestry, S. F. Austin State University, Nacogdoches, Texas. 552 p.

Kuusinen, M. 1996. Cyanobaterial macrolichens on *Populus tremula* as indicators of forest continuity in Finland. Biological Conservation 75: 43-49.

Lambeck, R.J. 1997. Focal species define landscape requirements for nature conservation. Conservation Biology 11: 849-856.

Lammerts van Buren, E.M. and Blom, E.M. 1997. Hierarchical framework for the formulation of sustainable forest management standards. Principles, Criteria, Indicators. The Tropenbos foundation. Backhuys Publisher.

Liaison Unit in Lisbon 1998. Third ministerial conference on the protection of forests in Europe. General declarations and resolutions adopted. Ministry of Agriculture.

Linder, P. and Östlund, H. 1998. Structural changes in the three mid-boreal forests Swedish forest landscapes, 1885-1996. Biol. Conserv. 85: 9-19.

Lugo, A.E., Brown, S. and Brinson M.M. 1990. Forested wetlands. Ecosystems of the world. Volume 15. Elsevier Scientific Publications, Amsterdam, The Netherlands. 504 p.

Mahmood, S., Finlay, R.D. and Erland, S. 1999. Effects of repeated harvesting of forest residues on the ectomycorrhizal community in a Swedish spruce forest. New Phytologist 142: 577-585.

Majewski, P., Angelstam, P., Andrén, H., Rosenberg, P., Swenson, J., Hermansson, J. and Nilsson, S.G. 1995. Differences in the structure of the pine forest on deep sediment in pristine and managed taiga. *In* Angelstam, P., Mikusinski, G. and Travina, S. (eds.). Research in Eastern Europe to solve nature conservation problems in the Nordic countries. Swedish University of Agricultural Sciences, Department of Wildlife Ecology, Uppsala, Report 28.

Margules, C.R. and Pressey, R.L. 2000. Systematic conservation planning. Nature 405: 243–253.

Martikainen, P., Kaila, L. and Haila, Y. 1998. Threatened beetles in white-backed woodpecker habitats. Conservation Biology 12: 293-301.

Masters, R.E., Lochmiller, L., McMurry, S.T. and Bukenhofer, G.A. 1998. Small mammal response to pine-grassland restoration for red-cockaded woodpeckers. Wildlife Society Bulletin 26: 148-158.

Mayer, H. 1984. Die Wälder Europas. Gustav Fischer Verlag, Stuttgart. 691 p.

McNeely, J.A. 1994. Lessons from the past: forests and biodiversity. Biodiversity and Conservation 3: 3-20.

Mikusinski, G. 1997. Winter foraging of the Black Woodpecker *Dryocopus martius* in managed forest in south-central Sweden. Ornis Fennica 74: 161-166.

Mikusinski, G. and Angelstam, P. 1998. Economic geography, forest distribution, and woodpecker diversity in Central Europe. Conservation Biology 12: 200-208.

Mikusinski, G. and Angelstam, P. 1999. Man and deciduous trees in boreallandscape. Pp. 220-224 *in* Kovar P. (ed.): Nature and Culture in Landscape Ecology. Experience for the 3rd Millenium. The Karolinum Press, Prague.

Mikusinski, G., Gromadzki, M. and Chylarecki, P. 2001. Woodpeckers as indicators of forest bird diversity. Conservation Biology 15: 208-217.

Mönkkönen, M. and Reunanen, P. 1999. On critical thresholds in landscape connectivity - management perspective. Oikos 84: 302-305.

Nilsson, C. and Götmark, F. 1992. Protected areas in Sweden: is natural variety adequately represented? Conservation Biology 6: 232-242.

Nilsson, S., Sallnäs, O. and Duinker, P. 1992. Future forest resources of western and eastern Europe. Parthenon, Carnforth, United Kingdom.

Nilsson, S.G., Niklasson, M., Hedin, J., Aronsson, G., Gutowski, J.M., Linder, P., Ljungberg, H., Mikusinski, G. and Ranius, T. Densities of large living and dead trees in old-growth temperate and boreal forests. Forest Ecology and Management *(in press).*

Noss, R.F. 1999. Assessing and monitoring forest biodiversity: A suggested framework and indicators. Forest Ecology and Management 115: 135-146.

Noss, R.F., O'Conell, M.A. and Murphy, D.D. 1997. The science of conservation planning, Washington, D.C. The Island Press.

Ohlson, M., Söderström, L., Hörnberg, G., Zackrisson, O. and Hermansson, J. 1997. Habitat qualities versus long-term continuity as determinants of biodiversity in boreal old-growth swamp forests. Biological Conservation 81: 221-231.

Peterken, G.F. 1993. Woodland conservation and management (2nd edition). Chapman & Hall, London. 374 p.

Peterken, G.F. 1996. Natural woodland. Ecology and conservation in northern temperate regions. Cambridge University Press, Cambridge. 522 p.

Plentovich, S., Tucker, J.W. Jr., Holler, N.R. and Hill, G.E. 1998. Enhancing Bachman's sparrow habitat via management of red-cockaded woodpeckers. Journal of Wildlife Management 62: 347-354.

Prendergast, J.R., Quinn, R.M. and Lawton, J.H. 1999. The gaps between theory and practice in selecting Nature Reserves. Conservation Biology 13: 484-492.

Pressey, R.L., Ferrier, S., Hager, T.C., Woods, C.A., Tully, S.L. and Weinman, K.M. 1996. How well protected are the forests of north-eastern New South Wales? – Analyses of forest environments in relation to formal protection measurements, land tenure, and vulnerability to clearing. Forest Ecology and Management 85: 311-333.

Punttila, P., Haila, Y., Pajunen, T. and Tukia, H. 1991. Colonisation of clearcut forests by ants in the southern Finnish taiga: a quantitative survey. Oikos 61: 250-262.

Rackham, O. 1976. Trees and woodland in the British landscape. J. M. Dent, London. 234 p.

Röhrig, U. (ed.). 1991. Temperate deciduous forests. Elsevier, Amsterdam.

Sader, S.A. and Joyce, A.T. 1988. Deforestation rates and trends in Costa Rica, 1940-1983. Biotropica 20: 11-19.

Salem, E. and Ullsten, O. 1999. Our forests our future. Cambridge University Press, Cambridge. 228 p.

Samuelsson, J., Gustafsson, L. and Ingelog, T. 1994. Dying and dead trees: a review of their importance for biodiversity. Swedish Threatened Species Unit, Uppsala, Sweden. 110 p.

Scott, J.M., Csuti, B., Jacobi, J.D. and Estes, J.E. 1987. Species richness: A geographic approach to protecting future biodiversity. BioScience 37: 782-788.

Scott, J.M., Csuti, B., Smith, K., Estes, J.E. and Caicco, S. 1988. Beyond endangered species: An integrated conservation strategy for the preservation of biological diversity. Endangered Species Update 5(10): 43-48.

Scott, J.M., Csuti, B., Estes, E. and Anderson, H. 1989. Status assessment of biodiversity protection. Conservation Biology 3(1): 85-87.

Scott, J.M., Davis, F., Csuti, B., Noss, R., Butterfield, B., Groves, C., Anderson, H., Caicco, S., D'Erchia, F., Edwards, D.T.C. Jr., Ulliman, J. and Wright, R.G. 1993. Gap analysis: A geographic approach to protection of biological diversity. Wildlife Monographs 123: 1-41.

Selander, S. 1957. Det levande landskapet i Sverige. Göteborg: Bokskogen.

Shugart, H.H., Leemans, R. and Bonan, G.B. (eds.). 1992. A systems analysis of the global boreal forest. Cambridge University Press, Cambridge, Massachusetts, USA. 542 p.

Siitonen, J., Martikainen, P., Punttila, P. and Rauh, J. 2000. Coarse woody debris and stand characteristics in mature managed and old-growth boreal mesic forests in southern Finland. Forest Ecology and Management 128: 211-225.

Spies, T.A., Franklin, J.F. and Thomas, T.B. 1988. Coarse woody debris in Douglas-fir forests of western Oregon and Washington. Ecology 69: 1689-1702.

Stahl, H.H. 1980. Traditional Romanian Village Communities: The Transition from the Communal to the Capitalist Mode of Production in the Danube Region Cambridge University Press, Cambridge. 227 p.

Storch, I. 1999. Auerhuhnschutz: Aber wie? Wildbiologische Gesellschaft München. 43 p. (In German).

Sykes, M.T., Prentice, I.C. and Laarif, F. 1999. Quantifying the impact of global climate change on potential natural vegetation. Climatic Change 41: 37-52.

Thirgood, J.V. 1989. Man's impact on the forests of Europe. Journal of World Forest Resource Management 4:127-167.

Thompson, R.L. (ed.). 1971. The ecology and management of the red-cockaded woodpecker. Bureau of Sport Fisheries and Wildlife, U.S.D.I., U.S. Government Printing Office. 188 p.

Thompson, I.D. and Angelstam, P. 1998. Special species. Pp 434-459 *in* Hunter, M.L. (ed). Maintaining biodiversity in forest ecosystems. Cambridge University Press. Cambridge.

Tilman, D. and Kareiva, P. 1997. The role of space in population dynamics and interspecific interactions. Monographs in Population Biology 30. Princeton University Press, Princeton. 368 p. ISBN 0-691-01652-6

Tucker, G.M. and Evans, M.I. 1997. Habitats for birds in Europe. BirdLife International, Cambridge. 464 p.

Turner, B.L., Clark, W.C., Kates, R.W., Richards, J.F., Mathews, J.T. and Myer, W.B. 1990. The Earth as Transformed by Human Action: Global and Regional Changes in the Global Biosphere over the Past 300 Years. Cambridge University Press, New York. 713 p.

UNCED. 1992. Report of the United Nations Conference on Environment and Development. Rio de Janeiro, 3-14 June 1992.

UN-ECE/FAO. 2000. Forest resources of Europe, CIS, North America, Australia, Japan and New Zealand (Industrialized Temperate/Boreal Countries). UN-ECE/FAO Contribution in the Global Forest Resources Assessment 2000: Main Report. Geneva Timber and Forest Study Papers, No. 17. United Nations. New York and Geneva. 445 p.

Walters, C.J. 1986. Adaptive management of renewable resources. New York: McGraw Hill. 374 p.

Walters, J.R., Doerr, P.D. and Carter III, J. H. 1992. Delayed dispersal and reproduction as a life-history tactic in cooperative breeders: Fitness calculations from red-cockaded woodpeckers. American Naturalist 139: 623-643.

Whitmore, T.C. 1990. An introduction to tropical rain forests. Clarendon Press, Oxford.

With, K., Gardner, R.H. and Turner, M.G. 1997. Landscape connectivity and population distributions in heterogenous environments. Oikos 78: 151-169.

Wood, D.A. (ed.). 1983. Red-cockaded woodpecker symposium II. Proceedings. State of Florida Game and Fresh Water Fish Commission. 112 p.

Young, T.P. 2000. Restoration ecology and conservation biology. Biological Conservation 92: 73-83.

Zwicker, S.M. and Walters, J.R. 1999. Selection of pines for foraging by red-cockaded woodpeckers. Journal of Wildlife Management 63: 843-852.

5.5   OPERATIONS WITH REDUCED ENVIRON-MENTAL IMPACT

P. Hakkila

5.5.1   Nutrient content of stem and crown

The greatest concentration of plant nutrient elements occurs in the parts of the tree where essential life processes take place, i.e. foliage, cambial zone, inner-bark and root tips. The stemwood proper is less rich in minerals than the tree components normally left in the forest after harvesting.

When leaf buds break in the spring and new foliage begins to grow, the leaf tissues have high concentrations of N, P and K. As the foliage matures and accumulates carbohydrates, the initial concentrations of these elements are diluted. On the other hand, concentrations of Ca, Mg and Fe usually increase with leaf age. After leaf maturation, the content of N, P, and K remains relatively constant during the growing season until a rapid reduction takes place in the fall. Such losses are a result of active withdrawal of nutrients from foliage for reuse during the following year. Leaf nutrient content is also affected by rainfall that leaches some elements, particularly K, from the leaf surface. Leaching rates often increase as foliage undergoes senescence before abscission. Losses of nutrients by leaching follow the decreasing order K, P, N, Ca (Waring and Schlesinger 1985). The same loss pattern seems to occur after harvesting during summer storage of whole-trees and residual crown mass. This means that a proportion of the mineral element content is leached or moved from foliage to the woody components of the branches before the foliage is shed through transpiration drying.

Table 5.5-1. Average concentration of macro-nutrient elements in dry biomass of young softwoods and hardwoods in Maine (Young and Carpenter 1967).

Category	Tree component	Concentration in dry mass, g $100g^{-1}$				
		N	P	K	Ca	Mg
Softwoods	Stem	0.15	0.02	0.06	0.33	0.03
(5 spp.)	Branches	0.31	0.07	0.14	0.53	0.05
	Foliage	0.93	0.15	0.42	0.69	0.10
Hardwoods	Stem	0.25	0.03	0.13	0.49	0.03
(3 spp.)	Branches	0.43	0.05	0.17	0.58	0.06
	Foliage	1.35	0.19	0.75	0.78	0.19

The content of mineral elements in woody biomass varies considerably. Table 5.5-1 gives an example of the macro-nutrient element content of young softwoods and hardwoods in Maine, USA. Concentrations in foliage are 6-7 times as high as those in the stem.

It is thus inevitable that the extraction of crown mass will mean an increase in nutrient loss from the forest, certainly more than the increase in biomass yield suggests. Yet, particularly in managed forests, crown mass represents such a large proportion of forest residues that in most cases bioenergy production would not be feasible without it. This is demonstrated in Figure 5.5-1 for Finland, where most of the technically-harvestable forest fuel in managed forests consists of crown mass rather than residual stem mass. Furthermore, large-scale production of forest fuel in conjunction with conventional forestry is in many cases only profitable if all above-ground biomass is harvested simultaneously in one integrated operation.

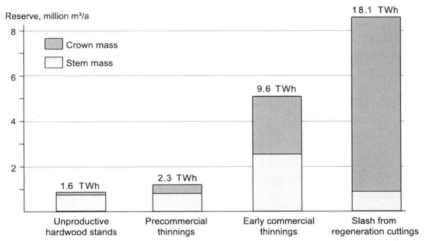

Figure 5.5-1. Technically-harvestable forest residues from conventional forestry in Finland.

5.5.2   Loss of nutrients in conventional logging

The harvesting of timber disturbs the natural development of the ecosystem. Removal of organic matter and nutrients is accelerated. Nutrients are lost not only through direct extraction of biomass but also through leaching, exposure of the forest floor, soil disturbance and even erosion associated with logging (Mann *et al.* 1988). The negative impact of harvesting on the productive capacity of the soil is dependent on site characteristics and the technology applied.

Table 5.5-2 presents examples of nutrient loss after conventional stem-only extraction during thinning of young Scots pines, and birch and after regeneration-cutting of an over-aged Norway spruce stand. In softwood stands, the loss of N is about 400 g m^{-3} and that of P about 40 g m^{-3}. Losses from birch stands are about three times as much. Species differences for K and Ca loss are smaller. Before World War II, when de-barking was often carried out at the stump, nutrient removal was about one third less.

Table 5.5-2. Average nutrient losses in conventional stem-only harvesting in Finland (data from Mälkönen 1974, Kubin 1977, Mälkönen and Saarsalmi 1982).

Tree species	No. of stands	Average stand age, years	Nutrient loss, g m^{-3} of timber			
			N	P	K	Ca
Scots pine	3	40	406	44	222	423
Norway spruce	1	135	396	44	308	1012
Birch	2	30	1149	121	367	734

In a typical thinning operation in a Scots pine stand yielding 70 m^3 of unbarked timber per ha, approximately 30 kg N, 3 kg P, 15 kg K and 30 kg Ca are removed. For spruce and birch the losses are higher. These levels of nutrient loss from stem-only logging of unbarked timber are thought to be acceptable when the whole rotation period is considered. Losses caused by logging and leaching are compensated by atmospheric precipitation, chemical weathering and biological fixation of nitrogen. The nutrient balance of the site and sustainability of the ecosystems are not endangered by stem-only harvesting (Nykvist 1977).

5.5.3   Additional losses from crown mass recovery

When harvesting is extended to the nutrient-rich crown mass, the situation changes. Losses of nutrients, particularly N and P, are much greater. Figure 5.5-2 shows the relative distribution of dry mass and macro-nutrient elements in above-ground components of a 22-year-old radiata pine plantation in New Zealand. Crown mass contains more N and P than stem mass and a smaller but still substantial proportion of other nutrient elements.

Data from European forests showing relative amounts of macro-nutrient elements in components of the above-ground biomass are presented in Table 5.5-3. If all the crown mass were recovered, increases in loss of N and P would be approximately 50% for leafless beech, 150% for pine and as much as 300% for spruce, due to its large foliage mass.

As a rule of thumb, and using conventional stem-only harvesting for comparison, each percentage increase in biomass recovery represented by crown mass with foliage can be expected to incur increased nutrient losses amounting to 2-3% for pines, 3-4% for spruces and 1.5% for leafless hardwoods. In the case of Ca, the relative loss is smaller in all species.

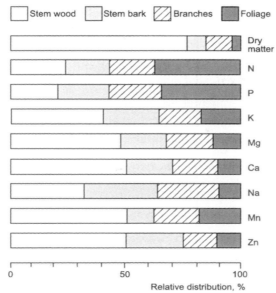

Figure 5.5-2. The relative distribution of dry mass and the primary nutrients in the above-ground components in a 22-year-old radiata pine stand in New Zealand (redrawn from Madgwick et al. 1977).

Table 5.5-3. Ratio of whole-tree nutrient content to nutrient content of merchantable stem. Sources of data: Binnis 1975, Kreutzer 1975 and Mälkönen 1975).

	N	P	K	Ca
Pine	2.4	2.8	2.3	1.8
Spruce	3.9	4.5	3.1	2.1
Beech	1.4	1.6	1.4	1.3

In practice, the recovery of crown mass is never complete. The loss of nutrients is not as great as theoretical calculations suggest, since no technology is able or intended to remove all of the crown mass. For example, the harvesting of logging residues from regeneration areas in northern Europe, using forwarders for off-road transportation, extracts only 60-80% of the crown mass. The percentage of needles removed is even lower, particularly if residues are stored over the summer (Hakkila *et al.* 1998).

5.5.4   Reducing the loss of nutrients

The effects of crown mass removal on subsequent growth of remaining trees in a thinned stand, or on growth of young seedlings in a regeneration area, are being studied in field experiments in many countries. Stem-only logging is compared with whole-tree logging in permanent sample plots. These experiments do not correspond to reality in the following respects:

- In control plots representing conventional stem-only logging, residual crown mass is distributed evenly across the whole area. This is not the case in current mechanized harvesting operations.

- In plots representing whole-tree logging, all crown mass is removed. This can never be achieved operationally.

A high degree of experimental precision is necessary for the determination of relationships between residue retention and productivity in the succeeding rotation. Any negative impacts from operational residue harvesting would be intermediate between those resulting from experimental stem-only and whole-tree removal. Differences between experimental treatments and residue retention associated with actual harvesting practice must be taken into account when interpreting the results. Unfortunately, reference to the results of scientific experiments does not always include allowances for these differences.

Of all tree components, foliage is the richest in nutrients. Accelerated removal of nutrients caused by biomass harvesting is therefore dependent on the extent to which needles are included. Methods suggested below will reduce the amount of foliage removed from the site without interfering with the rationalization of current harvesting systems.

Active separation of nutrient-rich biomass components during fuelwood harvesting is rare, but minor adjustments can be made to operating methods and procurement logistics. In general, adjustments made to supply systems incorporating processing or storage at the logging site are more effective than those made to systems based on processing at roadside or terminal (Danielsson 1993).

Transpiration drying of small trees

Summertime transpiration drying (also known as biological drying or leaf seasoning) is an effective and economical way of achieving simultaneous reduction of the moisture and foliage components of small felled trees. The tree is left on the ground with its crown intact. Drying takes place through transpiration as long as the foliage retains its vitality, and progresses from crown base to apex. The foliage gradually turns brown but may still remain attached. Eventually leaves or needles start to fall off. As with all natural seasoning methods, transpiration drying is dependent on weather conditions and is affected by temperature, rainfall, air humidity and wind. Drying is very slow in a humid, rainy climate, and during winter.

Transpiration drying is a simple procedure which leaves foliage in the logged area and improves the quality of the fuel. Nutrient loss from the site is reduced, although a proportion of the nutrients is translocated into branches and stem before the foliage is shed.

Risk of insect damage is a negative feature of transpiration drying of early thinnings in softwood stands. Small trees left in the forest may become a habitat for harmful bark beetles which damage living trees and unbarked timber. In Sweden and Finland, storage of felled pine and spruce trees in the forest is prohibited during the bark beetle swarming season in late spring and early summer. Transpiration drying of coniferous trees is only permitted after mid-summer. In Denmark, large-scale transpiration drying associated with early thinning operations in young spruce stands has not created insect problems. No insect risks are attached to transpiration drying of hardwoods.

Whole-tree chipping of young trees with breast height diameter less than 20 cm has been practiced in Denmark for more than ten years. In 1998, at least 15% of fellings in conifers were whole-trees chipped for fuel. This practice promotes tending in young plantations, where the primary species is usually Norway spruce, but it greatly increases nutrient loss. To reduce this loss and improve fuel quality, transpiration drying during summer has become a common practice in small-tree harvesting (Heding 1999). Guidelines for state forests in Denmark advise that coniferous trees should be left to dry at the site for at least two months in spring or summer.

When Norway spruce trees were harvested fresh in a whole-tree harvesting experiment in Denmark, nutrient loss, especially of N and P, was very high. When trees were left to dry at the site for a period of six months, most of the foliage and some thin twigs were shed. Drying reduced the biomass yield by 15% and nutrient removal by 20-45% (Møller 2000).

Transpiration drying of logging residues

In the mechanized cut-to-length system based on the use of one-grip harvesters and forwarders, logging slash accumulates in heaps and along strip roads (Jacobsson and Filipsson 1999). In regeneration areas of Norway spruce, approximately half of the residue accumulates in layers thicker than 35 cm, an area of 100 m^2 sometimes having a weight of several tonnes. On the other hand, a large percentage of the ground surface remains free from residual biomass (Nurmi 1994). This uneven distribution makes released nutrients available to only some of the young trees in the new generation. It may also increase nutrient leaching into ground water. When residues are extracted from regeneration areas using the Nordic system, 20-40% is left at the site. Only heaped material is removed and remaining crown mass is fairly evenly distributed. Figure 5.5-3 shows the distribution of logging residues in successive sample plots. Even distribution makes the released nutrients available to more trees and reduces the risk of leaching. It also facilitates site preparation and planting.

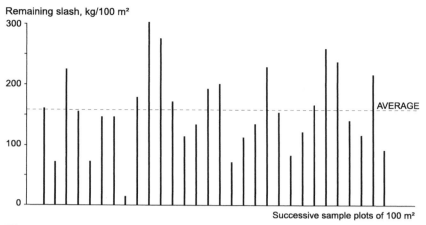

Remaining slash, kg/100 m²

Figure 5.5-3. Amounts of logging residue remaining after forest fuel harvesting with a terrain chipper in a clear-cut Norway spruce stand in Finland. About 80% of the original residue was removed overall (Hakkila et al. 1998).

Since foliage contains more than a half of the nutrients in the crown mass, it has been recommended that crown mass should be recovered without foliage (Egnell *et al.* 1998). Although transpiration drying and the subsequent shedding of foliage is technically simple, it affects the procurement chain and product quality in several ways:

- The yield of fuel is reduced. If all foliage is left in the forest, the yield of crown mass is approximately 20% lower for pine and 30% lower for spruce. In practice, the shedding of needles is only partial. On the other hand, seasoned twigs tend to become brittle and break off.

- Since the yield of chips is reduced per heap, per hectare and per site, work productivity decreases and the cost of recovery and overheads of procurement increases. On the other hand, the cost of trucking may be reduced due to the drier product.

- Since transpiration drying requires storage of residues during summer at the logging site, the flow of fuel from logging site to energy plant slows down. This interferes with the logistic system of fuel procurement and may delay subsequent site preparation and establishment of a new stand.

- Transpiration drying reduces the moisture content of chips and results in an increase in the effective heating value. This compensates for the loss in energy yield associated with foliage removal.

- Removal of foliage improves the storage properties of chips by reducing microbial activity and chemical reactions in the biomass. The rate of decomposition is slower, the number of micro-organisms causing allergic reactions is reduced and the risk of fire during long storage periods diminished (see Section 3.4).

- When the proportion of foliage is decreased, the fuel properties of chips improve. Compared with the other biomass components of a tree, needles contain more chlorine which causes corrosion in boilers. They contain more nitrogen, resulting in an increase in NO_x emissions, and they cause more sulfur emission. Needles tend to cause bridging of chips in truck loads and fuel silos particularly in winter. Furthermore, they produce excessive ash and lower its melting point, which may give rise to problems in combustion.

The effects of transpiration drying on the nutrient loss from forest sites and on fuel quality are thus beneficial. However, the effectiveness of transpiration drying depends on the tree species, season, weather and the method of storage. In a seasoning experiment of logging slash from a typical Norway spruce stand in Finland, the proportion of foliage was 28% at the time of clear-cutting in September. When the slash was seasoned at the site, no obvious reduction occurred in the winter, but by mid-summer the proportion had been reduced to 20% and by September to 7% (Nurmi 1999). Shedding proceeds much faster in hardwoods, but is slower in pines.

Whether residues are removed from the site unprocessed, in the form of chips or in the form of tight bundles, handling always starts with a grapple loader. If transpiration drying is at an advanced stage, a simple way of loosening and removing part of the remaining foliage is by vibrating the crane. This spreads the needles over a wider area than that covered by the pile made by a harvester.

Topping of trees

In whole-tree harvesting of early thinnings, the amount of foliage can be reduced by topping the trees at the stump. Figure 5.5-4 shows the vertical distribution of foliage in Scots pine and Norway spruce trees removed in selective thinning and regeneration cuttings in Finland. It also illustrates the relative effectiveness of bucking a 3m section from the top of a tree.

Figure 5.5-5 shows the proportion of the total crown mass and needle mass left at the site in whole-tree logging if the crown is cross-cut at a distance of 3 m from the top of the stem. Topping of single trees or bunches of trees can be performed motor-manually by a chainsaw operator, by a chainsaw mounted on the grapple of a forwarder or terrain chipper crane, or by a guillotine-type bucker at the rear of the load space of a forwarder. From the standpoint of nutrient loss, this compromise with the principle of whole-tree logging is inefficient for large spruce trees but efficient for small pine trees. In early thinning of Scots pine plantations in Finland, it is possible to reduce foliage removal by 52% if a 3-m top from each tree is left at the site. The loss of crown mass is considerably less (30%) and the loss of above-ground biomass only 8%. In mature pine stands and all spruce stands, the method is not effective.

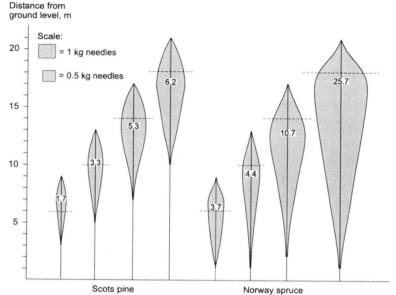

Figure 5.5-4. Vertical distribution of dry foliage mass in 9-, 13-, 17-, and 21-m-high Scots pine and Norway spruce trees removed in selective thinning and final cutting in southern Finland. The average dry mass of foliage (kg) and the location of 3 m top (broken line) is indicated for each height category (Hakkila 1989).

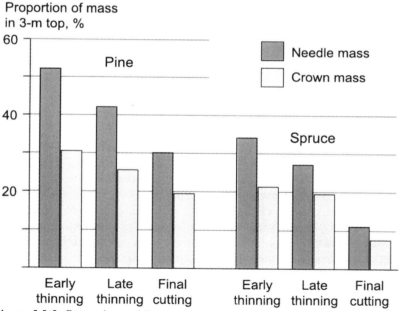

Figure 5.5-5. Proportions of live crown and foliage mass attached to a 3 m top bucked from whole Scots pine and Norway spruce trees in southern Finland (Hakkila 1991).

Removal of foliage by mechanical means

Previous sections demonstrated that transpiration drying is a simple and natural way of separating foliage from small-sized trees and logging residues. However, its use is often problematic as the requirement for summer on-site storage interferes with system logistics. Interruption of the procurement process necessitates additional movement of machines from site to site, extra costs of overheads and interest, and delay in subsequent stand establishment. There is consequently interest in the development of methods for removing foliage by mechanical means, in conjunction with the harvesting of green crown mass.

In Canada, Bjerkelund (1989) has introduced the trimmed tree concept which involves mechanical removal of foliage at the stump before undelimbed trees are transferred to the roadside. In Sweden, Filipsson and Andersson (2001) have listed a large number of technical solutions for partial removal of foliage in conjunction with chipping or bundling at the logging site, but no proven technology has been developed. The following possibilities have been explored in different countries:

- Combined motor-manual rough-delimbing and topping of small-tree bunches with a chainsaw. The use of manual labor for this purpose is becoming a prohibitive cost factor.

- Rough-delimbing with one-grip harvesters to integrate production of stemwood for fibre and large branches for fuel. Most of the thin twigs and foliage are removed.

- Equipping the load space of a forwarder with a light chain flail device for removing foliage from whole-trees or tree sections in conjunction with off-road transport.

- Equipping the infeed system of a terrain chipper with a light chain flail device for removing foliage from whole-trees or logging residue prior to reduction of biomass to chips.

- Removing the foliage component of chips in conjunction with terrain chipping using vibrating tractor loaders or wind sorting as the product is moved from chipper to bin.

None of these techniques has become operational. Another alternative is to separate foliage from green chips at the processing terminal, return it in chip trucks to the logging area, and spread it back over the site. Spreading could be carried out with forwarder-mounted equipment similar to that used for fertilizer application. However, when net profit from foliage return is weighed against the total cost of separation and return, and the fuel value of foliage is taken into account, return of foliage does not seem feasible. Nutrient return (with the exception of N) can be achieved at lower cost with ash from wood-based fuels.

5.5.5   Recycling of ash

With the exception of nitrogen and other volatile elements such as chlorine, nutrients removed from the forest in logging residues remain in the ash after combustion. Compensation for loss of nutrients from the harvesting site could therefore be achieved by recycling the ash as fertilizer or as a soil improving agent. In Sweden, the recycling of wood ash is strongly recommended in areas where whole-tree harvesting is practised more than once during a rotation period, and in areas subject to high levels of acid rain and nitrogen deposition (Lundkvist *et al.* 1999). Some producers of wood ash in Europe and Northern America have started to recycle ash on an operational scale. In Finland, more than 10% of wood and bark ash produced by forest industry is returned to the forest.

Recycling of ash offers environmental and in some cases economic benefits. However, no guidelines have been developed to indicate where and when ash should be applied in order to meet requirements for ecological sustainability and the best economic result. Several questions arise:

- If ash is recycled, should its destination be forest sites? Due to a high content of Ca and Mg, wood and bark ash is a low-cost alternative to commercial lime and could be used to counteract acidity in agricultural soils. Assuming that the content of cadmium and other heavy metals is acceptable, would spreading of ash on agricultural land be an easier operation, providing a better and faster economic result?

- On mineral soils, where availability of nitrogen is the growth-limiting factor, ash may not improve tree growth unless nitrogen is also returned. Are the principles of ecological sustainability compromised if wood ash containing nutrients derived from mineral soils is applied to nitrogen-rich peatland forests where it gives a better and longer-lasting growth response and may improve the health of trees exhibiting nutrient deficiency?

- Are ecological expectations met if ash derived from a mature tree stand is spread not on the original clear-cut site, but under a young stand at the stage when a better growth response can be expected?

- Is it feasible and acceptable to delay wood ash application to a regeneration site for 10-20 years, when the new stand will utilize the nutrients more effectively?

For technical reasons, wood ash is spread at high rates. In Finland 3-5 t ha^{-1} and in Sweden 1-3 t ha^{-1} are applied. It follows that the amount of ash produced will not be sufficient for spreading in all stands from which crown mass has been removed. On the other hand, ash from industrial residues (bark and sawdust) have equal nutritional value and can be used as a supplement.

To make ash recycling economically and environmentally acceptable, ash producers have to pay attention to the following aspects of quality control:

- Mixing of fuelwood with soil should be avoided during harvesting, transport and storage to keep the concentration of useful nutrients high and to avoid the transport and spreading of useless material. Load-carrying forwarders keep biomass cleaner than load-dragging skidders.

- Woody biomass should be burnt as completely as possible. Charcoal increases ash volume and incurs extra transport and spreading costs. If ash is to be hardened, crushed and screened in order to avoid dust problems during spreading and also impact shock to flora and fauna, the carbon content should be less than 10-20%. Ash with a high carbon content should be reburned before recycling (NUTEK 1994).

- Ash cooling to reduce fire risk and dust whirling must be done with the minimum amount of water to avoid extra weight and freezing during winter.

- Ash storage facilities at the plant must be designed for easy and dust-free loading of trucks.

- To reduce dust problems at all stages of the recycling process, and to avoid shock effects to flora and fauna, ash should be hardened or granulated.

- Wood and bark ash should be kept separate from fossil fuel ash to avoid problems with heavy metals and to prevent dilution of important nutrients. Mixed combustion of biomass, peat, municipal waste and fossil fuels (co-firing) is in many cases a constraint to ash recycling. Co-firing of wood with other solid fuels may be an efficient way of reducing chlorine, sulfur or nitrogen emissions.

Ash is transported from combustion plant to the forest site in trucks, for instance returning chip trucks, and tipped on to the ground. Later it is loaded into a tractor bin. Storage areas at roadside landings may be uneven, resulting in ash loss and extra cost. Stones picked up with the ash may cause breakage of the spreader. To avoid these problems and also the wetting of ash in roadside storage, as well as to improve the logistical recycling flow of ash back to the site, a delivery system based on a single on-off-road vehicle has been devised in Finland.

Since ash may be used at rates which are 10-20 times as great as those used for artificial fertilizers, spreading costs are crucial. The cost of aerial spreading would be prohibitive since spreading from the air is only competitive when small quantities are spread over large areas. In the case of ash, large quantities are spread over small areas.

Ash recycling systems are based on the use of load-carrying farm or forest tractors (Hakkila and Kalaja 1981). In agricultural application, ash can be dropped directly behind the tractor and the width of the spreading swath is not critical. In the forest, spreading is done from logging strip roads, and the ash must be spread evenly to a distance of 10-15 m on either side of the tractor. Because spreading is impeded by trees, particularly those with long crowns such as spruces and firs, technical demands are greater in forest application.

The tractor must be capable of carrying the ash load over rough terrain. Forwarder-mounted equipment is therefore most suited to large-scale operations. Usually a 10-12 m^3 ash bin is mounted in the load space of the forwarder. The crane of the forwarder is adapted for loading ash into the bin by replacement of the timber grapple with an ash scoop. The bottom of the ash bin is equipped with a conveyor for transferring the ash to one or two slings at the rear of the spreader. These throw out the ash on both sides. In Finnish operations, the spreading output varies between 7 and 10 t h^{-1}. The total cost of recycling is USD 25-35 t^{-1}, and includes handling at the power plant, trucking, spreading and overheads. On sites with a low load-bearing capacity, spreading is feasible from technical and ecological points of view, but the cost of the operation is considerably higher.

The driver and the engine of the spreading machine are both subjected to the dust hazard. The tractor cabin must therefore be tightly closed and pressurized and air supplies to cabin and engine must be filtered. The dust problem can be reduced but not removed by increasing the moisture content of the ash. If the amount of water in the ash is high, trucking costs are increased, spreading becomes difficult and freezing problems may occur in the winter.

Dust problems can be largely avoided by stabilizing the ash (NUTEK 1994, Väätäinen *et al.* 2000). A simple form of ash treatment involves addition of water and crushing after *self-hardening*. In large-scale operations *granulation* techniques are used. Here the moistened ash is aggregated into granules in drums or on rotating disks. Addition of binding agents may be necessary to produce stable granules. Powdery wood ash has a density of 250-500 kg m^{-3} while the density of granulated ash may exceed 1100 kg m^{-3}. This difference is very important from a logistical point of view. It is also possible to *pelletize* ash although the material causes heavy wear to the pelletizing equipment. Pre-treatment of ash allows a lower moisture content and decreases mass per unit volume. It also reduces dust problems, allows more uniform spreading, reduces the clogging of machinery, makes storage easier and lessens shock to the ground vegetation.

5.5.6   Effect of residue removal on soil compaction

Logging machines are known to cause soil compaction and tree damage, especially on soft soils. Reduced tree growth and timber quality may result. The risk is high in stands of spruce and many hardwoods which grow on soft soils have shallow root systems, and are sensitive to decay. Pines, which grow on relatively dry soils, have deeper root systems and are resistant to decay, incur less risk from harvesting machinery.

In the Nordic cut-to-length system, timber is hauled to the roadside using load-carrying forwarders. On soft soils and at thinning sites, directed felling accumulates crown mass on the strip roads, providing a protective layer which reduces rut formation. As shown in Figure 5.5-6, even a thin layer of slash reduces the depth of ruts.

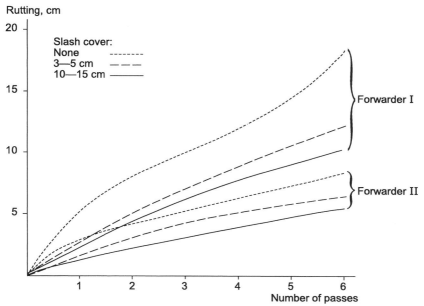

Figure 5.5-6. The effect of slash layers on rut formation as a function of the number of passes with two different forwarders on soft soil (Hallonborg 1982).

The avoidance of soil disturbance and root damage requires correct timing of logging, good work organization and careful selection of equipment. Soil, season, strip road width and tractor behavior all influence the amount of damage. In the Nordic countries, when timber is harvested from spruce stands for industrial processing, regeneration cuttings are carried out in the summer season and thinning in the winter season to reduce damage to trees growing in soft soils. This affects the scheduling of chip harvesting, since the harvesting of fuel is integrated with the harvesting of industrial raw material.

In most cases, ground damage can be prevented through the training of workers, careful planning and the use of proper equipment and work methods. If this is not possible, the entire logging site or sensitive parts of it should be excluded from crown mass harvesting. In Finland, it is estimated that 20% of the area logged annually and 15% of the area with fuelwood potential should be excluded because nutrient losses would be unacceptable or because the load-bearing capability of the soil would be compromised without a protective residue layer (Hakkila *et al.* 1998).

5.5.7   Summary

Because of the environmental, social and silvicultural advantages, use of woody biomass for fuel is attracting increasing global attention. Residues from conventional forestry which are currently left at the site of a logging or tending operation, will be increasingly utilized for fuel in the future. However, intensive biomass harvesting involves potential ecological risks.

When removal of biomass is intensified, the sustainability of wood production must not be endangered through excessive nutrient loss from forest sites. Just as productivity can be increased by fertilizer addition, it may also be decreased by improper management practice. Positive and negative effects of biomass harvesting must be understood and management practice and technology applied accordingly.

The following conclusions can be drawn from current knowledge:

- Use of residual forest biomass from regeneration cuttings and small-sized trees from thinning is increasing in many countries because the resource is renewable and offers opportunities for the reduction of CO_2 emissions as well as other social and silvicultural benefits.

- Where industrial demand for wood is high and forests are managed accordingly, residual biomass available for energy production consists largely of crown material.

- Compared with tree stems, crown material and particularly the foliage component are rich in nutrients. It follows that crown mass removal increases the loss of nutrients from a forest ecosystem when it is combined with conventional stem-only logging. Consequently, when the use of crown mass for fuel becomes a common practice in conventional forestry, certain restrictions will be needed to control the harmful effects of intensive harvesting (Rekommendationer 1998).

- Studies in many countries show that crown mass removal may endanger the sustainability of production capacity, depending on site conditions and the amount and composition of biomass removed (Jacobson and Kukkola 1999, Mattsson 1999). However, field experiments usually incorporate uniform distribution of material after logging in control plots and complete removal of crown components from whole-tree logging plots. Since this degree of precision is impossible in operational forestry, experimental results tend to over-estimate the negative impact of fuelwood harvesting on the growth potential of the site. This should be taken into account when research results are interpreted and applied.

- Biomass harvesting technology should not be developed for improved productivity and lower procurement costs only. Depending on site and stand conditions, it should also be used to control loss of nutrients and to minimize ground compaction.

- Negative ecological impacts can be reduced by careful planning and the adoption of appropriate technology. Examples of available methods are the appropriate timing of operations, transpiration drying and foliage shedding, topping of undelimbed small-sized pine trees, development of foliage trimming techniques, and recycling of ash from forest and industrial residues. These methods will not completely compensate for nutrient losses, but they will greatly reduce them.

- Procedures for reducing negative ecological impacts will affect the process of procurement of forest fuel in many ways: biomass yield will be reduced; the cost of recovery increased; storage properties of biomass improved; fuel quality enhanced; procurement logistics altered; and the flow of biomass from forest to energy-producing plants may be slowed down.

- There is an urgent need for development of technology and guidelines to ensure that the use of forest bioenergy has optimal environmental and economic results. The main issues will then be reduction of CO_2 emissions through replacement of fossil fuels, and conservation of the production capacity of forest soils.

5.5.8   References

Binnis, W.O. 1975. Whole-tree utilization - consequences for soil and environment. Experience and opinion in Britain. Elmia 75 Konf Sk2: 18-25.

Bjerkelund, T. 1989. Development of conceptual forest operations system based on a trimmed tree form. IUFRO meeting on Harvesting and Utilization of Tree Foliage. Riga, Latvia.

Danielsson, B-O. 1993. Harvesting methods and available quantities in the light of ecological constraints. *In* Mattsson, J.E. and Mitchell, P. Environmental issues in supply of biomass for energy from conventional forestry. Sveriges Lantbruksuniversitet. Uppsatser och Resultat 252.

Egnell, G., Nohrstedt, H-Ö., Weslien, J., Westling, O. and Örlander, G. 1998. Miljökonsekvensbeskrivning (MKB) av skogsbränsleuttag, asktillförsel och övrig näringskompensation. Skogsstyrelsen. Rapport 1.

Filipsson, J. and Andersson, G. 2001. Teknik för barr- och rISspridning. Skogforsk. Arbetsrapport 474.

Hakkila, P. 1989. Utilization of residual forest biomass. Springer Verlag, Berlin.

Hakkila, P. 1991. Hakkuupoistuman latvusmassa (Crown mass of trees at the harvesting phase). Folia Forestalia 773.

Hakkila, P. and Kalaja, H. 1981. The technique of recycling wood and bark ash. Folia Forestalia 552.

Hakkila, P., Nurmi, J. and Kalaja, H. 1998. Metsänuudistusalojen hakkuutähde energialähteenä. Metsäntutkimuslaitoksen tiedonantoja 684.

Hallonborg, U. 1982. Ristäckets inverkan på spårbildningen. Skogsarbeten, Results 3.

Heding, N. 1999. Sustainable use of forests as an energy source. *In* Lowe, A. and Smith, T. (eds.). Developing systems for integrating bioenergy into environmentally sustainable forestry. New Zealand Forest Research Institute. Forest Research Bulletin 211.

Jacobson, S. and Filipsson, J. 1999. Trädresternas rumsliga fördelning efter slutavverkning. Skogforsk. Arbetsrapport 422.

Jacobson, S. and Kukkola, M. 1999. Skogsbränsleuttag i gallring ger kännbara tillväxtförluster. Skogforsk. Resultat 13.

Kreutzer, K. 1975. The effect of economizing on nutrient economy in central European forests. Elmia 75, Konf Sk2: 37-54.

Kubin, E. 1977. The effect of clear cutting upon the nutrient status of a spruce forest in northern Finland (64°28'N). Acta Forestalia Fennica 155.

Lundqvist, H., Eriksson, H.M., Nilsson, T. and Arvidsson, H. 1999. Ecological effects of recycling of hardened wood ash. New Zealand Forest Research Institute, Forest Research Bulletin 211: 91-92.

Madgwick, H.A.I., Jackson, D.S. and Knight, P.J. 1977. Above-ground dry matter, energy and nutrient contents of trees in an age series of *Pinus radiata* plantations. New Zealand Journal of Forestry Science 7: 445-468.

Mann, L.K., Johnson, D.W., West, D.C., Cole, D.W., Hornbeck, J.W., Martin, C.W., Reikerk, H., Smith, C.T., Swank, W.T., Tritton, L.M. and Van Lear, D.H. 1988. Effects of whole-tree and stem-only clearcutting on post-harvest hydrologic losses, nutrient capital, and regrowth. Forest Science 42: 412-428.

Mattsson, S. 1999. Tillväxtförluster ger "dolda kostnader" vid uttag av skogsbränsle-framförallt i gallring. Skogforsk Resultat 14.

Mälkönen, E. 1974. Annual primary production and nutrient cycle in some Scots pine stands. Communicationes Instituti Forestalis Fenniae 84:5.

Mälkönen, E. 1975. Whole-tree utilization-consequences for soil and environment. Experiences from Finnish research and practice. Elmia 75, Konf Sk2: 26-30.

Mälkönen, E. and Saarsalmi, A. 1982. Biomass production and nutrient removal in whole tree harvesting of birch stands. Folia Forestalia 534.

Møller, I. 2000. Calculation of biomass and nutrient removal for different harvesting intensities. *In* Richardson, J., Lowe, A., Hakkila, P. and Smith, T. (eds). Conventional systems for bioenergy. New Zealand Journal of Forestry Science 30(1/2): 29-45.

Nurmi, J. 1994. Työtavan vaikutus hakkuukoneen tuotokseen ja hakkuutähteen kasautumiseen. Folia Forestalia 1994(2): 113-122.

Nurmi, J. 1999. The storage of logging residue for fuel. Biomass and Bioenergy 17: 41-47.

NUTEK. 1994. Askåterföringssystem. Tekniker och möjligheter R1994:3.

Nykvist, N. 1977. Skogliga åtgärders inverkan på storlek och tillgänglighet av ekosystemets näringsförråd. Sveriges Skogsvårdsförbunds Tidskrift 75: 167-178.

Rekommendationer vid uttag av skogsbränsle och kompensationsgödsling. 1998. Skogsstyrelsen. Mimeograph.

Väätäinen, K., Sikanen, L. and Asikainen, A. 2000. Rakeistetun puutuhkan metsäänpalautuksen logistiikka (Logistics of forest spreading granulated ash). University of Joensuu, Faculty of Forestry. Research Notes 116.

Waring, R.H. and Schlesinger, W.H. 1985. Forest ecosystems. Concepts and management. Academic Press, London.

Young, H.E. and Carpenter, P.M. 1967. Weight, nutrient element and productivity studies of seedlings and saplings of eight tree species in natural ecosystems. Maine Agricultural Experiment Station, Technical Bulletin 28.

5.6   CONCLUSIONS

Forests will be under greater pressure to satisfy environmental, economic and social needs of society in the future. A challenge facing forest managers will be the development of management systems that satisfy local and global demands today, while maintaining and improving the ability of forests to provide for the needs of future generations. This challenge is central to the concept of sustainable forest management (SFM), and has a major influence on conceptual developments related to bioenergy system planning and implementation. One premise underpinning bioenergy programs at the international level is that they will be *sustainable* in the broadest sense of the term.

This chapter addressed the major environmental implications of utilizing biomass from naturally regenerated and plantation forests for bioenergy production, focusing specifically on soil, long-term productivity, hydrology, biodiversity and forest habitats. The primary objective of each section was evaluation of the effects of management on these environmental components at all levels, and to identify guiding principles and recommendations that will facilitate sustainable bioenergy production from naturally regenerated and plantation forests.

The following recommendations can be made as a contribution to sustainable forest bioenergy production:

- Adaptive Forest Management (Figure 5.1-1) offers a structured approach to planning, operations, monitoring, and improvement activities. It should increase the likelihood that social, economic and environmental management goals will be achieved and provides opportunities for continuous improvement. This approach is central to certification programs, and should increase public acceptance of forest operations.

- Site-specific management is critical to the achievement of sustainable forest management. Site specific information systems, codes of forest practice, and local management prescriptions are essential.

- Soil quality and site productivity can be significantly increased by sustainable forest management practices.

- Soil organic matter should be conserved. It is possible to maintain soil organic matter levels through management practices that increase forest productivity, and to minimize losses by the reduction of site preparation disturbance. Soil organic matter conservation is important at both local and global levels.

- Soil nutrient supply and availability should be managed carefully. Loss of nutrients from forest soils can be minimized through a variety of operational techniques (Sections 5.2.6 and 5.5) including retention of foliage at the harvesting site. Nitrogen management is especially important for long-term site productivity. Some essential nutrients can be returned to forest sites in wood ash or wastewater biosolids. Equipment has been developed for wood ash and biosolids application in natural and plantation forests, and its use has been demonstrated operationally in Scandinavia, New England, the Pacific Northwest USA, and New Zealand.

- Soil physical properties should be conserved or improved operationally. Machine traffic damage to forest sites should be minimized or eliminated. Harvest residues can be placed so that machine travel during harvesting and forwarding operations is facilitated and rut formation reduced. Seasonal management guidelines should be developed and implemented on sensitive sites with poorly-drained soils. Damaged sites should be repaired, and amelioration of soils with physical limitations should be considered. Specialized equipment may be required for prevention of site damage and avoidance of costs associated with site restoration.

- Soil erosion should be prevented. Harvesting, forwarding, site preparation, and road building operations can be managed to minimize the erosion potential if codes of practice developed for specific sites and regions are followed.

- Forest harvesting and road-building operations can be managed according to regional codes of practice so that stream water quality is not impaired by increased levels of dissolved ions, nitrate nitrogen, and sediment.

- Peak streamflow responses to harvesting are variable and depend on site characteristics. Large increases are rare, except as a result of catastrophic events. Regional codes of forest practice should take potential post-harvest peak streamflow into account.

- Features important to forest biodiversity include maintenance of balanced natural habitat processes (e.g. nutrient supply, browsing, predation), old growth and old, large trees, dead trees (both standing and fallen) and a diversity of tree species.

- Forest managers can sustain forest biodiversity and maintain habitats by creating structurally diverse stands, arranging stands so that species interactions are preserved, and ensuring connectivity between habitat types. Management planning should include consideration of outcomes at both stand and landscape levels.

- Forest managers are encouraged to seek continuous improvement in the implementation of practices that promote sustainability. Through the monitoring of progress towards desired goals, Adaptive Forest Management can facilitate site-specific decisions, the adoption of locally-relevant codes of forest practice based on current knowledge, and recognition of the need for new information.

CHAPTER 6

SOCIAL IMPLICATIONS OF FOREST ENERGY PRODUCTION

N. W. J. Borsboom, B. Hektor, B. McCallum and E. Remedio

Forest management and use are largely dictated by human interests and concerns. The discussion of sustainability of forest ecosystems must therefore consider public perceptions and values relating to forests and their use, including their use for energy production. This chapter discusses firstly the importance of wood-based energy in the lives of people in developing countries, and then the particular role of forest energy production in remote northern regions with aboriginal peoples. Public perceptions and values of urban people are examined because these can have a strong influence on forest management and use for energy production, even though they are not based on direct experience of the forest. Finally, some insight is provided about the design of fuelwood operations in urbanized societies and the relationships between forest energy production and rural employment.

6.1 SOCIAL IMPLICATIONS OF BIOFUEL USE IN DEVELOPING COUNTRIES

Use of biofuels appears to be a cultural, as well as an economic activity, and public attitudes can be a barrier to understanding the advantages of bioenergy use. Methods of production depend not only on economics but on the purposes for which biofuels are intended and the extent to which efficiency of production is balanced against levels of employment and earnings.

Values and attitudes in Asian and African cultures are discussed in the following section: they appear to differ from those of Western societies in that biofuel collection and use are an accepted part of everyday life. Issues

Richardson, J., Björheden, R., Hakkila, P., Lowe, A.T. and Smith, C.T. (eds.). 2002. Bioenergy from Sustainable Forestry: Guiding Principles and Practice. Kluwer Academic Publishers, The Netherlands.

of labor and earnings, mechanization and environmental considerations are recurring themes in this examination of the subject.

6.1.1   Biofuel use in Asian countries

Among Asian countries, wood and other types of biomass are widely used as domestic and industrial fuel because they are locally-available and cheaper than other energy sources.

Biomass fuels may consist of woody or non-woody material derived from trees, shrubs, crop residues and other vegetation. All can be converted into charcoal. Among Asian countries, important biomass fuels are wood and residues from coconut, rubber and oil palm trees. By-products such as sawdust, bagasse, husks and rice straw are also included in this category.

These materials do not all come from forest areas. In fact, few or none may come from forest zones. A study in the Philippines showed that woodfuels produced, traded and consumed in 1993 originated in private woodlots, small woodlots, agroforestry areas, tree fallow areas, secondary regrowth, shrublands, idle lands and tree plantations in combination with food and cash croplands (Bensel and Remedio 1993).

Contemporary perceptions of wood energy

Use of fuelwood is fairly well incorporated into the energy policies of India, Nepal, Sri Lanka, Thailand, Malaysia, Cambodia, China and the Philippines. Issues that surround this source of energy are its importance for the national economy and employment; the fact that fuelwood is not being phased out; lack of a sound basis for the 'fuelwood-gap theory'; the potential for modern applications of wood energy; and benefits to global climate.

The main problems associated with use of wood energy are poor combustion techniques; restrictive government policies that tend to deprive poor people of possible benefits; and wasted opportunities for economic and social development. Fuelwood supplies the needs of more than one billion people in the region – close to 40% of the total population. In most countries, it is still the largest single energy source, used mainly in poor rural households. Consumption patterns are complicated and linked with socio-economic, environmental and institutional factors that are site-specific. Fuelwood is important to food security and the quality of life in rural communities.

The *'fuelwood-gap theory'* formulated during the 1970s led to the belief that fuelwood use (including charcoal production) was the major cause of deforestation. It was assumed that all fuelwood came from forest areas and that consumption would increase at the same rate as population growth. These assumptions ignored the fact that a substantial amount of fuelwood comes from non-forested areas, and the fact that users have turned to alternative supply sources, different fuels, and diffent cooking habits.

Legal, infrastructural, institutional and other factors place serious constraints on wood energy development. Energy departments tend to focus on electricity and fossil fuels and to ignore traditional fuels. This is often due to a lack of understanding. Forest departments regard fuelwood as a marginal by-product. The majority of foresters have little up-to-date knowledge about the effects of tree growth and bioenergy use on greenhouse gas concentrations in the atmosphere.

Volume of fuelwood flows

A number of misconceptions surround the use of wood energy. In fact, there is no general link between fuelwood use and deforestation; fuelwood use does not trap people in poverty; fuelwood is not being phased out; it is not a marginal product; and it is not just a traditional commodity serving poor people. Fuelwood flow studies help to provide a more accurate understanding of local situations and assist in the formulation of appropriate policies (Hulscher 2000).

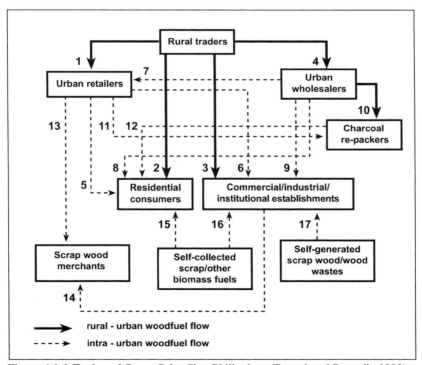

Figure 6.1-1 Fuelwood flows, Cebu City, Philippines (Bensel and Remedio 1993).

Figure 6.1-1 describes fuelwood flows into and within Cebu City in the Philippines. Slight variations may occur in other provinces, but the pattern is similar throughout. Flows 1, 2, 3 and 4 represent deliveries of fuelwood from rural traders to retailers, residential consumers, commercial establishments and wholesalers respectively. Approximately 53,400 tonnes of fuelwood and 13,500 tonnes of charcoal are shipped annually into the 49 urban barangays of Cebu City in this way. Only 2.6% of the fuelwood and 5.7% of the charcoal goes directly from rural traders to residential consumers (Flow 2). Another 17.8% of fuelwood and 12% of charcoal pass from rural traders directly to commercial and industrial establishments in the city (Flow 3). The remainder (79.6% of the fuelwood and 82.3% of the charcoal) goes to urban retailers and wholesalers.

Rural income and employment

Studies have shown that wood and other bioenergy sources create more local employment within the national economy than other forms of energy (Table 6.1-1). Rural non-farm incomes are generated from the manpower (unskilled labor) required for harvesting, processing, transporting and trading of the fuels.

Table 6.1-1. Estimated local employment associated with six household fuels.

Fuel type	Amount of fuel per terajoule	Employment per TJ of energy[a], person-days[b]
Kerosene[c]	29 m³ TJ	10
LPG[c]	22 TJ	10-20
Coal[d]	43 TJ	20–40
Electricity[e]	228 MWh	80–110
Fuelwood[f]	62 TJ	100–170
Charcoal[f]	33 TJ	200–350

[a] *Source UNDP/WB ESMAP. 1992.*

[b] *Employment includes growing, extraction, production, transmission, maintenance, distribution and sales, including reading meters. It excludes employment outside the country.*

[c] *Assumes that crude oil (for refining), kerosene and LPG are imported.*

[d] *Values vary according to capital intensity of the mine, seam thickness, energy value of the coal and distance from demand centers.*

[e] *Values vary according to production methods, which range from hydro to traditional oil- or coal-fired units and according to the efficiency of electricity generation, transmission and distribution.*

[f] *Values depend on site productivity, production efficiency and distance from markets.*

Fuelwood trading is mostly small-scale and informal. There are no comprehensive statistics from which its scale, magnitude or total number of people employed can be assessed. Nevertheless, several studies conducted in recent years can be combined to give a general picture of fuelwood flows.

In Phnom Penh, Cambodia, wood energy supply from rural areas involves a multilayered system of traders which can be described as informal and unregulated. Urban demand by households and industry is high, since wood is the main energy source (FAO RWEDP 1998). This is especially true of areas in which fuelwood has been treated as a commodity and assigned a monetary value within the local market system. In times of crisis, e.g. extreme drought, subsistence households resort to the selling or bartering of fuelwood. When harvests are inadequate for subsistence, the opportunity to generate income in fuelwood business provides a safety-net for the people affected (FAO RWEDP 1997).

Bioenergy use in the domestic sector

The domestic or household sector is the primary user of bioenergy in Asian countries. Fuelwood is used mainly for cooking and space heating. Almost all village cottage industries use bioenergy to produce goods and services intended for sale. In urban centres, this type of use is limited to certain goods, for instance food items such as barbecues, or services such as tire re-treading.

6.1.2   Gender influences

Gender roles are important to the consideration of bioenergy use. They differ according to the country, area and ethnic group. Asian women generally do the cooking, and are often considered to be responsible for gathering fuel. Rural and urban cottage industries based on biofuel are also mostly operated by women (FAO RWEDP 1999). Women tend to work longer hours than men.

Men usually find work outside the home. They have control over land and money matters, and rarely see the value of spending on improved cooking stoves that may save effort in collection of fuel and in cooking.

Women generally do not have time for planting and care of trees and other bioresources. In cases where women do plant trees, men usually harvest and sell the products. Men tend to select species for timber, while women prefer fast-growing species that will provide fuel. Fruit trees are a compromise.

Trade and commerce in biofuels are mostly carried out by women, either as a full time job, a source of extra income, or a survival strategy in times of crisis.

Gender influences the quality of the data collected on biofuels. Even if women are present during a survey interview, they usually allow a man, as head of the family, to provide the information. Responses are not always accurate because men are not directly involved in gathering, using and managing biofuels and may not be aware of all the related issues.

In policy-making, little attention has been paid to gender roles, except in matters related to the improved cooking stoves program.

6.1.3   Examples of socio-economic and cultural views

Majalaya sub-district, West Java: habit and necessity

People living in villages and small cities like Majalaya, West Java, use wood energy because they are poor and it is cheaper than kerosene. Wood is a traditional cooking fuel which does not require the purchase of a special heater and accessories. In rural areas, many sources of bioenergy are directly accessible whereas in large urban centers, a wider range of commercially-supplied energy is available.

The fuelwood energy business in Majalaya involves both lower and middle socio-economic levels. A chain of activity extends from primary production processes through to retail sales. Production takes place in rural areas and is dominated by economically weak groups, landless laborers and very small landholders. Members of the middle socio-economic level of society provide capital and organizational skills. They usually reside in urban areas and act as middlemen or retailers (FAO RWEDP 1991).

Laguna Province, Philippines: festivals and celebrations

In every village and town of Laguna Province, an annual religious festival is held in honor of the patron saint. This custom is part of the cultural heritage derived from early Spanish settlers. The town 'fiesta' is held on the last Friday of August. Seven other religious festivals are regularly observed, three of them throughout the whole town.

During these celebrations, large amounts of bioenergy (fuelwood, charcoal, coconut fronds, sawmill waste, coco shells, and others) are used in food preparation. The amount of extra fuelwood consumed during these socio-cultural activities is considerable, but has never been quantified (FAO RWEDP 1991).

Pokhara, Nepal: ceremonial uses

Hindu and Buddhist rites dictate that a dead body must be cremated. Fuelwood is arranged in criss-cross layers to make a funeral pyre. An immediate relative, usually male, sets the pyre alight. In Pokhara, 1-3 corpses are cremated every day on the banks of the Seti River. Approximately 200-300 kg of fuelwood are required on each occasion, and a committee of the Gurung community manages a depot which supplies the necessary material. Some communities bring the fuelwood required for cremation from their homes to share the cost.

Among ethnic Gurungs, fire is used on auspicious social and ceremonial occasions. In one religious ceremony, Aughaun, the Gurungs worship Buddha and eat a meal cooked in a community kitchen. About 15-20 ceremonies of this type are organized annually at one particular Buddhist temple, and for each approximately 500 kg of fuelwood is consumed (FAO RWEDP 1991).

6.1.4 Attitudes to environmental issues

Fuelwood

It is now widely accepted that the cause of deforestation is not fuelwood utilization but conversion of land use from forestry to agriculture or urban development. Population increase leading to a growing demand for food is a major influence. It is accepted that fuelwood use contributes to deforestation and forest degradation, but it is not a major cause. Commercial fuelwood trading can have a positive effect on the environment, as the market mechanism provides incentives for producers to plant and care for the trees which are their main income source.

Global climate change

Recent concern over global climate change and its relationship with use of fossil fuels has stimulated interest in wood as a renewable, sustainable and environmentally-benign source of energy. Wood energy, carefully used and produced, is carbon neutral. Although wood emits CO_2 during burning and natural decomposition, growing trees absorb carbon from the atmosphere during photosynthesis. Natural decomposition also emits methane, which is not a product of complete combustion. From the environmental standpoint the burning of logging and industrial processing residues is beneficial. Unlike coal and oil, fuelwood does not emit SO_2.

In Asia, most households use traditional stoves fired by biofuels. Due to incomplete combustion these stoves have a low efficiency and emit pollutants such as carbon monoxide, methane and oxides of nitrogen. These pollutants have a negative effect on human health and increase greenhouse gas concentrations in the atmosphere. Effort is needed to develop stoves that conserve fuel, enhance user convenience, minimize health hazards and reduce greenhouse gas emission.

At present, electricity in Cambodia is generated from oil which, like LPG, is a finite energy source. Wood, on the other hand, would be a renewable and indigenous energy source and could provide other forest products and environmental stability (FAO RWEDP 1998).

Sustainable use of bioenergy assists the sustainable management of the environment. If wood and other biomass materials are valued by local populations as an important energy source, trees are more likely to be cared for and protected. Sustainable use of biomass benefits the global climate because it is carbon-neutral. Any replacement of bioenergy by fossil fuels will add to the greenhouse effect. (FAO RWEDP (no date)).

6.1.5   Biofuel use in African countries

Botswana

Fuelwood is gathered by household members, mainly children, indicating that the opportunity cost of labor for children is low. Women are also frequent collectors. Contrary to the usual assertions, men, whose opportunity costs are high, also participate in fuelwood collection. Some households supplement gathered fuelwood with cow dung and commercially-available energy sources. The most common means of transport for fuelwood is headloading, used by 67% of households in Khakhea and 62% in Mnangkgodi. Donkey carts are the second most common method of transport, used by 14% of the households in both villages. Bicycles, pick-ups and trucks are also used.

Those with trucks and pick-ups collect logs, whereas headloaders tend to collect twigs and small branches. Fuelwood collection is a selective process and households have detailed technical knowledge regarding tree species, their qualities and economic benefits. Species providing good fuelwood ember are preferred; smokeless burning and ease of cutting are also desirable. Main end uses are cooking, space heating and water heating; other uses include brewing beer, ironing and lighting (Karenzi 1994). Level of demand is influenced by household size, household wealth and the availability of wood.

Most Mnangkgodi households cook three meals a day, but in Khakhea fewer meals are cooked. Khakhea is in the semi-arid part of the country where agricultural productivity is much lower. Average household size is 6.2 persons. A decrease in the amount of available fuelwood indicates that harvesting exceeds the limits of sustainable land use. Availability of wood will depend on distance from remaining forests and the availability of household labor.

In wealthier households, there will be more use of gas and paraffin as fuelwood becomes scarce and more expensive. Drought, the rapidly-growing urban population, increasing demand for poles, and land clearances are the main causes of depletion of biomass resources (Fraser 1991).

Mali

About 75% of the fuelwood supply for Bamako, the largest city, comes from as far away as 175 km. Most is transported by motorized road vehicles and trucks; the rest by donkey carts, bicycles and pedestrians. Fuelwood is obtained from natural savanna woodlands and is collected 3-5 km on either side of six radial roads in a network that brings fuel to the city. As it becomes scarcer there is a tradeoff between consumption and the extra time and effort required for collection; as a result, consumption is reduced. Alternative materials including previously-rejected small wood and branches as well as millet stalks, other agricultural residues, and dung, come into use. Bamako, unlike other big metropolitan areas in neighboring countries, seldom uses charcoal.

Because a six-year fallow rotation practice is functioning successfully, it is expected that there will be opportunities for the expansion of woodfuel collection during the next two decades (Foley 1987). Supply sources include communal fallow land, village woodlots, hedgerows and shelterbelts. In a number of villages located 120 km north of Bamako, between two tributaries of the Senegal River, socio-economic activities consist mainly of semi-nomadic herding of zebu cattle, sheep and goats in the wooded savanna and the tending of livestock. Farmers live in small villages in extended families. Cultivation is done by hand, with hoes and axes as the main tools. When fallow land is cleared for recultivation, at least part of the tree and shrub growth is harvested as fuelwood. An eighth-year savanna fallow may provide 3-4 m^3 ha^{-1} of fuelwood. Cultivators have free access to the land to graze livestock or to gather material other than tree products. This results in localized overgrazing and annual fires that hinder regeneration. Existing farming systems such as use of fallow periods and plowing with animal traction play an important part in maintaining a balance between land and people vulnerable to the vagaries of agroclimatic conditions (Ohler 1985).

Rwanda

The Rwandese rely mainly on wood and agricultural residues for their energy needs. Sources of fuel are woodlots, agricultural residues (mainly sorghum stems) and bushes, and vary from region to region. In the Ntyazo region, residues of the tallest species are used whereas in the Cyimbo area, weedy grass rhizomes are the main fuel source. Twigs and small branches collected from woodlots and neighboring fields are used more than wood from cut trees. Greater fuelwood use means less use of agricultural or weed residues, and *vice versa*. End uses include cooking, water heating, lighting, beer preparation and banana ripening.

Cooking in the city of Kigali is carried out mainly on traditional three-stone fires and ordinary charcoal stoves, two or three meals being cooked per day. There are no fuel shortages; fuel conservation methods are known but not often used. City residents use cut wood in contrast to rural households which use twigs. As Rwandese towns develop a higher standard of living, there is a shift from fuelwood consumption to charcoal consumption (Karenzi 1994).

Senegal

Some fuelwood is collected and some is purchased. The availability of cash determines choice of fuel. Refugees rely on collection because of their lack of cash. Wealthy local women have access to charcoal, whereas women refugees are likely to use animal dung. Because most of the charcoal is produced inefficiently, local people have a greater impact on the environment than do the refugees (Black and Sessay 1997).

South Africa

Food preparation is the main end use for fuelwood, although beer brewing is also practised, and requires more fuelwood. The average distance travelled to gather fuel is 4.1 km, but in areas where there are woodlots, about 40% of all households travel more than 6 km. Most wood collection is done by women (who have considerable knowledge about woody species), some by children and some by hired carters. As household size increases, wood consumption *per capita* decreases. The cost of fuel per household is approximately 15% of total income. Very few households use stoves as the sole method for cooking; the majority use the traditional open wood fire, because of lower cost and socio-cultural considerations – an open fire is often preferred to an efficient stove (Bembridge 1995).

Zaire (now the Democratic Republic of Congo)

In Kinshasa, fuelwood is mainly used for cooking, although other uses include local beer brewing, handcrafts and small industries. People try to adapt to limited fuel supplies by reducing the number of cooked meals per day, abandoning traditional meals that take longer to cook and resorting to rewarmed meals (Tsibangu 1996).

Zambia

Fuelwood and charcoal are used daily for cooking, heating and lighting, and occasionally during social events like funerals, initiations and weddings. Non-household users include bakeries and potteries, brick producers, social centers and churches (Elmazzoudi 1986).

Zimbabwe

In the four provinces (Eastern, Central, West Mashonaland and Midlands) where most of the flue-cured virginia tobacco is grown, there are 10 034 ha of tree plantations, of which 92% are eucalyptus. Forests and woodland cover about 40% of the land area. The total annual increment of wood is estimated to be approximately 29 million tonnes, and the total annual consumption of fuelwood approximately 5.1 million tonnes. Half of the population lives in communal lands where the annual wood increment is only 1.1 million tonnes, and the demand 2.8 million tonnes. The availability of coal from Zambia offers tobacco farmers an alternative fuel in times of local shortage or high prices. A total of 1304 registered producers of flue-cured tobacco cultivate 54,000 ha of land. Tobacco has contributed positively and significantly to the economy. The tobacco industry accounts for 8% of the national fuelwood consumption, but wood is only used to cure 30% of the tobacco, the remainder being cured using coal. The low value of wood in Zimbabwe, compared with coal, means that supplies are not transported for long distances. Wood used by the tobacco industry comes from areas where there is a surplus of sustainable yield over demand (Fraser 1987).

6.2   SOCIAL IMPLICATIONS OF BIOENERGY USE IN DEVELOPED COUNTRIES

Attitudes of the Western public toward 'nature' and the consequences for bioenergy utilization are examined in the following section. The discussion draws on scenarios from two distinct and contrasting cultures: the remote aboriginal communities of Northern Canada and the urbanized populations of Western Europe.

6.2.1   Development of bioenergy utilization in remote communities

Remote communities in Canada

The social benefits of development of bioenergy utilization in rural communities are perhaps more immediately evident than benefits to urbanized societies, which have less direct experience of and involvement with forests. The use of biofuels by aboriginal communities in Northern Canada brings practical and social advantages, although its adoption remains a challenge to traditional and institutional practices.

In Canada, interest in bioenergy has ebbed and flowed with world prices for oil. There was considerable activity between 1973 and 1985 when oil prices were high and expected to rise further. The rationale for bioenergy development during that period was mainly economic and related to security of energy supply. When oil prices decreased, government and industry interest in bioenergy waned.

Taxation of fossil fuels in Canada is higher than that in the United States, but low compared to that in Europe. The need to remain competitive with the USA makes it difficult for Canada to increase energy taxes beyond current levels. Carbon taxation has been strongly resisted by the province of Alberta, which has produced most of the oil in Canada. For these reasons, bioenergy has not been attractive to industry or institutions over the past decade and net usage has actually declined in Atlantic Canada. Advocates have been unsuccessful in selling bioenergy in Canada as the 'green alternative' which it is in Europe. The public perspective in Canada generally is that cutting trees for any reason is a bad thing.

One area where bioenergy is economically attractive is in forested regions of Canada's North, where energy costs are high because of the remote location. First Nations Reserves are the responsibility of the Federal Government Department of Indian and Northern Affairs. The Canadian Forest Service of the Department of Natural Resources has been delivering forest management programs to First Nations communities since the early 1980s. In 1995, the Minister of Indian and Northern Affairs asked the Service to explore ways of increasing economic development based on community-owned forest resources. The rationale was clearly to create social benefits through forest management and development. The Canadian Forest Service has been exploring opportunities for silviculture contracting, timber and sawmilling operations, and biomass energy, with a number of northern communities.

Fuel supply situation

Canada has more than 150 remote communities located to the north of the natural gas pipelines, the electricity grids and the year-round road networks. These communities have access to significant biomass and timber resources. In addition, there are approximately 200 semi-remote communities located in forested areas to the south which are connected to the electric power grids and year-round roads, but not to low-cost natural gas. The size of populations in First Nations Reserves is variable. Most consist of 400-600 people, and a few have as many as 1500-2000. The birth rate among aboriginal people in Canada is high, and all native communities are growing rapidly. The need for social services is also growing. Poor housing, primitive water and sewer systems, high unemployment, alcoholism and suicide are common problems. Government and community band councils are struggling to deal with these issues.

Remote northern communities have many things in common. Most public buildings are heated by oil-fired hot air or hot water furnaces. The cost of oil is high. Heating oil is transported to the reserves over winter-ice roads between January and March and stored in local tank farms. In recent years, the winters have been quite warm and winter road access has been limited or nonexistent. As a result, most of the heating fuel has been flown in at even higher cost.

Most of the remote communities are located in the vast boreal forest which spans Northern Canada. Reserve forests are not usually managed or harvested, although small volumes of firewood are used to heat private homes. Firewood is generally procured as needed. Woodlands near the reserves are often overmature and may constitute a serious fire hazard to the community.

Case studies have been carried out in a number of communities to examine opportunities for biomass conversion. Most of the communities have large, under-utilized forest resources and clusters of public oil-heated buildings that could be heated with woodchips. There is an acute need for sustainable local employment.

Biomass harvesting system options

Two harvesting system options have emerged as the most practical for remote areas. In larger communities and those already possessing some forest machinery such as skidders, interest has focused on the use of conventional forest machines. Following motor-manual (chainsaw) felling, a skidder or a forwarder would be used to extract wood to a roadside landing. Chipping could take place at the landing, and existing dump trucks with raised sides could be used to transport chips to a storage facility at the heating plant. Alternatively, whole-trees or trees cut into two or three sections could be transported to a wood yard located near the community and chipped during the heating season.

For small communities with no existing commercial forestry equipment, a system based on tractor-powered forestry equipment is a practical, relatively low-cost option. After chainsaw felling, a 4WD, 75 kW farm tractor could be used to extract material for chipping as well as roundwood. The tractor could also be used to transport woodchips from a forest chipping site to a heating plant, or tree sections to a wood yard located near to the community (McCallum 1999). Farm tractor-based woodchip supply systems are used extensively in Denmark and Finland (Hakkila 1984).

A tractor could also be used to power a chipper at the wood yard, to transfer woodchips to the heating plant and to feed woodchips into the combustion unit using a front-end loader. It could also handle sawlogs and power a sawmill and a firewood processor producing blocked and split firewood.

Community-owned tractors could be used for other municipal purposes such as blowing or plowing snow, pulling a drag on winter roads, pulling a road grader, running a standby generator, transporting equipment and powering water pumps for firefighting. Some grapple loaders can be equipped with a backhoe or ditch-digging attachment. Combined forestry and municipal uses could make a heavy-duty 4WD tractor an economic investment for a remote community.

6.2.2 Benefits of bioenergy utilization for remote communities

The use of bioenergy would benefit remote communities in direct and indirect ways. Direct benefits would be both economic and social. They include:

- Lower cost heating. Forest fuel would be cheaper than heating oil. Since the heating cost of public buildings in remote communities is generally paid by outside governmental agencies, conversion would only benefit the community if specific financial arrangements were made to pass the savings on to them. This incentive would be critical to any community decision involving a switch to a less convenient energy source.

- Creation of long-term employment. Job creation would be the most important direct benefit of bioenergy utilization in remote communities. In small-scale chipping operations, labor accounts for over 50% of the cost of woodchips (Hakkila 1984).

- Improved economy of forest operations. Fuelwood harvesting could enhance the economic viability of any existing forestry operations, because equipment and labor could be more fully utilized.

The indirect benefits are mostly social in nature. They include:

- Retention of extra money in the community. Over 90% of the money spent on heating oil is lost to the local economy because it is paid to outside companies and very little local employment is generated. Because labor is the major cost factor in fuelwood production, a large proportion of the money spent on fuelwood extraction and processing would be circulated in the local community.

- Greater community self-reliance. Reduced dependence on government funds means that unforeseen oil price increases (such as the one experienced in 2000) have a less severe effect on the local community.

- Enhanced self-esteem. People with steady employment related to an obvious need in the community have a much higher sense of self-worth and fewer social problems.

- Reduction of negative environmental impacts. Automated biomass combustion systems burn very cleanly, and the sustainable use of forest biomass is CO_2-neutral. If fuelwood replaces heating oil, net production of greenhouse gases will be reduced. Oil spills and contamination of soil and water, which are common problems in remote communities, will be reduced.

6.2.3 Complementary forestry activities for community development

Commercial timber operations are unlikely to be economically viable as stand-alone operations in remote regions of Canada. Distance and the fact that road access is limited to a few months in winter severely restricts the export of timber to the South. The market for products from most reserve woodlands is limited to the reserve itself and possibly other nearby reserves.

The development of community-based bioenergy production is seen as a complementary forest activity that could increase cash flow and justify investment in forestry equipment and the training of forest workers. Bioenergy use has been shown to complement conventional timber harvesting in other regions such as Prince Edward Island (McCallum 1995). Fuelwood production for domestic purposes could be added to other commercial forest operations. In most remote communities, firewood is produced privately and transported on toboggans pulled by snow machines. This work is arduous, inefficient and expensive. Many residents are currently switching to oil for domestic heating and this reduces the self-sufficiency of the community.

Bioenergy harvesting offers silvicultural benefits that can deliver long-term social benefits. Mature overstories could be removed and young stands could be thinned for high-quality timber production. Bioenergy utilization would provide a market for mature, non-merchantable wood and for thinnings. Harvesting logging residues and non-merchantable trees from patch cuts would facilitate replanting and other types of stand regeneration.

The two harvesting options discussed in Section 6.2.1 are both based on motor-manual, rather than mechanical harvesting. Motor-manual methods are more appropriate for small-scale operations and involve a minimal amount of capital investment. They are also more labor intensive than mechanical systems. Experience in Prince Edward Island and Finland has shown that woodchips can be produced very economically when motor-manual felling is combined with mechanical extraction and an efficient chipping operation (McCallum 1995, 1997; Hakkila 1984). However, the creation of inefficient work must be avoided. If cost savings are not substantial and sustainable, other communities are unlikely to perceive the benefits of adopting new bioenergy technology.

To date, bioenergy systems have only been installed in semi-remote communities with permanent road access. Examples can be found in Ouji Bougamou, Quebec, Grassy Narrows in Ontario and in several communities in the Northwest Territories. The heating plants use sawmill waste rather than woodchips from harvested forest residues. They produce low-cost heat and create some employment, but do not provide additional forestry work or contribute to long-term forest management. Bioenergy utilization by remote communities has yet to be demonstrated in Canada.

Potential applications in other remote regions

Alaska and Siberia are regions which are similar to Northern Canada in terms of geography, climate and forestry, and which also have a need for space heating during the long cold winters. The benefits of forest bioenergy utilization described in Section 6.2.2 could well apply to remote communities in these countries. In the Southern Hemisphere, the end-use would probably be heating for small-scale industrial processing rather than space heating.

6.2.4   Opportunities and challenges for remote communities

Bioenergy utilization can bring about significant economic savings in almost all remote communities. It offers long-term social benefits such as the creation of stable employment, enhancement of self-esteem for workers, retention and circulation of scarce income within the communities, greater community self-reliance and reduction of the risks associated with sudden oil price fluctuations. While these benefits are highly desirable, they often fail to fit within the specific mandates of existing decision-making bodies, both in government and on the reserves.

An additional problem is that although bioenergy systems make logical sense, people in remote communities have yet to be convinced that their community should assign priority to local biomass utilization schemes. Successful demonstration projects are needed to illustrate the benefits to local people and to government agencies responsible for investment on reserves in Canada. Convincing demonstrations will require strong leadership from forestry and energy institutions to facilitate collective action by the many parties involved, and to find the funds necessary for successful implementation.

6.2.5 The relationship between urban communities and the forest

Current public attitudes of Western society to nature and forests are based on two perceptions: on the one hand, nature is regarded as beautiful and in need of protection; on the other, nature is unknown and even threatening. The public sees and admires nature on television and the Internet, or enjoys the reality during Sunday afternoon walks or short holidays, appreciating the silence and occasionally spotting the wildlife.

Although for most urban people nature consists of beautiful scenery, a smaller group is deeply interested in how nature works and in analyzing its value. People in this group often have a special interest in the wellbeing of certain species or groups of species. Some focus on relationships among species. Whatever the interest, many are members of non-governmental environmental protection organizations such as Greenpeace, the World Wildlife Fund, Friends of the Earth, the British National Trust, or groups dedicated to the protection of particular species. Collectively, these organizations have considerable lobbying power and can influence the management of nature. The standard of living in Western society is so high that protection and even extension of components of the natural environment may be achieved at the cost of other land uses.

Most members of the general public have some kind of relationship with environmental organizations. Individuals who are not active members may be financial contributors or may sympathize with specific goals. Environmental organizations are highly appreciated for their educational work, but are sometimes resented by business managers who are perceived to pay too little attention to environmental sustainability.

Most professional members of nature conservation and forestry organizations do not have a large influence on public attitudes. In the absence of other information, people have to rely for their knowledge on interpretation by environmental organizations. There is a marked contrast between the general acceptance of environmental conservation and knowledge about our dependence on natural processes. The public accepts the beauty of nature but does not understand how it works. Thus, it is not accepted that animals must die, that trees compete with each other for essential resources, or that human survival depends on the work of farmers, fishermen and foresters. One explanation for this contradiction is that in the past, professionals tended to advocate exploitative management of natural resources for short-term gains. Now, however, entrepreneurs close to nature are becoming wiser. They seek certification to show that their management methods are compatible with environmental sustainability, and they present themselves to the public as stewards of our natural heritage.

People who earn their living directly from natural resources have not communicated well with the general public. The urban public is not directly dependent on nature, and its members are not well-educated about the ways in which the natural environment can be influenced by human intervention.

Forestry is not an important business in urbanized regions because the cost of land is high and the financial return is low. As a consequence, wood and timber production are not well-understood and have a low political profile.

Public reaction to use of wood for fuel

Collection and recyling of newspaper to avoid unnecessary cutting of trees is popular in Western countries; some have an excess supply and export used newsprint. There is a general attitude that raw materials should be reused as much as possible. Burning of 'waste materials' is not generally acceptable because burned material cannot be reused, and burning is associated with negative effects (odors, dust, disease-causing organisms and chemicals).

In modern urban communities few people depend directly on wood for home heating; electricity and gas are more comfortable, and cheaper and easier to use. In some communities the use of fireplaces is banned to reduce air pollution.

It is difficult for lay people to understand that utilization of wood rather than fossil fuels reduces CO_2 levels in the atmosphere. For this reason, most politicians when talking about CO_2 reduction programs prefer to discuss photovoltaic systems, windpower, waterpower and solar heating systems. The use of bioenergy is often ignored or presented as a second choice (Bootsma 1998). Nevertheless, some energy producers are developing a policy designed to improve the relationship between consumers and the environment. This is exemplified in a statement by Alwin Schoonwater of Nuon International/Renewable Energy:

"We intend to find locations for new bioenergy plants which answer the criteria of high energy efficiency and the relation with the local environment. This means that the installed capacity of new bioenergy plants depends on the amount of local collectible raw materials from forests, parks or energy crops (from an average radius of 50 km) and on the demand for energy by the local customers (district heating or industry). When we have a new location which has for example plenty of raw materials but a limited energy demand, then the bioenergy plant will be designed to fulfill the energy demand to achieve a high energy efficiency. In this way, we create a balanced relation between the environment and the customers".

The need for standards

Wood as a substitute for fossil fuels is generally accepted by environmental organizations, most of whom understand the benefits of forest management. Their strongly-held and generally accepted view is that wood should be used and reused for as long as possible. The rationale for the *'timber cascade effect'* is that wood production fixes carbon and as long as it is not burned or decomposed, wood is a means of preventing the return of carbon to the atmosphere as CO_2 gas. To this is added the view that humans should be careful not to exhaust nature's ability to supply raw materials. Decisions about the use of certain types of timber must be based on life cycle analysis. The question for the forest manager is not only 'What product gives me the best economic returns?' but also, 'What product gives the best environmental returns?'

Utilization of energy from wind, water and sun has a much higher public profile than bioenergy production (Bootsma 1998). Use of bioenergy is often regarded as the burning of waste and there is a fear that combustion products will increase air pollution. Governments have responded with regulation of energy plant emissions (Stichting Platform Bio-Energie 2000). Strict standards are imposed on the inputs and outputs of biomass-burning facilities that produce heat, electricity or both. These standards require the installation of equipment which increases investment levels and reduces the opportunities for use of fuelwood in small-scale energy plants (Stichting Platform Bio-Energie 2000). Different countries have different standards, and these affect trade on the international market. Sweden, for example, is importing fuelwood over distances of more than 1000 km. The European Community is developing a system of quality standards for biofuels in order to create a transparent market.

The adoption of combustion standards is not only important for the international market. It also signals to the public that wood is an acceptable and legitimate energy source. The development of standards for application of life cycle analysis will enable appropriate decisions to be made about the use and ultimate destination of all forest products.

Public reaction to harvesting and storage of wood in the forest

The reaction of the urban public to forest harvesting is generally negative. 'Why disturb the peacefulness, why disturb the animals, why disturb the birds, why disturb the darkness?' Few members of the public recognize the need for balance between nature and the profitable use of nature. The urban population wants free access to forests for recreation, and does not appreciate forest management operations that interfere with this. There is no doubt that forest machinery is noisy, that it compacts and disturbs the soil, or that harvesting operations kill breeding birds (Bijlsma 1999, Fopma 1999, Joapik de Witkop van Roennos 2000).

Forests in densely-populated prosperous countries are becoming increasingly dedicated to recreation, and harvesting activity is being reduced both in time and space. Forestry professionals are paying increased attention to park management, public accessibility, reduction of size and number of storage and loading areas, and restriction of commercial vehicle access and noise.

In response to public pressure, the National Forest Service of the Netherlands has limited harvesting operations during the bird-breeding season. This policy change was communicated to the bird protection society (van Tuyll van Serooskerken 1999, Boon 1999) and all forest operations are now monitored by birdwatchers. Approximately 80% of all forest operations must be carried out during the seven months of autumn and winter. This means that delivery of fuelwood to energy plants has become a more complicated and more expensive procedure.

6.2.6   Design of fuelwood operations in urbanized societies

Design elements required for optimum supply

Components of the supply chain of fuelwood from forest to combustion plant are often determined by the economic scale of the operation. The aims are usually maximization of productivity and reduction of the number of employees. Regulations relating to air pollution impose high costs for exhaust filtering systems, which can only be borne by large-capacity plants.

Because of the nature of the market, the plant must run continuously. Heating plants must operate at their highest capacity during the colder seasons. Energy plants in urban areas do not have excess storage capacity for fuelwood, because of the high cost of storage space. Production plants developed by the energy sector usually obtain raw material from the market shortly befure use.

Forests in densely-populated countries are often small and scattered. They may have numerous owners. Restrictions to forest operations include limitations on storage capacity, access to the forest and noise. The forest is not the only supplier of wood to energy-generating plants. Large quantities are also obtained in the form of cuttings from city parks and roadsides, and as industrial process residues from sawmills.

Design of the fuelwood supply chain should take account of the following:

Production

- Quality and dimensions (moisture content; particle size; contaminants such as sand, clay, stones).
- Supply sources (forests; sawmills; other woodworking industries; municipalities; parks and roadsides).
- Location of chipper.
- Costs associated with sustainable forest management.

Transport

- Distances.
- Mode (road; rail; water).
- Vehicle type.
- Restrictions on supply route to plant.
- Cost of forest road repair.
- Cost (sometimes more than 25% of total production costs).

Storage

- Drying conditions.
- Location of piles.
- Capacity of facilities at combustion plant.
- Opportunities for integration of production and storage (in the forest, in the transport chain, at sawmills, at municipal sites).

Environment

- Costs associated with environmental sustainability.
- Regulations relating to seasonal activity, harvesting area, access, noise, emissions.
- Ash utilization.

Details of the design of the supply chain must be discussed with suppliers (forest owners, mill owners, municipal authorities), transport operators, energy plant operators, local and national government agencies, and the local community. The points mentioned above could serve as a checklist for the planning process.

Design elements for forestry operations in densely-populated countries

Forest managers have to pay attention to both sustainability of production and sustainability of the forest environment. For example, forestry operations should minimize the removal of nutrients from the forest site, especially on poor sandy soils such as those of the heath belt of north-central Europe. Extraction of logging residues from clearcuts, as practised in Sweden, would not be acceptable on the poor soils of the Netherlands. In the latter country there is a general understanding that removal of whole-trees should be restricted to first and second thinnings. Fertilizer application in forests is not acceptable, because of high cost and the general attitude that fertilizers should be used only where absolutely necessary. The age of the forest plantation is an important factor in the design of the forest operation for energywood.

The following elements should also be taken into account:

- Forest composition: stand age classes and expected tree growth rates; need for silvicultural treatment; forest health.
- Physical accessibility of stands at different seasons.
- Restrictions on operational access.
- Noise.
- Wood storage (methods; capacity; public acceptance; phytosanitary restrictions). See Section 3.4.
- Duration of operations.

To prevent criticism by the general public and environmentalists, the following rules of thumb should be applied:

- The duration of the operation should be as short as possible. This requires strict planning and a high degree of mechanization.
- Harvesting operations should be carried out during autumn and winter.
- Chipping should be done outside the forest if possible.
- Chipping within forest stands should be restricted to first and second thinnings.
- The production chain should be used as intensively as possible.
- Drying piles should be located along forest roads only if generally acceptable.
- Transportation in the forest should be properly planned, taking road use by the public into account.

Non-forest fuelwood

Use of forests as the sole source of fuelwood is costly, because of distance from the energy plant, and because production may not be continuous. In densely-populated areas, the demand may exceed forest productivity. On the other hand, forest fuelwood is sometimes highly valued for domestic use because of its 'clean and natural' image. Where this perception is not justified, value can be added through processing, e.g. by modification into pellets or briquettes.

At present wood from the forest costs twice as much as coal on the world market. To meet competition from other fuels, sources of fuelwood other than the forest must be available. In densely populated areas of countries with a high standard of living, other readily-available sources are woodchips from amenity plantations (parks and roadsides) and woodworking industry residues. Selection of uncontaminated material, the application of standards and objective life-cycle analysis are important to the image of bioenergy production as a 'green energy' business. Wood from outside the forest is relatively cheap. The forest owner can contribute to the energy supply of the country and generate a small profit if forest residues are included in the fuelwood mixture.

When energy generation from wood becomes a common practice, supplies in a densely-populated country are unlikely to satisfy the demand for fuelwood. Conversion to wood will depend on prices, and imports from other countries may be necessary.

6.2.7   Changes in urban culture and economics

Strengthening of the relationship between public and forest

Technological developments and efforts to reduce atmospheric CO_2 levels offer opportunities for greater public understanding about forest ecosystem processes. The economy of Western countries is based on a secure and plentiful supply of energy. At present, this is derived mainly from fossil fuels, although more use is being made of sun and wind power. Greater utilization of biomass from the forest can increase the public need for forests. It would encourage urban people to appreciate the forest as a source of energy as well as a source of inspiration and recreation. Understanding of forest processes will be strengthened if the public knows where fuelwood is coming from and wants assurance that the forest is well managed. This assurance can be provided through the certification of forest management, forest products (including fuelwood), and energy plants. The certification process is slowly changing the timber market. Already, there are circumstances under which small-scale bioenergy production based on currently-available technology is becoming competitive with generation from other sources.

Cost and profits

In prosperous and densely populated areas of Western Europe, the high cost of forest fuel is due to high land costs and a complex ownership structure. In countries with a large forestry sector (the Nordic countries, USA, Canada) land and operational costs are lower because of the efficiency of scale. This affects fuelwood production and forest management as a whole. It has been shown that production of fuelwood from the forest gives the forest owner a slightly better return than the traditional market (Rietkerk and Verhagen 2000). The return is sufficient to enable the owner to thin tree stands where such work has been postponed for lack of funds. In economic terms, the possibility of fuel production places the forest owner in a better position in the market for low-value bulk material (such as wood for particleboard). In addition, there is an opportunity for higher output in terms of volume: undersized stems and logging residues become marketable.

If wood enters the homes of urban populations not only in the form of construction material, furniture and floorboards, but also in the form of energy, forest owners will be able to strengthen their position and show the general public how useful the forest is in both economic and environmental terms. They will be able to demonstrate that they are professional forest managers who care for nature and the environment.

6.3   FOREST ENERGY AND EMPLOYMENT

Increased employment and earnings in rural areas are included among the many benefits attributed to forest bioenergy production. Politicians use the term 'job creation' whereas economists and planners refer to 'income formation' and 'employment'. In a rural society, the main issue is: 'Will bioenergy production provide earnings that are high enough to make the mobilization of local resources worthwhile?'

It is assumed (and probably true in rural areas) that some of the required resources are currently under-utilized. Such resources include labor, machinery, forest residues, land, infrastructure, and management capacity. Moreover, the work is performed not for wages but by self-employed farmers, forest owners and local contractors. They are interested in adequate returns or 'earnings', in the form of working hours, use of machines, sales of fuel, or other benefits.

In spite of the fact that the term 'earnings' describes the benefits of bioenergy projects to a rural community more accurately, most studies express benefits in terms of 'employment'. This is because 'employment' data fit better into general economic planning models and political paradigms. In this section data on both employment and earnings are provided where possible.

6.3.1 Methods for measuring employment and earnings

When measuring earnings and employment, a number of practical problems are encountered. It is often difficult to separate the production costs for fuelwood from those of other products of integrated activity. Fuelwood is often regarded as a by-product and only the variable costs are accounted for; in other cases fuelwood harvesting may be regarded as a nuisance by the logging entrepreneur, who will then allocate a high proportion of the fixed operating costs to the fuel product. Even where data on employment and earnings from fuelwood are available, their quality and the accounting principles applied may lead to misinterpretation.

Two methods are used for the analysis of employment and earnings. The direct method is applicable where detailed data are available. There may be a need for evaluation of the quality of the data and adjustments may be required if relevant and reliable information is to be obtained.

The indirect method is useful when statistical information is scarce or covers aggregated activities. It is based on the assumption that total cost data for fuelwood production are available, which is normally the case. From the total production costs, standard costs are deducted for activities that are not related to the rural economy. These include all external inputs, e.g. machinery and equipment, gasoline, interest on borrowed money, and various services. The remainder represents the earnings of the local rural community, and forms the basis for further analysis of employment.

The second method demonstrates the effect of the use of existing under-utilized resources. Individuals active in the rural economy may only be interested in fuelwood production if it utilizes spare capacity in terms of time, machinery, land, trees, skill, etc.

6.3.2 Effects of production systems on employment and earnings

In this section three groups of case studies are presented. The first represents conditions in Sweden and focuses on the effects of a rapid increase in demand for, and production of, bioenergy and a parallel decrease in biofuel cost and prices. It illustrates differences in employment and earnings between various production systems. The second group relates to biofuel production under tropical conditions, and presents data for highly-mechanized and less highly-mechanized operations. The third group consists of selected European studies of multiplier effects.

Sweden: the effects of rapid increase in production

Information on Swedish conditions is based on three studies (Danielsson and Hektor 1992, Ahlgren 1998, Stridsberg 1998) in which a similar methodological approach was used. The latter two can be regarded as revisions of data from the first study.

Danielsson and Hektor (1992) reviewed employment in several different biofuel production and distribution chains. They evaluated the jobs generated by production of additional quantities of biofuels, and results therefore illustrate the effects of an increase in demand and production. Analysis included the direct employment effects, but not indirect employment or multiplier effects.

Table 6.3-1. Job creation in biofuel production in Sweden (person-years per petajoule).

Biofuel type	Harvesting	Comminution	Transport	Combustion	Administration	Total
Industrial processing residues						
Internal use[1]	–	1	–	–	–	1
Sold to market[1]	–	1	5	1	2	9
Sold to market[2]	–	–	–	–	–	8
Sold to market[3]	–	–	–	–	–	11
Demolition wood[1]	0	5	3	1	4	13
Logging residues[1]	13	8		1	4	32
Logging residues[2]	9	7	5	0	3	24
Logging residues[3]	–	–	–	–	–	25
Tree-sections from integrated operations[1]	2	11	15	1	4	34
Small trees						
Manual work[1]	58	5	5	1	4	73
Mechanized[1]	20	5	5	1	4	35
Mechanized[2]	20	3	5	1	3	32
Firewood for own use[1]	58	5	0	0	0	63
Willow short rotation fuelwood						
Local heat contract[1]	24	23	25	37	4	113
Local heat contract[2]	11	9	10	36	4	69
Local heat contract[3]	–	–	–	–	–	44
Mechanized, sale[1]	11	2	6	1	4	25
Canary reed (*Phalaris arundinacea*)[1]	18	0	6	1	1	26
Straw[1]	12	0	8	1	2	23

[1]*Danielsson and Hektor 1992;* [2]*Ahlgren 1998;* [3]*Stridsberg 1998.*

The study used both direct and indirect methods of analysis. The first step was application of the indirect method, using theoretical relationships to calculate machine capacity and work output, and labor inputs. The formulae were developed for traditional forestry, agriculture and transport operations. Because they were developed for products other than biofuel, the validity of the results for biofuel production was checked. Selected representatives of the bioenergy sector were interviewed and provided information on practical operations. At that time very few data on bioenergy were available to the public. The survey results were compared with data from theoretical analysis. Differences were found to be small, and only a few minor adjustments were needed.

It was assumed in the 1992 study that the technological level chosen to represent the best present method and technique would be the approximate average level during the decade 1990-2000. Within a few years it became apparent that the development of technology and methods was much faster than had been expected. Costs (in real terms) decreased by approximately 40% between 1990 and 1997, continuing the trend that had prevailed in the previous decade. Revision of the data was required so that results could be related to actual performance levels. In independent studies, Ahlgren (1998) applied the direct method, and Stridsberg (1998) used mainly the indirect method to rework the data. Both concluded that the relative employment effects decreased with increased efficiency. Table 6.3-1 summarizes the results of all three studies. Different levels of mechanization were found to result in wide differences in labor input. Thus, employment was influenced substantially by the choice of technology level in the operation.

Developing and tropical countries: the effects of mechanization

This section summarizes and draws conclusions from studies carried out for public and private organizations in Southeast Asia and Latin America. The Master Plan for the Dendro Thermal Power Programme of Thailand (NEA 1988) is an example of the material reviewed.

In all the studies, it was assumed that wood-based energy production would be competitive on its own merits and thus satisfy the following four criteria:

- The price of fuelwood must be higher than its production cost.

- The income of farmers or forest workers must be kept at an acceptable level.

- The price of fuelwood must be competitive with that of other locally-available fuels.

- Wood-based energy generation must be competitive with generating systems using other energy sources.

These criteria ensured that there would be strong pressure for efficiency in fuelwood production, and no specific subsidies were included. Traditional economic factors were considered to be the leading factors. Social factors such as employment were regarded as positive side effects. The level of earnings for individuals and rural communities was one of the primary conditions.

Systems for fuelwood production are grouped here into three categories: (1) Intensive production on marginal farmland. (2) Fuelwood production with intercropping. (3) Large-scale production on cleared, previously-forested land. Prices and production costs for fuelwood at the power plant site before combustion were of the same order of magnitude irrespective of the system.

Table 6.3-2 shows that for more labor-intensive systems, much of the work consists of growing and tending activities. Manual harvesting operations also generate many jobs. For intercropping systems, only jobs related to wood production were included, so values for employment are somewhat lower. Total employment in wood and crop production would probably have been higher. Large-scale mechanized systems generate fewer jobs than small-scale farm systems. Even so, the large-scale systems generate three times as many jobs per energy unit as mechanized systems in northern Europe (Table 6.3.2).

Table 6.3-2. Job creation and earnings in biofuel production from plantations grown under tropical and subtropical conditions

	Estab-lishment	Weed-ing, etc.	Harvest-ing	Trans-port	Chipp-ing	Admin-istration	Total
Job Creation (person-years petajoule^{-1})							
Intensive production, farmers	112	338	248	70	13	19	799
Intensive intercropping	71	196	251	71	13	19	620
Large-scale 'energy forestry'	34	59	85	51	13	11	252
Community Earnings (000 USD petajoule^{-1})							
Intensive production, farmers	82	206	257	69	14	69	696
Intensive intercropping	55	127	257	69	14	69	590
Large-scale 'energy forestry'	17	27	38	21	14	34	151

Earnings in the local community are highest for the manual systems. Most production is carried out by manual labor; transport and related work with locally-available trucks and tractors. In mechanized systems much of the revenue from fuel production is used for system costs, and trickles down to the local community.

Indirect effects, induced effects and multipliers

In the examples given above, employment and earnings were calculated for fuelwood production in isolation from other activities. In the real world conditions are much more complex. When fuelwood production is introduced or increased, there is an effect on other activities. Woody biomass harvested for energy production cannot be used for other purposes. Labor and machine time spent on fuelwood production cannot be simultaneously applied to other activities. On the other hand, increased fuelwood production generates increased earnings, which gives more purchasing power to the local community. This may have a positive secondary effect on earnings, provided that some of the money is spent locally. Higher purchasing power may also create opportunities for new secondary jobs, thus encouraging people to stay in the community or to join it. These latter effects are referred to as '*induced effects*'.

The BIOSEM study, sponsored by the European Union, was carried out in 1997-1998 to examine these effects in bioenergy projects in several European countries (EU/AEAT 1999). Although the size and type of the projects varied, the results serve as a basis for comparisons and conclusions (Table 6.3-3).

Table 6.3-3. Employment and earnings per petajoule of annual fuel consumption in a range of European bioenergy projects

Project type	Size (MWth)	Direct jobs	Indirect jobs	Induced jobs	Total jobs	Labor earnings, (000 USD)	Other earnings, (000 USD)	Multi-plier	Comment
Short rotation crop, gasifier	2	51	11	36	98	1 004	1 003	1.25	UK
Miscanthus, heat	0.13	321	0	214	534	6 349	3 728	1.21	Belgium
Forest residues, combined heat and power	40	52	33	30	115	1 409	204	1.30	France
Triticale, process heat	2	134	60	28	222	3 472	-426	1.33	Germany
Artichoke, heat	1	269	19	93	380	1 571	-430	1.53	Greece
Short rotation crop, gasifier	5	36	21	23	80	909	360	1.29	Ireland
Industrial residues, combined heat and power	17	41	11	13	65	877	-237	1.46	Italy
Waste, etc., combined heat and power	5	13	2	27	42	216	2 205	1.18	NL
Logging residues, heat	10	52	2	21	76	652	925	1.26	Sweden

Projects based on agricultural crops generate more employment and earnings, but their economic feasibility depends on the assumption that fuel production is subsidized under the Common Agriculture Policy for projects using set-aside land.

Large projects tend to have a lower impact on employment and earnings than small projects. This observation runs counter to the generally-accepted theory that the economy-of-scale effect applies to all energy plants and projects.

Values for the multipliers calculated for projects in Table 6.3-3 are lower than most appearing in the general literature. Multipliers are slightly underestimated, as only the effects of the first loop in the money flow have been included. As the study progressed it became obvious that European social security systems buffer the multiplier effects. For instance, in many cases, unemployment compensation payments are directly or indirectly made from national or regional sources, and not by the local community. This reduces the local induced effects of the bioenergy projects.

The reliability and relevance of these analyses depend on the quality and structure of available data, and on the application of the methodology to individual cases. The calculations are easier and probably provide more realistic results for communities with little contact with the outside world. Most modern European local communities have a multifaceted pattern of regional, national, and international interactions. The extreme complexity of the mechanisms to be examined makes it difficult to carry out detailed analysis.

6.3.3   Implications for employment and earnings

Local employment and earnings resulting from increased production of fuelwood vary according to wage level, the price paid for the fuelwood, the availability of under-utilized resources, and productivity. Changes in these factors alter the magnitude of employment and earnings. For example, if prices decrease and productivity increases, the relative earnings and employment will be lower. However in this case, an increase in volume of production would normally counterbalance the effects on number of jobs and total earnings.

Within the limits set by revenues from the sale of fuelwood, among other factors, the number of jobs and net earnings can be influenced by choice of production method and organization. In some cases, manual methods and mechanized systems lead to similar production costs, but different levels of employment and earnings.

The variety of influences makes it difficult to apply simple standard methods for general appraisal of employment and earnings. Lack of relevant data often prevents detailed analysis. This is especially true when more sophisticated theories are to be examined, for instance those including induced and multiplier effects. Analysis of case-specific models based on available data and focused on the relevant issues will give more reliable results although the possibility of comparison between cases will be reduced.

The precision of appraisal of employment and earnings associated with fuelwood production cannot be very high. Values that are applicable in practice probably lie within the ranges presented in this section.

6.4   CONCLUSIONS

The social consequences of forest energy production are complex and vary from region to region. It is clear that they can be significant and merit close consideration. Design and implementation of production systems should take account of the following points:

- Tradeoffs must be made between mechanization of forest energy production operations and employment levels. Increasing mechanization is likely to reduce job opportunities.

- Although stakeholders hold a variety of values and opinions regarding forest energy production, they must be involved in decision-making related to bioenergy use if implementation is to be successful.

- Better communication among stakeholders may be required if traditional attitudes towards forest energy production are to be modified.

- There will be a need for up-to-date information and guidance as economic and socio-cultural factors affecting energy production and utilization continue to evolve.

6.5   REFERENCES

Ahlgren, K. 1998. Bioenergins sysselsättningseffekter (The employment effects of bioenergy). Länsstyrelsen i Värmland. Rapport 1998:12.

Bembridge, T.J. 1995. Woodlots, woodfuel and energy strategies for Ciskei. South African Forestry Journal 155: 42–50. University of Fort Hare. Alice, Ciskei.

Bensel, T.G. and Remedio, E.M. 1993. Patterns of commercial woodfuel supply, distribution and use in the City and Province of Cebu, Philippines. FAO RWEDP Field Document No. 42. Food and Agriculture Organization of the United Nations, Bangkok, Thailand. 123 p.

Bijlsma, R.G. 1999. Zomervellingen desastreus voor Broedvogels. Nederlands Bosbouw Tijdschrift 71(2): 42–46.

Black, R. and Sessay, M.F. 1997. Forced migration, environmental change and woodfuel issues in the Senegal River Valley. Environmental Conservation. 24(3): 251–260.

Boon, C. 1999. Reactie AVIH, Bosbeheer: da's werk in uitvoering. Nederlands Bosbouw Tijdschrift 71(2): 51.

Bootsma, M.H. 1998. Gemeenschappelijk standpunt van de Stichting Natuur en Milieu en de 12 provinciale Milieufederaties. Stichting Natuur en Milieu, November. 12 p.

Danielsson, B-O. and Hektor, B. 1992. Biobränslens sysselsättningseffekter (The employment effects of biofuels). Uppsats 45. ISSN 0284-6187, ISRN SLU-SIMS-UPP–45–SE. Sveriges lantbruksuniversitet, Institutionen för Skog-Industri-Marknad Studier.

Elmazzoudi, E.H. 1986. Statistical analysis of the wood consumption and forest resources in the Republic of Zambia. Wood energy consumption and resources survey. Ministry of Land and Natural Resources and FAO. 26 p.

EU/AEAT European Commission. 1999. FAIR CT96-1389. Socio-economic multiplier technique for rural diversification through biomass energy deployment: BIOSEM. Final Report.

FAO RWEDP. 1991. Woodfuel flows: rapid rural appraisal in Asian countries. Field Document No. 26. Food and Agriculture Organization of the United Nations, Bangkok, Thailand.

FAO RWEDP. 1997. Regional study on wood energy today and tomorrow - Executive Summary. Field Document No. 50. Food and Agriculture Organization of the United Nations, Regional Wood Energy Program for Asia, Bangkok, Thailand.

FAO RWEDP. 1998. Woodfuel flow study of Phnom Penh, Cambodia. Field Document No. 54. Prepared by the Woodfuel Flow Study Team, Regional Wood Energy Program for Asia, Food and Agriculture Organization of the United Nations, Bangkok, Thailand.

FAO RWEDP. 1999. Wood energy and gender. Food and Agriculture Organization of the United Nations, Regional Wood Energy Program for Asia, Bangkok, Thailand. Website http://www.rwedp.org

FAO RWEDP. (no date). Biomass: more than a traditional form of energy? *In* Biomass management in ASEAN Member Countries. Food and Agriculture Organization of the United Nations, RWEDP/ASEAN-EC Energy Management Training Centre and EC-ASEAN COGEN Program.

Foley, G. 1987. Exaggerating the Sahelian woodfuel problem? Synopsis. Ambio 16(6): 367–371.

Fopma, A. 1999. Houtoogst en verstoring van vogels. Nederlands Bosbouw Tijdschrift 71(2): 52–56.

Fraser, A.I. 1987. The use of wood by the tobacco industry in Zimbabwe. International Science Forest Science Consultancy, Edinburgh, UK. 32 p.

Fraser, A.I. 1991. A study of energy utilization and requirements in the rural sector of Botswana. Energy for sustainable rural development projects. Case studies. Training Materials for Agricultural Planning 23/2. FAO Rome.

Hakkila, P. 1984. Forest chips as fuel for heating plants in Finland. Folia Forestalia 586, Finnish Forest Research Institute, Helsinki, Finland.

Hulscher, W. 2000. FAO RWEDP Report.

Joapik de Witkop van Roennos van d'n Ugchelse Barg. 2000. Sniebonunholtmoll' n herrie in de bos ' anklach an Staatsbos, Stamblad, personeelskrant van staatsbosbeheer, 30 maart, 14 p.

Karenzi, P.C. 1994. Biomass in Botswana. Pp. 68-130 *in* Hall, D.O. and Mao, Y.S. (eds). Biomass energy and coal in Africa. Zed Books, in association with African Energy Policy Research Network – AFREPREN, Gaborone.

McCallum, B. 1995. Case studies of small commercial chipping operations in PEI 1989–1995. Natural Resources Canada, Ottawa.

McCallum, B. 1997. Handbook for small scale fuelwood chipping operations. IEA Bioenergy, Vantaa, Finland.

McCallum, B. 1999. Woodchip supply system options for remote communities. Natural Resources Canada, Ottawa.

NEA (National Energy Administration) Thailand 1988. Master Plan. The Dendro Thermal Power Programme of Thailand. Bangkok

Ohler, F.M.J. 1985. The fuelwood production of wooded Savannah fallow in the Sudan zone of Mali. Agroforestry Systems 3: 57–62.

Rietkerk, F. and Verhagen, S. 2000. Megawatts uit brandend sortiment. Studie voor de AVIH, Zeist, juni. 56 p.

Stichting Platform Bio-Energie. 2000. Samenvatting van reacties op het conceptvoorstel 'nieuwe emissie-eisen voor energiewinning uit biomassa en afval, De Haag. 3 p.

Stridsberg, S. 1998. Biobränslens sysselsättningseffekt (The employment effects of biofuels). Vattenfall AB Projekt uthålliga energilösningar. Rapport 1998/1, Stockholm.

Tsibangu, T.W. 1996. Résultes d'une enquête sur la consommation des combustibles ligneux à Kinshasa, Zaire. Tropicultura 14(2): 59–66.

UNDP/WB ESMAP (United Nations Development Programme/World Bank, Energy Sector Management Assistance Programme). 1992. Philippines: Defining an Energy Strategy for the Household Sector. Results of a Joint Study by ESMAP and the Philippines Office of Energy Affairs. Vol I: Main Report.

van Tuyll van Serooskerken, F.W. Baron. 1999. Reactie Staatsbosbeheer. Nederlands Bosbouw Tijdschrift 71(2): 47–50.

CHAPTER 7

POLICY AND INSTITUTIONAL FACTORS
AFFECTING FOREST ENERGY

A. Roos

The production of forest energy is highly influenced by policies covering energy, the environment and forestry, with varying emphasis in different countries. The main reason for this is that energy production and forest use both have important effects on society and the environment. Fossil energy sources still dominate the energy market in the industrial world. Supportive policies or policies that account for the total costs of different energy sources are therefore required for the implementation of bioenergy technologies.

Policies affecting forest biomass harvesting and utilization have shifted over time. Energy policy was not a controversial issue in the 20th century until the oil embargoes of 1973 and 1978. The practice of nature conservation has improved and forest regulations have been implemented since the late 1800s. These have received renewed interest since the 1960s. The direct effect of energy production on the environment became clear during the 'acid rain' debate of the 1970s and 1980s. Global climate change has been discussed by the world community since the 1980s and increased use of forest energy is seen as one way of reducing greenhouse gas emissions.

An explicit 'forest energy policy' scarcely exists in any country. Forest biomass represents only one (often dominant) component of biomass energy, accounting for 34% of total biomass energy production in the IEA countries in 1995 (IEA 1997a). Other sources are wastes, agricultural residues and energy crops. Bioenergy accounts for 90% of the non-hydro renewable energy production in IEA countries and the latter represents 6.1% of the total energy production. Policies influencing forest energy use may therefore be specifically designed for forest fuel, biomass energy, or renewable energy, or they may address energy use in general (IEA 1997a).

In the policy debate, the supply of forest bioenergy is also connected with forest policy. Forest policies in many countries relate primarily to timber production and multiple use of forests but do not explicitly address modern fuel harvesting activities. Environmental policy and agricultural policy may also be important

Richardson, J., Björheden, R., Hakkila, P., Lowe, A.T. and Smith, C.T. (eds.). 2002. Bioenergy from Sustainable Forestry: Guiding Principles and Practice. Kluwer Academic Publishers, The Netherlands.

in forest energy development. Indirect influence comes from planning laws, industry policy, and regional policy. The conclusion is that forest energy issues may be affected by several different policies on forest use, environmental protection or energy involving legal measures, information and economic incentives. Furthermore, support for forest energy production can be motivated by regional and employment policies. The range of possible policy influences on forest energy use at different stages of forest fuel production is shown in Figure 7.1-1. The diagram illustrates how different policies influence the whole sequence of fuelwood production from forest site to end-consumer. Information, and research and development can be important at all stages.

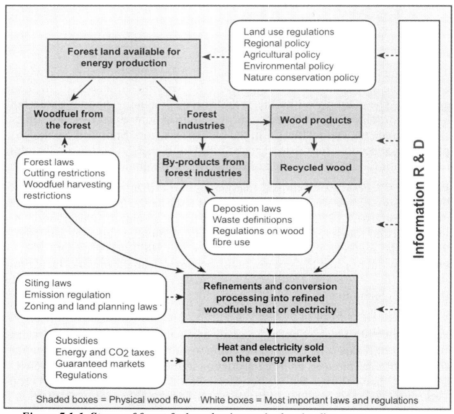

Figure 7.1-1. Stages of forest fuel production and related policy areas.

Policies on natural resources ideally relate to prevention of conflicts over resources, allocation of public resources, and the achievement of satisfactory multiple objectives. In the forest energy arena, policies resolve conflict of interest along the energy production chain from forest to consumer. Stakeholders need to understand the relevant policies. Better awareness is also beneficial for assessment of ways in which laws and institutions might influence forest energy production. Insights into interactions between policy and industry will improve the ability of the forest energy industry to influence policy, e.g. by better communication with policy makers and the public. Knowledge of current or potential future policies is also necessary for identification of research needs.

This chapter describes important policies relating to forest energy production from forest to market and into the recycling phase. The aim is to present a general picture of how forest energy-use and policy interact. Differences between countries and regions are discussed where possible. Unfortunately, all aspects of forest energy use in countries with varying natural conditions and policy requirements cannot be considered fully. It is hoped, however, that the points made will help the understanding of the policy context of forest biomass energy use in most countries.

7.1   POLICIES INFLUENCING FOREST ENERGY

7.1.1   Policies regulating forest fuel harvesting

Forest energy harvesting practices

Experience in Finland, the USA and Sweden shows that forest bioenergy can be harvested at relatively low marginal cost because the traditional forestry sector has already depreciated a large share of the investment cost of roads, labor and machinery development. Forest fuel is often transported and bought within the forestry sector, e.g. forestry companies may form forest fuel enterprises, or logging contractors may also harvest fuelwood (Connors 1994, Roos *et al.* 1999). Costs of forest fuel harvesting are discussed in Chapter 4.

Forest energy harvesting on an industrial scale occurs in only a few regions where a developed forest sector, consisting of large-scale forestry operations and a forest industry, is combined with a large market for forest fuel. Normally, the harvesting of energy assortments (tops, branches or low-value small trees) follows the extraction of sawlogs and pulp timber. In Sweden 3.5 Mt or 38 PJ of forest fuel was collected in this manner in 1997 (National Board of Forestry 1999). Since the early 1990s, tops, branches and cull trees have been collected for fuel in much in the same way in Maine, USA (Connors 1994). In Australia, industrial residues and harvesting residues could be an important source of energy and would help to reduce CO_2 emissions (Fung *et al.* 2000).

Figure 7.1 shows that wood-based fuels are produced at several stages of the traditional wood supply chain: as logging residues and trees with low or no industrial value at the logging site; as low quality logs that cannot be used for pulp or sawing; or as saw- or pulp-milling residues and residues from secondary forest products industries. The production of forest fuel for energy is discussed in Chapter 3. Fuelwood is also derived from recycled wood products and wood wastes resulting from demolition.

Forest fuel harvesting, if done thoughtlessly, can result in machine damage to soil and uncut trees, and decreased site productivity due to nutrient removal. In the worst instances water quality may be compromised, landscape quality degraded and habitats may disappear. Environmental issues surrounding forest energy systems are discussed in Chapter 5. However, as pointed out in Chapter 3, fuelwood harvesting can have beneficial side-effects. Forest regeneration can be improved and insect attacks on standing trees reduced (Egnell *et al.* 1998). Removal of dead and dry wood may reduce the risk of uncontrolled forest fires (Graham *et al.* 1998, Fung *et al.* 2000). Furthermore, extraction of residues may have a positive effect on soil quality in areas where nitrogen deposition from atmospheric pollutants is a problem.

Forest and environmental policy

Although there is general consensus that forest policy should be aimed at sustainability, multiple use and consistency with other policies (Hummel 1991), formulation and content of *forest legislation* both vary greatly. Several countries have a long history of forest legislation, extending as far back as the Middle Ages. Regulations have been implemented in several countries since the 19th century to stop excessive cutting and deforestation and to ensure responsible and sustainable use of the forest resource by the evolving forest industry (Buttoud 1995, Schmithüsen 1999).

During recent years Finland, Germany, Great Britain, Spain and Sweden have reviewed and modified their forest legislation. Some of the changes involve multifunctional objectives, putting more emphasis on ecosystem management, biodiversity and landscape management (Schmithüsen 1999). Aspects of nature conservation, wildlife protection, landscape variety and biodiversity are considered although they are often included as recommendations without accompanying sanctions for non-compliance. In the Swedish Forest Law environmental goals and forest production goals are rated equally. Trends in new forest legislation include more delegation of policy formulation and implementation to regional governments or state entities; fewer regulations; more joint management systems with more negotiation between authorities and forest owners; more information transfer and communication with non-governmental organizations; and more flexible silvicultural practices (Schmithüsen 1999). Forest laws do not often create a barrier to forest fuel harvesting, partly because wood has always been used as a source of energy.

Because forests have a role in the carbon cycle, forest policy is becoming a major component in environmental climate policies outlined in the United Nations Framework Convention on Climate Change in Rio de Janeiro in 1992 and the Kyoto Protocol in 1997 (UNCED 2000).

Bioenergy harvesting, like most forest activities, is affected by *environmental policies* for the protection of forest ecosystems. These policies include delineation of natural reserves, restriction of forest management operations in sensitive areas, and the protection of certain species and water bodies. Conservation laws may state that specified cutting methods must be used in sensitive areas to prevent erosion, protect ecosystems and maintain biodiversity. These restrictions sometimes mean that bioenergy assortments cannot be removed from some specified areas.

Forest management techniques and conservation policies are constantly upgraded on the basis of the latest research results, so that optimal productivity can be achieved without harmful effects on the environment. Future research on forest ecology, including the role of the carbon cycle, will probably influence future policies on forestry activities.

Forest fuel harvesting instructions

Specific regulations for forest fuel harvesting have been implemented in some countries, most extensively in Sweden. Since small branches and foliage contain more nutrients than other parts of the tree, concerns were raised that repeated harvesting of fuelwood would deplete soil nitrogen and other nutrients and reduce site productivity. Removal of dead wood could threaten biodiversity (Löfstedt 1998). These issues are discussed in Chapter 5.

Rapid increase in the use of forest bioenergy by the Swedish district heating sector led to a thorough environmental impact assessment describing the consequences of increased removal of tree tops and branches from forest sites (Egnell *et al.* 1998). Effects investigated were acidification,and changes to soil nutrient status, soil compaction, greenhouse gas emission, soil temperature, water balance, forest productivity and the activity of plants, animals, insects and other organisms, including those living in soil and on dead wood (Egnell *et al.* 1998). Current knowledge about the return of wood ash to the soil was also reviewed. It was concluded that forest fuels should be harvested under a specified set of conditions, e.g. that wood ash is returned to the forest site, that leaves and needles are left in the forest, and that sensitive areas are exempted from fuelwood harvesting.

Similar impact assessments of whole-tree harvesting in Maine, USA, suggested that care must be exercised during harvesting on coarsely-textured, shallow or wet and poorly drained soils. It was also concluded that biomass harvesting should be conducted carefully in order not to damage standing trees. This can be achieved through appropriate education and good management instructions (Connors 1995). Several public initiatives have promoted appropriate harvesting in Maine. A series of biomass harvesting demonstrations for loggers was established, and information was distributed to field personnel through conferences, information packages and extension activities (Connors 1995).

The findings of the Swedish impact studies were transformed into the following recommendations by the central forest authority on forest energy harvesting (National Board of Forestry 1998):

- Most needles should be left in the forest and distributed as evenly as possible, except in areas where there is high nitrogen deposition from the air.

- Trees with environmental or landscape value, or rare tree species should not be harvested.

- Harvesting is not appropriate on biotopes with high environmental value, e.g. wetland forest.

- Dead wood or tree tops and large branches from broadleaved trees should be left in the forest.

- A significant proportion of the forest residues should always be left on the ground. The site should not be completely cleared.

- Damage to soil and standing trees should be reduced.

- Insect damage from stacks of fuelwood should be minimized.

Instructions for use of fertilizer

Several recommendations specify how and when fertilizer application should be legally required. Nutrients, including those in ash from wood burning, should be returned to the soil to avoid long-term productivity loss. The ash should be distributed in slowly-soluble pellet form and should not contain larger quantities of heavy metals and other pollutants than were removed in the biomass. (National Board of Forestry 1998) Removals and additions should always be documented (SLU 1999). Ash from wood burnt with oil, coal or treated wood should not be returned to the forest.

7.1.2   Energy policy

Reasons for energy policies

In many countries, e.g. the United States, the earliest energy policies were aimed at reduction of the negative effect of grid-based so-called 'natural monopolies'. New energy policies were issued after the oil embargoes of 1973 and 1978 to develop cheap alternative energy sources. At this time the industrial base of the rich countries and increasing living standards were under threat. Policy goals were the creation of alternative technologies and fuels to reduce dependency on oil imports from the OPEC countries. For safety and defence reasons the nuclear industry in most countries is regulated by the government.

The polluting effects of fossil fuels, including coal, have become a matter of greater concern. Acid rain and its effects on soil water and forest health have been discussed since the late 1970s. Energy policy has been influenced by government ambition to reduce energy prices by the restructuring and deregulation of electricity markets. Early movers in this process were the UK and Scandinavian countries. Several states in the USA are currently undergoing electricity market restructuring. Utilities are losing their favored status and monopolies are being privatized.

The latest energy issue is global climate change, which has been the subject of international negotiation and documentation in the Kyoto Protocol to the UN Framework Convention on Climate Change (UNFCCC), adopted at the third session of the Conference of the Parties to the UNFCCC in Kyoto, Japan, on 11 December 1997 (see also Chapter 1).

Energy policy may influence fuel choice, costs and technologies. In most national energy policy documents, the main goals relate to energy security, production and supply of low-cost energy, the environment, employment, social and regional production issues, the link between energy policy and foreign and defence policy, international obligations, trade in technology and expertise, development, regional development, and public awareness (IEA 1992).

Renewable energy sources are supported by governments for a wide range of reasons. Commitments on the reduction of greenhouse gases under the UNFCCC state that renewable sources, including bioenergy, play an important role in substitution for fossil fuel use and hence reduction of greenhouse gas emissions. Renewable sources also reduce energy import dependency, increase energy diversity, and provide export opportunities for national industries (IEA 1997a). Other benefits include the contribution of bioenergy to development programmes, agricultural income and employment, cheaper energy supply to rural populations, and the implementation of more sustainable energy systems (IEA 1997a).

Several countries have included renewable energy in their national energy programs (IEA 1997b). In a White Paper issued by the European Commission in 1997, targets were set for an increase in the renewable share of total energy production from 5.3% in 1995 to 12% in 2010. Biomass energy production in the European Union is expected to increase from 44.8 Mtoe in 1995 to 135 Mtoe in 2010 (EC 1997). In the United States the Energy Policy Act and the Climate Change Action Plan announced in October 1993, still represent the basis for renewable energy policy. An important shift in American bioenergy policy was reflected in the Executive Order of August 11, 1999, which encouraged further research into bioenergy with the goal of tripling US use of biobased products by 2010. In Japan the Law on Special Measures to Promote Use of New Energies was passed in 1997. Policies on forest energy in Canada were presented in the 1996 document on Renewable Energy Strategy and further developed in 1998 in the Renewable Energy Development Initiative and Renewable Energy for Remote Communities (IEA 1999a). Energy policies in Australia were outlined in 1998 in the document 'Sustainable Energy for Australia'. The Australian Government has funded policies for a national target of 9.5 TWh of electricity from renewable sources in 2010. Targets for bioenergy use have been formulated in several countries, e.g. Belgium, Denmark, Finland, Italy, the Netherlands and Bavaria in Germany (IEA 1999a).

Capital and tax subsidies

Financing is often a serious problem in bioenergy production. Financial incentives, subsidies, taxes or tax grants for bioenergy are available in some IEA countries to assist the development of new technologies. Subsidies are generally available for a limited time because of budget restrictions, or because they are used to support the development of new bioenergy technology during an initial phase. Specific measures vary between countries.

Capital subsidies are often designed as a capped percentage subsidy. Shifts in subsidy levels may depend on government shifts. In some countries, special funds have been created, e.g. the Sustainable Energy Fund in New South Wales, Australia (IEA 1999b). Grants for bioenergy use have been offered in Austria, Belgium, Denmark, Finland, France and many other countries (AFB-Nett 1995, Hakkila 1999).

Swedish energy policy has presented a long series of support programs for different types of investment through which, for instance, district heating networks and renewable energy investments have been supported. Between 1981 and 1986 subsidies were given for investments reducing oil dependency through development of district heating boilers which used solid fuels, e.g. peat, coal and forest fuel. Since 1991, incentives have been in place to encourage installation of combined heat and electricity plants which burn biomass (Hillring 1996). These programs have contributed to the present high proportion of forest fuel in the Swedish energy system.

In Austria small biomass-fired heating plants, ranging from a few hundred kW to 8 MW received federal, state and EU support in the 1980s and 1990s. The objectives were stimulation of social and economic activity in rural communities and increase in the use of local and clean energy sources. The first biomass district heating plant was built in 1979 and by 1997, 300 small district plants were built. An evaluation concluded that success depended on early community cooperation, the social position of the operators, support by municipal authorities, public opinion, information, and avoidance of conflicts in the village (Rakos 1995).

Tax incentives such as federal income tax credits, advantageous depreciation schemes, or incentive payments are used in the United States to support use of bioenergy. These include alcohol fuel credits, alternative fuel credits, credit for the sale of electricity ('closed-loop'), and renewable electricity production incentives. At the state level different tax incentives, direct project grants, subsidy payments, income tax credits, property tax exemptions, and sale or use tax exemptions are available for bioenergy projects (Rotroff and Sanderson 1998). Tax exemption and lower rates of VAT for renewable sources and bioenergy use exist in some IEA countries. Incentives are available in the United States for feasibility studies of bioenergy projects (IEA 1999a).

Energy tax policy

Energy taxation policy is a much-debated and powerful tool in energy policies. *Energy taxes* can be purely fiscal, designed to change consumption, or to account for environmental external effects. In many countries there is increasing interest in the use of taxes and fees that account for the external effect of an activity.

In Sweden an energy tax is levied on electricity and all fuels except those used in industry. Fuels used for electricity production are not taxed. Sweden has taxes on nuclear power. The carbon dioxide tax was introduced in Sweden in 1991 (Swedish National Energy Administration 1999). Industry pays only 50% of this CO_2-tax. A sulfur tax was also introduced in 1991 and there has been a levy on NO_x since 1992 (Swedish National Energy Administration 1999). The taxes on fossil fuels increased consumption of biofuels by 100% between 1991 and 1997. Examples of energy/environmental and CO_2 taxes in some European countries are shown in Table 7.1-1. Several countries, including Finland, Denmark, Sweden, the Netherlands and Norway, tax CO_2 emissions. Despite repeated recommendations from the Commission, no energy tax exists at the EU level.

There have been attempts to introduce federal energy (Btu) taxes in the United States that would internalize the negative externalities of energy use. However, a proposal by the President in 1993 was rejected by the Senate and the introduction of different environmental taxes on energy use in the United States is unlikely.

Taxing energy may involve unintended negative effects. Tax differences between countries may lead to economic inefficiency and wasteful allocations and trade flows. Policy differences between countries may reduce confidence in energy policies. High CO_2 taxes may lead to high energy costs in some countries, whereas in others the governments still subsidize fossil energy use. As a consequence, polluting industries may choose to invest where energy and carbon taxes are low. This increases costs for the country concerned and for the world community, distorting economic efficiency and increasing CO_2 emissions (IEA 1999b).

Table 7.1-1. Examples of energy/environmental and CO_2 taxes (Holm 1999, Schlegermilch 1998)

Country	Environmental taxes	CO_2 taxes
Sweden	Energy taxes Sulfur: 27 SEK m^{-3} per 0.1% S-content in oil NO$_x$: 40 SEK kg^{-1} Fuelwood: Nil	0.368 SEK kg^{-1} emitted CO_2. Industry pays 50%
Denmark	Energy tax, higher for natural gas, gasoline/diesel Sulfur: 20 DKK kg^{-1}	100 DKK t^{-1} CO_2
Netherlands		CO_2 tax only for SME[1]
Finland	Energy taxes Environmental fees on oil products	102 FIM t^{-1} CO_2, lower for natural gas and peat
Norway	Sulfur tax on oil	CO_2 tax on oil and coal
	Environmental/ CO_2 taxes	
Italy	Excise taxes on gasoline, diesel, coal and mineral oils to reduce GHG emissions	
Austria	Tax on natural gas and electricity	
France	Mineral oil tax on diesel: 0.01 Euro l^{-1}	

[1]*Small- and medium-sized enterprises*

Guaranteed markets

Institutional measures may create markets in which alternative, renewable energy sources can develop. Such measures include competitive bidding procedures within ring-fenced markets, e.g. non-fossil fuel obligation (NFFO) in the UK and possibilities created by the Electricity Feed Law in Germany. Similar policies have been implemented in Australia, Denmark, France, and Portugal. Experience of the NFFO in the UK shows that this policy is successful in attracting bids, and that these bids have decreased over time, indicating that technology is progressing (IEA 1999a).

In Germany, a 1998 law (Energiewirtschaftsgesetz) obliges distributors to include a certain percentage of renewable components in their energy mix. Minimum prices are based on the average consumer price for electricity.

Regulations

Regulations have been used extensively in several countries to promote new energy alternatives, including those based on forest biomass. The US Public Utility Regulatory Policies Act (PURPA) of 1978 (PL 95-617) guaranteed non-utility generators a market for power by mandating that utilities pay 'avoided costs' rates for any power supplied by a qualifying facility, e.g. small power producers using a certain percentage of renewable energy sources. The implementation of PURPA resulted in a rapid increase in use of bioenergy in several states (EIA 1993). However, decreasing electricity production prices and continuing deregulation of electricity markets have decreased the competitiveness of renewable energy in the United States.

The current trend in many countries is towards deregulation of natural gas and electricity markets. Price pressure on electricity and energy will reduce the competitiveness of biomass energy, but the new environment will also attract new stakeholders, a greater variety of energy sources and perhaps more innovation (IEA 1999a).

Administrative rules, i.e. the large number of different permits required when starting a bioenergy production operation, can be a barrier to forest-based bioenergy projects such as the introduction of co-firing in coal-fired power plants in the United States.

Voluntary actions

Efforts to create partnerships between industry and national or local authorities can remove obstacles and coordinate efforts to implement bioenergy production. Voluntary action may involve an increase in the exchange of information and better coordination. They can also lead to more binding agreements concerning renewable energy production (IEA 1999a).

Industry-state partnerships help to solve several problems concerning bioenergy. Cooperation between the US Department of Energy and industry enabled several issues to be resolved. These included information and education, environmental assessment and valuation, Energy Policy Act implementation, development of co-firing opportunities, expansion of the resource base, repowering, and the development of technology export. Other US initiatives are the National Biofuels Roundtable and the Coalition of Northeastern Governors Biomass Policy Roundtable. These bodies helped to streamline policies, develop bioenergy use and implement sound harvesting practices.

Research and development

An important component of energy policy is the creation of alternatives to traditional energy sources. National research programs on forest energy include investigation of new technology, as well as the analysis of economic and resource systems in the bioenergy sector. Of the total research and development expenditure on energy in IEA countries in 1995, non-hydro renewable resources accounted for 8.3%, and bioenergy for 1.3%. This is less than the amounts spent on wind, solar or photovoltaic technology (IEA 1997a). Countries placing most emphasis on bioenergy research are the United States, Spain, Sweden and Italy. Studies of the ecological effects of forest fuel harvesting and use are carried out in several countries. The Finnish research and development program includes conventional energy technology, biomass fuel production, harvesting, combustion, emission reduction, and transportation fuels (Hakkila 1999). The Biofuels Promotion Board in Sweden supports long-term development of commercial biofuels. In Sweden, research and development focuses on wood-based fuels, forest ecosystems, bioenergy systems research, energy crops, waste and biogas, alternative motor fuels, and ethanol development (Hillring 1996).

The European Commission White Paper on renewable energy stresses the importance of research and development in providing the best solutions for renewable energy introduction (EC 1997, IEA 1999a). Information is still required about the use and properties of bioenergy and the public should also be made more aware of environmental effects of bioenergy production and use (AFB-Nett 1995).

Trials with crops of *Salix* spp. and *Populus* spp. are conducted in several European countries: Sweden, UK, Finland, Denmark, Ireland, the Netherlands. Demonstration projects based on short rotation crops have been established in the Netherlands, Denmark, Ireland, Belgium, Austria and Germany (Venendaal *et al.* 1997).

Research on bioenergy harvesting and forest ecology takes place in most countries through several international organizations such as the Center for International Forest Research (CIFOR), International Centre for Research in Agroforestry (ICRAF), and the European Forestry Institute (EFI). Research on forest biomass has also been coordinated and undertaken for many years through the IEA Bioenergy Agreement. Regional efforts to promote renewable energy development have been made by the Asia-Pacific Economic Cooperation Expert Group on New and Renewable Energy Technologies. The European research programs Altener, Thermie, and the Fifth Framework program for research, fund renewable energy projects including production of bioenergy in agriculture and forestry.

Eco-labelling and green pricing

Several market-based instruments such as the green pricing of energy have an influence on opportunities for the development of forest fuel production. These systems shift power from governments to consumers and avoid the negative associations of regulations and taxes. They may be initiated by governments or by market players. Green pricing of electricity has been introduced in Denmark, Germany, Sweden, The Netherlands, Switzerland and the United States (IEA 1999a).

The issuing of *Green Certificates* is a method by which value can be added to energy from renewable sources, making it competitive in the open market. Production of electricity from renewable sources is rewarded with certificates which can be traded later. This market is created either by quotas or by voluntary agreements. Certification procedures currently being planned or implemented in the Netherlands and Denmark correspond to the Federal

Renewable Portfolio Standard (RPS) in the United States. Here, the first phase is planned for 2000-2004. The RPS sets targets for the purchase of electricity derived from renewable sources. It involves credit trading, which ensures that the energy is produced at lowest cost. Legislation or regulations containing the RPS have already been passed in several states, e.g. Arizona, Maine, Massachusetts, Nevada, and Connecticut.

According to polls undertaken in the United States, members of the public rate energy from forest biomass lower than solar or wind power, but higher than oil and coal. Investigations show that bioenergy is ranked higher as people receive more information (Paulos 1998). If the forest fuel industry can avoid negative environmental effects and communicate better with the public, it may in the future be regarded as more environmentally-friendly.

Forest fuel harvesting and utilization interests must consider *forest certification systems* that already exist, or are soon to be introduced. In Europe, the two main systems are administered by the Forest Stewardship Council (FSC) and the Pan European Forest Certification (PEFC) authority (EC 1999). In 1994, the American Forest and Paper Association produced a document to guide forest operations and planning for sustainable forestry, responsible practice, forest health and productivity, and the protection of special sites (Cantrell 1998). The main goal is enhancement of the environment by visible changes in the practice of forestry. Many bioenergy producers are aware of the value of an environmentally-friendly image and it is likely that the requirement for 'certified' forest fuel will increase.

Another market-based system involves the trading of emission rights. Indirectly, bioenergy producers will benefit from the Acid Rain Program established by the US Environmental Protection Agency (EPA), especially the SO_2 trading program, since similar trading for CO_2 is likely to be realised. A voluntary program for reporting of greenhouse gases is in place in the United States (Rotroff and Sanderson 1998).

The Kyoto mechanisms allow countries to reduce greenhouse gas emissions either by reductions in their own countries or through joint implementation, clean development mechanisms or emissions trade. Many major practical issues remain to be solved. Problems in emissions-trading relate to the initial distribution of emission rights, sanctions, and emission control.

Information and education

Lack of information is often regarded as a barrier to the implementation of forest bioenergy production. More than half of the IEA countries have included information and education measures in their energy policies, focusing on environmental awareness, education in schools, and targeted information for planners (IEA 1997a).

Provision of information to home-owners about residential wood heating (pellet stoves, wood stoves or central heating systems) and its environmental performance could improve consumer awareness of alternative methods (Roos *et al.* 1999). In a similar way suppliers, forest owners and farmers should be informed about forest fuel production. Energy from forests or agricultural land is a new concept to many people and prospective producers may be suspicious at the outset. Information must be clear and practical and must include comparisons with the environmental effects of fossil energy alternatives (IEA 1999a).

7.1.3   Policies on non-wood energy sources

Regulations relating to other energy sources will influence fuelwood production. If policies reduce the supply of other energy sources or increase their cost, the competitiveness of forest bioenergy will increase. The effect of taxes has already been discussed (Section 7.1.2). Other policies can also change the competitive balance between bioenergy and other types of energy. Changes in security standards for nuclear power affect the cost of power production in several countries. Political decisions made to phase out nuclear energy production capacity in Austria, Sweden and Germany, could create opportunities for use of alternative energy sources.

Restrictions on peat exploitation may reduce the use of peat as a fuel. Additional use of hydropower in Sweden is limited by restrictions concerning use of the four remaining major rivers (IEA 1996). Similar restrictions exist in other countries with potential for increased hydro-power production, e.g. the United States, Norway, and Austria. Reduced support for the coal industry could also increase opportunities for use of forest energy. Wind power has good prospects although wind turbines are restricted in some areas by the NIMBY (not-in-my-backyard) syndrome.

7.1.4   Other policies

Range of policies affecting woodfuel use

An inventory of policy areas that influence forests and forest industries (ECE/FAO 1997) describes the wider policy situation for energy from the forest. This document lists energy, forest and environmental policies, as well as several others that affect energy use: demography and social affairs, economy, land use, rural and regional development, agriculture, industry, trade, construction and the role of the public sector. These may influence the supply of timber, the structure of trade within forest industries, and the market for forest products. They may also influence the supply of forest fuel, the energy industry, fuelwood trading and markets (ECE/FAO 1997).

Agriculture

Policies influencing energy crop production are often linked to agricultural policy. Many countries in Europe have set objectives for increasing the use of forest energy as a means of increasing farmer income, stimulating regional development, and maintaining rural employment (EC 1997). If changes in agricultural policies were to reduce support for the agricultural sector within the European Union, these incentives would gain importance.

Regulations for energy-generating plants

Energy production for heat and/or electricity can produce a range of external effects. Energy plants may emit greenhouse gases and other atmospheric pollutants, or create other nuisances.

Normally, several environmental regulations are applied when new installations are commissioned. The Swedish Environmental Agency sets levels for emission control of particles, volatile organic compounds (VOC), acidifying chemicals, greenhouse gases, and NO_x. Stack height is also regulated. Restrictions are placed on dust, water pollution, and noise. The handling of process residues is controlled since ash may contain heavy metals, salt, or organic chemicals.

Zoning and planning laws govern the location of new energy plants. Several factors (e.g. heavy traffic, aesthetics, dust and noise) contribute to requirements for location of wood-fired energy plants at a distance from residential areas.

In Europe, stricter legislation for waste deposition will provide an incentive for use of recycled wood waste as a biofuel. Sweden imports large quantities of wood waste from other European countries. If this material is contaminated, special pollution control measures must be used and the ash must be stored in prescribed ways. Trade in wood wastes is regulated by authorities in the exporting and importing countries and by local authorities in Sweden (SLU 1999).

Industry laws

Several industrial regulations have influenced the Swedish bioenergy sector. Changes in the Building Act implemented between 1975 and 1987 (SFS 1975:1321), replaced by the Wood Fibre Act (SFS 1987:588) between 1987 and 1993, regulated the use of wood resources to ensure a supply of wood fibre to the forest industry. The rules were made in response to concern that increased competition for industrial residues from the sawmilling industry would raise the price for wood and wood residues and threaten the future of the board industry (Hillring 1996).

7.2   LAND USE

The supply of wood-based fuels is influenced by the size of the total forest resource and restrictions placed on harvesting. Extraction of fuelwood in addition to conventional roundwood assortments means that the total pressure on forest resources will increase. Theoretically, the value of forest land would also increase, and eventually other land types would be converted to forest. In practice, this would probably be a slow process which could be offset by other trends.

Protection of endangered species in the United States may result in exclusion of large forest areas from energy harvesting. A different approach would be to initiate reforestation programs on marginal land. This has been done successfully in New Zealand where, since 1913, large plantations have increased wood supply and reduced the level of cutting in natural forests. These plantations are mainly intended for pulp but they could eventually be used, at least in part, for energy as well (Dekker-Robertson and Libby 1998).

The area available for bioenergy production, including short-rotation coppice, often depends on supra-national factors. In the Netherlands, the area available for forestry and biomass production is controlled by world market development, European agricultural policy, and agricultural production in Eastern European countries (Faaij *et al.* 1998). Planting of willow for fuelwood on arable land in Sweden between 1991 and 1996 was the result of a shift in Swedish agricultural policy aimed at the reduction of agricultural over-supply of food crops. Even in densely-populated countries, the national goal is an increase in forest area, which could eventually augment the supply of forest fuels.

Long-term *demographic changes*, for instance the rural exodus that is taking place in large parts of the world, often results in conversion of farmland to forest for timber production or forest energy supply. Statistics from FAO show that in Europe, the forested area increased between 1973 and 1988. Urban development, e.g. in North American cities, takes large areas of farmland and also forest for housing. The production of forest fuel increases as land is cleared, but will decrease over time. It is not clear whether this demographic process improves the competitiveness of fuelwoods.

Within OECD Europe, approximately 76% of the total land area is suitable for growing biomass for energy and about 30-40 Mha of agricultural land could be taken out of agricultural production to be used for bioenergy. In theory 17% of present primary energy consumption could be derived from biomass, using 10% of the usable land area and 25% of potentially-harvestable residues from forests, agricultural land and urban areas. This would require a shift of subsidies away from food crop production and towards energy production (Hall and House 1995).

Ownership regimes are likely to influence the supply of wood for energy production. In the European context, forest owners may be companies, the state, institutions, cooperatives or private concerns, the proportion being different in each country (Hummel 1991). This is likely to influence fuelwood supply since owners will have different goals. Large-scale owners are more likely to sell forest fuel or to undertake short-rotation energy crop production than small forest owners because they are better informed about new assortments and are prepared to take risks. Institutional differences between countries, including ministerial responsibility, personal and corporate taxation, and the structure of the forest service and extension services can also influence forest energy use (Hummel 1991).

7.3   INTERNATIONAL AGREEMENTS

Many international agreements and initiatives set targets and restrictions for national policies or influence conditions for governments in some other way. For example in 1987 the Brundtland Report focused on sustainable use of the resources of the Earth (ECE/FAO 1997). Several conventions cover long-range transboundary pollution of the air and water. The United Nations Conference on Environment and Development (UNCED) in Rio de Janeiro considered ways of combining forest management with long-term sustainability and biodiversity. Chapter 11 in Agenda 21 of UNCED entitled 'Combatting deforestation and forest principles', stressed the need for multipurpose and sustainable forest management practices.

The Intergovernmental Panel on Forests (IPF) and, since 1997, the Intergovernmental Forum on Forests (IFF) promote actions to develop and implement criteria and indicators for sustainable forest management. Other processes or forums for this purpose are the Inter-Governmental Seminar on Criteria and Indicators for Sustainable Forest Management, the Ministerial Conferences in Strasbourg and Helsinki on the protection of forests in Europe, and other forest initiatives by the European Parliament.

The effect of emissions on global climate motivated the World Meteorological Organisation to arrange the first World Climate Conference in Geneva in 1985. The Intergovernmental Panel on Climate Change (IPCC) was set up in 1987. The text of the United Nations Convention on Climate Change (UNFCCC) was adopted at United Nations Headquarters, New York on May 9, 1992; it was open for signature at the Rio de Janeiro Conference from June 4-14, 1992 and the Convention came into force on March 21, 1994. The text of the Protocol to the UNFCCC was adopted at the third session of the Conference of the Parties to the UNFCCC in Kyoto, Japan, on December 11, 1997. The Protocol includes an agreement on specified reductions of greenhouse gas emissions, e.g. 8% in the EU, 7% in the United States and 6% in Japan between 1990 and 2008-2012. The protocol includes flexible mechanisms which permit cost-efficient reductions through joint implementation and tradable emission rights. Possible outcomes of ongoing discussions on climate change are that forest bioenergy use may increase and that the supply of fuelwood could eventually increase if more areas are planted with forests.

International funding for activities promoting renewable energy and energy efficiency is provided by the Global Environment Facility and the Prototype Carbon Fund (PCF). Both are supported by a consortium of partners, including the World Bank.

The United Nations Economic Commission for Europe (UNECE) decided to reduce 1980 SO_2 emission levels by 30-80% between 1994 and 2010. A strategy has been elaborated in Europe which sets limits to emissions of SO_2, NO_2, and ammonia (Swedish National Energy Administration 1999). Other global processes that may influence world energy use include the world trade negotiations. Energy markets are becoming more international and diversified and could stimulate trade in biomass fuel and technology.

7.4   LOCAL VERSUS NATIONAL POLICY

National and regional policies are important for the success of bioenergy projects. Local authorities often provide for practical implementation of national policies. Implementation of the PURPA legislation had an important impact in Maine, California and Minnesota whereas it was less influential in the southeastern states of the USA.

Local policy is often more flexible and can be adapted for specific conditions. Local policy makers have a deeper knowledge of industry and of the resource base in their region. Often, forest energy development has strong local support because it is not oligopolized and because it is renewable, recyclable, and creates more local jobs and economic input than other energy sources, e.g. imported oil. If conditions are appropriate and there is local competence, community bioenergy projects are likely to succeed (Rakos 1995, Forsberg 1999). Where other energy sources are important, for instance in coal mining districts, local policy makers and the public are less likely to be interested in forest fuel.

The enthusiasm and dynamism of individuals is an important factor in the growth of a forest energy industry. Studies have shown that local extension organizations and local fuel dealers play an important role in increasing the supply of forest energy and the willingness to sell fuelwoods (Ling 1999). Adoption of short rotation energy crop production depends on the availability of local advisors and contractors. Dynamic advice and marketing can also promote increased use of wood pellet heating. Even in a deregulated environment, the fact that electricity is produced locally from renewable energy sources may constitute a selling argument.

7.5   POLICY CHANGE

Policies on forest energy are the result of a complex political and legal process. They reflect both general public opinion and the mechanisms of the political and legal systems. The outcomes of the formulation process (forest and environmental regulations, energy policies, taxes etc.) may not coincide with the hopes and expectations of stakeholders in forest energy (environmental groups, the bioenergy lobby, fossil fuel industry, energy consumers). Eikelend (1993) showed that the shaping of energy policy in the US is a complicated process in which many stakeholders put forward their own interests. The policy process for wood-based energy production in Sweden has been described by Löfstedt (1998) and Hektor (1990). Key players in this process were the forest industry, energy-using industries, the bioenergy lobby and environmental groups. In this context biomass energy interests must define themselves and work in contact with policy makers.

The implementation of wood-based power production in Maine, USA, demonstrates how policy influences the growth and decline of bioenergy industries. After implementation of the PURPA legislation, there was a considerable increase in forest energy use. This was expected to continue, but was interrupted when oil prices fell to USD10/barrel in the 1980s while simultaneously the deregulation of the electricity market increased market competition for bioenergy (Lee and Hill 1995).

The phasing out of nuclear power in Sweden is another policy-driven change that has altered conditions affecting choice of energy source by individuals and industry in that country.

Interest in biomass utilization is generally represented by trade organizations. The United BioEnergy Commercialization Association (UBECA) is a trade association which includes representatives from the US biomass power industry. This organization is interested in market outlook, work associated with utility restructuring, trade and global climate change. Business associations similarly concerned with national bioenergy production exist in many countries.

7.6   CONCLUSIONS

Forestry and energy sectors are both affected by policies which change as new knowledge is obtained and the political scene alters. It is likely that the energy markets of tomorrow will be less regulated than today and that the forest bioenergy industry will depend more on market forces. It is also likely that priorities in forest, environment and energy policies will change over time. Good energy policies should be transparent, cost-effective in the achievement of their objectives, and fair. This means that negative as well as positive external effects should be reflected in the energy price. In order to gain public support and legitimacy, support systems must not favor any particular user group. The requirement for efficiency means that policies should be designed to reach the intended target in the most cost-effective manner. Market instruments, e.g. environmental taxes and permit trading rather than subsidies, should be used to allocate resources where they are most cost-effective. If policies fulfil several of the above conditions, they have a better chance of stability and will avoid wasteful modification. Coordination between countries is important, particularly when global environmental problems such as S and CO_2 emission problems are addressed.

Some will see the above list as unrealistic. There is likely to be progress in some areas and inconsistency in others. In a reasonable setting where external effects are taken into account and frequent policy shifts are avoided, it is likely that forest energy use can increase and make a worthwhile contribution to energy supply in many regions. Members of the forest energy sector will find it useful to provide relevant information to policy makers and to the public, and to be prepared to participate in discussions in a constructive way. It will also become more important for the forest fuel industry to take advantage of its environmentally-friendly profile in the energy market.

7.7   REFERENCES

AFB-Nett (European Agriculture and Forestry Biomass Network). 1995. To Establish a European Network to Co-ordinate Information Exchange Between National Biomass Energy Programmes on Agricultural and Forestry Biomass (AFB-Nett). Phase 1 Final Report. EU ALTENER. ETSU, Harwell, Didcot, UK.

Buttoud, G. 1995. Forest Policy and Environmental Considerations in France - In Search of Coherence. Pp. 91-101 *in* Buttoud, G. and Solberg, B. (eds.). Forest Policy Analysis – Methodological and empirical aspects. Proceedings from tour workshop during IUFRO workshop.

Cantrell, R. 1998. AF & PA's sustainable forestry initiative - A bold new program that works for the U.S.A. Biomass and Bioenergy 14(4): 325-328.

Connors, J.F. 1994. Impact of biomass energy development on forest management practices in Maine. Proceedings from Bioenergy '94, "Using biofuels for a better environment", Reno/Sparks, Nevada.

Connors, J.F. 1995. The Wood Fired Electric Generating Industry in Maine. Maine State Planning Office, Augusta, Maine.

Dekker-Robertson, D.L. and Libby, W.J. 1998. American forest policy - global ethical tradeoffs. Bioscience 48(6): 473-477.

EC (European Commission). 1997. Energy for the future: Renewable energy. White Paper. Brussels.

EC (European Commission). 1999. A forest strategy for the European Union. (2000/C51/23) Brussels.

ECE/FAO. 1997. The Policy Context for the Development of the Forest and Forest Industries by T.J. Peck and J. Descargues. ETTS Working Paper ECE/TIM/DP/11. United Nations

Economic Commission for Europe, Food and Agriculture Organisation of the United Nations. Geneva.

EIA (Energy Information Administration). 1993. Renewable Resources in the U.S. Electricity Supply. U.S. Department of Energy, Washington, D.C.

Egnell, G., Nohrstedt, H-Ö., Weslien, J., Westling, O. and Örlander, G. 1998. Miljökonsekvensbeskrivning (MKB) av skogsbränsleuttag, asktillförsel och övrig näringskompensation. National Board of Forestry, Rapport 1:1998. Jönköping, Sweden.

Eikelend, P.O. 1993. US energy policy at a crossroads? Energy Policy October 21: 987-999.

Faaij, A., Steetskamp, I., van Wijk, A. and Turkenburg, W. 1998. Exploration of the land potential for the production of biomass for energy in the Netherlands. Biomass and Bioenergy 14(5/6): 439-456.

Forsberg, G. 1999. Assessment of Bioenergy Systems - an integrating study of two methods. Acta Universitatis Agriculturae Sueciae, Silvestria 123. Uppsala.

Fung, P., Kirschbaum, M.U.F., Raison, R.J. and Stuckley, C. 2000. Bioenergy production potential from Australia's forests, its contribution to greenhouse targets and developments in conversion. Pp. 183-194 *in* Krishnapilly *et al.* (eds.). Proceedings of the XXI IUFRO World Congress, Kuala Lumpur, Malaysia, Volume 1.

Graham, R.L., Huff, D.D., Kayfman, M.R., Shepperd, W.D. and Sheehan J. 1998. Bioenergy and watershed restoration in the mountainous regions of the West: What are the environmental/community issues? Pp. 1262-1271 *in* Wichert D. (ed.). Proceedings "Bioenergy '98. Expanding Bioenergy Partnerships", 4-8 October 1998, Madison, Wisconsin, Volume 2.

Hakkila, P. 1999. Finnish bioenergy goals and policy initiatives. *In* Lowe, A.T. and Smith, C.T (eds.). Proceedings of IEA Bioenergy Task 18 workshop "Developing Systems for Integrating Bioenergy into Environmentally Sustainable Forestry", Nokia, Finland.

Hall, D.O. and House, J.I. 1995. Biomass energy in Western Europe to 2050. Land Use Policy 12(1): 37-48.

Hektor, B. 1990. Aktörsystem för Biobränslen - Många medverkande men få kärnintressenter. Department of Forest Industry Market Studies, Swedish University of Agricultural Sciences, Uppsala.

Hillring, B. 1996. Administrative policy instruments for the biofuel market: Legislation of natural resource- and wood fiber use. Report 44, Swedish University of Agricultural Sciences, Department of Forest Industry Market Studies, Uppsala.

Holm, S. 1999. PM: Alternativa styrmedel. Statens Energimyndighet. Eskilstuna, Sweden.

Hummel, F. 1991. Comparison of Forestry in Britain and Mainland Europe. Forestry 64(2): 141-155.

IEA (International Energy Agency). 1992. The role of IEA governments in energy. OECD, Paris.

IEA (International Energy Agency). 1996. Energy Policies of IEA Countries – Sweden – 1996 Review. OECD, Paris.

IEA (International Energy Agency). 1997a. Renewable Energy Policy in IEA Countries, Volume 1: Overview. ISBN 92-64-15495-7. Paris.

IEA (International Energy Agency). 1997b. Renewable Energy Policy in IEA Countries, Volume 2: Country Reports. ISBN 92-64-16186-4. Paris.

IEA (International Energy Agency). 1999a. The Evolving Renewable Energy Market. publication by Rodney Janssen, NOVEM/IEA.

IEA (International Energy Agency). 1999b. World Energy Outlook - 1999 Insights. Looking at Energy Subsidies: Getting the Prices Right. ISBN 92-64-17140-1. Paris.

Lee, C.R. and Hill, R.C. 1995. Evolution of Maine's electric utility industry, 1975-1995. Maine Policy Review, October 1995.

Ling, E. 1999. Bioenergins nuvarande och framtida konkurrenskraft. Acta Universitatis Agriculturae Sueciae, Silvestria 123, Uppsala.

Löfstedt, R. 1998. Sweden's Biomass Controversy - A case study of communicating policy issues. Environment 44(4): 16.

National Board of Forestry. 1998. Rekommendationer vid uttag av skogsbränsle och kompensationsgödsling. Jönköping, Sweden.

National Board of Forestry. 1999. Statistical Yearbook of Forestry. Jönköping, Sweden.

Paulos, B. 1998. Green Marketing and Biomass Energy. Pp. 178-185 *in* Proceedings of Bioenergy '98, Madison, Wisconsin.

Rakos, C. 1995. The diffusion of biomass district heating in Austria. Institute of Technology Assessment, Austrian Academy of Sciences. Vienna.

Roos, A., Graham, R.L. Hektor, B. and Rakos, C. 1999. Critical factors to bioenergy implementation. Biomass and Bioenergy 17: 113-126.

Rotroff A.S. and Sanderson, G.A. 1998. Tax incentives for bioenergy projects. Bioenergy and watershed restoration in the mountainous regions of the West: What are the environmental/community issues? Pp. 212-219 *in* Wichert, D. (ed.). Proceedings "Bioenergy '98. Expanding Bioenergy Partnerships", 4-8 October 1998. Madison, Wisconsin.

SFS (Svensk Författningssamling) 1975: 1321.

SFS (Svensk Författningssamling) 1987: 588.

Schlegermilch, K. 1998. Energy taxation in the EU and some member states: Looking for opportunities ahead. Wuppertal Institute for Climate, Environment and Energy. Wuppertal, Germany.

Schmithüsen, F. 1999. The expanding framework of law and public policies governing sustainable uses and management in European Forests. Pp. 1-30 *in* Schmithüsen, Herbst and Le Master (eds.). Experiences with new forest and environmental laws in European countries with economies in transition. Forstwissenschaftliche Beiträge der Professur Forstpolitik und Forstökonomie, ETH Zürich, 21.

SLU (Swedish University of Agricultural Sciences). 1999. Energi från Skogen. SLU-9 Uppsala, Sweden.

Swedish National Energy Administration. 1999. Energy in Sweden. Eskilstuna, Sweden.

UNCED (United Nations Conference on Environment and Development). September 2000. United Nations Framework Convention on Climate Change. Webpage. *(http://www.unfccc.de)*.

Venendaal, R., Jørgensen, U. and Foster, C.A. 1997. European energy crops: A synthesis. Biomass and Bioenergy 13(3): 147-185.

CHAPTER 8

FRAMEWORK FOR CONVENTIONAL FORESTRY SYSTEMS FOR SUSTAINABLE PRODUCTION OF BIOENERGY

J. Richardson

The forests of the world serve many purposes. Without trying to arrange these purposes in any particular order, they protect soils from erosion; help ensure a steady supply of water; they provide wood for an enormous variety of structural, domestic and industrial products, including housing, furniture, paper, cardboard and panelboards; they absorb carbon dioxide from the atmosphere and release oxygen, thus helping to keep greenhouse gas emissions in balance; and they provide habitats for countless plants and animals, thus conserving biodiversity as well as options for the enjoyment of aesthetic, spiritual, cultural and traditional values.

Forests are also a source of energy. The woody biomass of trees in forests can be converted into convenient solid, liquid or gaseous fuels to provide energy for industrial, commercial or domestic use. About 55% of the 4 billion m^3 of wood used annually by the population of the world is used directly to meet daily energy needs for heating and cooking, mainly in developing countries. Of the remainder, 40% ends up as industrial process residues that are viewed either as waste material for disposal, or as a potential source of renewable energy. In total, 70-75% of the global wood harvest is used or is potentially available for bioenergy production. This does not include the large amounts of logging slash and other biomass residues left in the forest after conventional silvicultural and logging operations.

Conventional forestry systems are defined here as natural forests and plantations in which biomass for energy can be considered as a by-product alongside other benefits and values such as timber production, environmental conservation, and biodiversity. The planting of woody crops dedicated to energy use goes one step beyond this by making production of bioenergy the primary goal.

The aim of this book is the provision of guidelines incorporating general principles that must be considered during production of biomass for energy

Richardson, J., Björheden, R., Hakkila, P., Lowe, A.T. and Smith, C.T. (eds.). 2002. Bioenergy from Sustainable Forestry: Guiding Principles and Practice. Kluwer Academic Publishers, The Netherlands.

from conventional forestry. As preceding chapters have shown, there are many facets to be considered, including:

- Methods for producing fuel from the forest - silviculture, forest management, harvesting and transportation.

- Cost of forest fuel production, and the impact of various factors on the economics of the system.

- Need for environmental sustainability, and how forest fuel production can make positive or negative impacts.

- Relationship between forest fuel production and people - social and cultural aspects.

These facets are all interlinked. At least one principle is common to them all - the concept of sustainability. This has become increasingly important to society in recent years, in the management of natural and planted forests as well as in other areas of human interaction with the environment. Sustainability in relation to management and use of the forest involves assurance that what we do in the forest and the benefits we derive from it today in no way compromise opportunities for use and benefit by future generations. Forest biomass can be a sustainable source of energy; a valuable, renewable alternative to finite fossil-based energy sources. But it can only be truly renewable if the principle of sustainability is maintained as a guideline for each of the above four aspects of forest fuel production.

8.1   PRODUCTION OF FUEL FROM THE FOREST

The normal life-cycle of a forest stand in a conventionally-managed forestry system consists of several stages, beginning with regeneration or stand establishment, continuing through the sapling stage of rapid height growth, the intermediate stage of steady growth in diameter and height, and finally reaching maturity and harvest before returning to the regeneration phase. There are many variations within this basic cycle, most of them related to time and geographic scale. The full cycle or rotation may last 10 years or more than 200 years, but typically 30-80. The 'forest stand' may be as small as the area of a single mature tree in a selection-managed mixed temperate rainforest, or as large as several hundred hectares of uniform, single-species forest in the boreal region.

At each stage of the rotation forestry operations may be conducted, many of which present opportunities for recovering woodfuel as a by-product. In young dense stands, early thinning leaves more growing space for the remaining trees; the trees which are cut have no commercial roundwood value, hence the term 'pre-commercial thinning'. If they can be collected and removed, they can be used for energy production. Thinning in older stands normally yields conventional products such as poles or pulpwood, but the tops and the branches are a potential by-product which can be used for production of bioenergy. Similarly, at final harvest tops and branches are available for fuelwood production in even greater quantities. Other circumstances, such as stand mortality caused by severe insect attack, disease or fire, may provide opportunities for the recovery of fuelwood from conventional forestry systems.

'Forest residues', the tops, branches and other forms of woody biomass that are conventionally left in the forest after silvicultural operations, are a source of energy. Such residues constitute 25-45% of all biomass felled in conventional forestry. Although environmental, technical and economic considerations may limit the proportion of this biomass that might be recovered for fuel, the resource is of great interest in countries with a large forest area *per capita*. Silvicultural practices are generally governed by the primary use of the forest, and may be tempered by other constraints that limit choices. Thus, bioenergy production seldom has a major influence on decisions made in the forest, although operations can be modified to improve opportunities for residue recovery.

Forest residues, whether derived from the tree where it was felled, or from whole-trees after transport to the roadside, have very low density and very low value. As a result, wood-based fuels must be used close to their source, or their density must be increased for efficient handling and transportation. Comminution and compaction are two techniques employed to achieve this aim. Comminution involves reduction to small pieces with a chipper, grinder or flail device. The choice of method and the location of the operation in the supply chain are very important factors. An example of compaction is a recently-devised technique for compressing forest residues into tied bundles which are cut into uniformly-sized 'compact residue logs' (CRLs) for efficient handling.

Cost-effective handling of forest fuel requires careful harvesting, avoidance of contamination and drying. Fuel quality is important to both the supplier and the consumer. Harvesting of forest residues may be conducted at the same time as the harvesting of primary products. Integration of the harvesting operation greatly enhances efficiency. Alternatively, residue collection may be postponed to a later date, by which time the material will have lost moisture and foliage and gained in energy value. A wide variety of harvesting and transportation equipment has been developed for different conditions and scales of operation. The choice of technology and methods depends on ecosystem conditions, infrastructure, forestry traditions and the desired level of integration.

If the bioenergy is to be used for heating, as it commonly is in the Nordic countries, the residue material is usually stored at some point. This is to allow the moisture and leaf content to be reduced, and to ensure that an adequate supply is available during the peak heating season. Residues may be harvested almost year-round, but the peak heating season is winter.

Storage may take place at the stump, in piles (comminuted or uncomminuted) at the roadside, at a central terminal, or at the energy plant. The choice of storage location is governed by biological, economic and logistic considerations. Storage and drying are strongly interrelated and must be carefully controlled to avoid excessive heating within piles and possible fire risk, as well as dry-matter loss and mold development with concomitant health risks. Storage increases handling costs and ties up capital. Maintenance of fuel quality during storage is essential.

8.2   ECONOMICS OF FUELWOOD PRODUCTION

Fuelwood is a low-value product. Its commercial value is less than that of conventional forest products such as sawn timber, veneer, pulpwood and poles. On the other hand, compared to fossil fuels, fuelwood currently has relatively poor economic competitiveness, which means that the use of wood energy is more often justified on the basis of environmental and social benefits. For forest residue harvesting to be economically viable, cost factors must be clearly understood and very carefully controlled during the design and operation of procurement systems.

There are a number of elements in the cost structure of a fuelwood procurement system. Typical operational costs include those associated with cutting, stacking, chipping, forwarding to the roadside, truck transportation and administrative overheads. The cost structure may vary greatly from one system to another and specific elements may be present or absent from each. Harvest operations should target stands where conditions are favorable for recovery. The scale of operations should permit efficient use of expensive equipment. The scale of the demand for fuelwood has a considerable impact on the cost of procurement. Availability in a particular procurement area directly affects harvesting and long-distance transport costs, as well as the selection of technology.

As a counter to the direct costs of producing fuelwood and the low price obtained for the material, there may be offsetting benefits that can have direct or indirect economic value. Removal of residues makes the harvesting site cleaner, thus facilitating access for subsequent site preparation and planting operations. It also reduces the risk of fire and attack from insects and disease-causing organisms that may inhabit the residues. It is important to consider these potential benefits when assessing the economic sustainability of a fuelwood project.

A final economic consideration is the policy context within which fuelwood production is undertaken. There are many regulations, laws, policies, subsidies and taxes that may hinder or foster fuelwood production. These include:

- Policies relating to forest land availability.
- Land use regulations.
- Regional, agricultural, environmental and nature conservation policies.
- Considerations relating to fuelwood extraction from the forest.
- Forest laws.
- Site-specific restrictions on fuelwood harvesting.
- Considerations relating to forest industry.
- Waste disposal laws.
- Regulations on wood fibre use.
- Considerations relating to energy production from fuelwood.
- Siting, zoning and land use laws.
- Emission regulations.
- Considerations relating to the energy market.
- Subsidies and other financial incentives.
- Energy and CO_2 taxes.
- Guaranteed markets.

Policies and political priorities change continuously in the light of new knowledge and information, and with shifts in public opinion. In the future, energy markets are likely to be less regulated and the forest energy industry more vulnerable to market forces. Effective policies should be transparent, cost-effective in the achievement of objectives, and fair as regards renewable versus non-renewable energy systems. A stable policy environment and better coordination between countries, particularly in relation to global environmental issues, can improve efficiency and the potential for sustainability of forest energy use.

8.3   ENVIRONMENTAL SUSTAINABILITY OF FUELWOOD PRODUCTION

Environmental sustainability is of vital importance for conventional forestry systems. At an appropriate level, the forest ecosystem must remain unimpaired for future generations. There are many criteria to be considered, and these are being codified in several international processes such as the Helsinki and Montreal processes for criteria and indicators of sustainable forest management. The environmental criteria include forest health, productive capacity, biodiversity, soil and water, and carbon budgets. The international processes do not specifically focus on biomass production for energy. The principles apply equally to fuelwood production and to the generation of conventional forest products. Because more intensive harvesting practices are involved, fuelwood production may have a greater potential for reducing sustainability, but actual impacts can be minimized.

Currently, the most important issues relating to energy and the environment are those of greenhouse gas emissions and carbon balances. It is evident that bioenergy systems, including those involving conventional forestry, offer significant possibilities for reducing greenhouse gas emissions when they are substituted for fossil fuel systems. Biomass utilization for energy can be considered as part of a closed carbon cycle, and it is thus effectively neutral with respect to carbon balance. Recognition of this benefit is one of the strongest driving forces behind present interest in wider use of bioenergy. The practice of fuelwood production also fits with a related trend towards cleaner, greener, smaller and more decentralized energy production units. Forests are increasingly recognized as an important source of 'green energy' in developed countries.

A commonly-expressed environmental concern about fuelwood harvesting is that it may create nutrient deficiencies for future forest growth. Mineral nutrient elements and nitrogen are especially abundant in foliage and other parts of tree crowns, so that harvesting all the above-ground biomass of a tree greatly increases the loss of nutrients from a forest ecosystem. Careful studies in several parts of the world have shown that forest residues can be harvested without appreciable loss of site productivity if reasonable precautions are taken. In practical, economically efficient operations, forest residues are never completely recovered. Nutrient removals in fuelwood can be minimized by transpiration drying and foliage shedding if operations are timed appropriately. As well as promoting better nutrient cycling, storage practices can improve the quality and properties of the fuel. Avoidance of harvesting on particularly sensitive sites and limiting removal of crown material to once during a rotation will help to minimize nutrient loss. Swedish authorities have established a series of recommendations for forest energy harvesting that are based on such principles.

Much of the concern regarding potential nutrient removal during forest residue harvesting can be alleviated if nutrients are returned to the forest. Where fuelwood is used in a combustion system, most nutrient elements are concentrated into the ash. In boreal forest trees, the ash content of bark is about 6 times greater and that of foliage 6-12 times as great as that of bark-free wood. Ash from fuelwood combustion is commonly returned to the forest as a 'natural' fertilizer, usually after treatment to make the ash easier to handle and distribute. Ground-based spreading equipment must be able to maneuver in relatively dense forest stands. It is not advisable to spread ash from co-firing combustion systems in which fuelwood is burned with coal or other fossil fuels. Other appropriate wastes, such as industrial and municipal biosolids and effluents can be used to maintain soil quality. If necessary, inorganic fertilizers can be used to replace nutrients removed in biomass. Wherever the recovery of forest residues for energy is contemplated, careful and comprehensive cost/benefit analyses should be undertaken to ensure that environmental and economic sustainability will not be compromised.

Forest residues are a source of organic matter, which is essential for maintaining soil properties such as structure and aeration. Organic matter is important for preserving the water-regulating properties of forest soils, and can influence both the quantity and the quality of stream water. Residue harvesting practices remove only a portion of the branches and tops, leaving sufficient biomass in the forest to conserve soil organic matter content as well as nutrients. Soil compaction is also a concern when harvesting activity is increased, since it can have a negative effect on soil. Moisture, air, heat balance, and the extent and rate of root growth can be seriously disturbed, especially if fuelwood recovery is not fully integrated with the harvesting of conventional products. Traffic impacts should be minimized by operating only when soils are dry. Bare-soil erosion should be avoided.

The key to combining environmental sustainability with fuelwood production is recognition that improvement of harvesting technology to enhance work efficiency, and productivity, and reduce procurement costs is not enough. Operations must also be designed to conserve the productive capacity of the soil, to safeguard forest health, and to maintain biodiversity. In practice this means the implementation of measures such as minimizing soil removal with forest residues, and allowing time for leaves or needles to fall from branch material in the forest. Fuelwood harvesting must be no less environmentally sustainable than conventional harvesting. With a few simple restraints such as avoidance of fuelwood harvesting on sites of high environmental value, leaving dead trees and snags, and reducing damage to soil and remaining live trees, the majority of forest values and benefits, including biodiversity, recreation and aesthetics, can be sustained.

8.4   SOCIAL ASPECTS OF FUELWOOD PRODUCTION

Fuelwood production affects people in a variety of ways. At the most basic level, its purpose is the production of energy for the direct or indirect benefit of people. Because a fuelwood production system requires people to operate it, employment, and particularly rural employment, is involved. In many forested regions and in developing countries, fuelwood production and use are traditional activities, and so community and cultural aspects are important. People outside the immediate system of producers and users of energy may be influenced by, or have an interest in the system. Public perceptions and values relating to forests and their use must be considered. Sustainability has an important social dimension.

Woodfuel production provides both direct and indirect employment to an extent which varies with the scale of the operation. Sometimes this employment can be more effectively discussed as 'earnings'. A farmer with a wood-fired heating system for home and farm buildings will probably own a woodlot and extract fuelwood from it without employment of outside labor. No wages are involved, only 'sweat equity' being required. The same farmer may be interested in returns or 'earnings' from sale of fuelwood or from the hiring out of his equipment.

Most forest residue harvesting is done by contractors. One contractor with a few employees may supply the entire fuel requirements of a small district heating plant. A large power generating plant will normally obtain fuelwood from a number of contractors with a larger number of employees, creating a multiplier effect on other employment. As efforts are made to achieve efficiencies of scale in larger operations, the level of employment may not be directly proportional to the amount of material harvested or the energy produced. Tradeoffs must be made between mechanization and employment. Integration of systems for harvesting energywood and conventional forest products may not increase the number of employees beyond that required by the conventional system, but working time may be longer.

Employment levels have a major effect in rural areas. This is an important consideration if rural employment and rural populations are decreasing. In many countries fuelwood production and use for heating and cooking are part of the rural culture. When distribution systems for oil, gas and electricity were less universal, there was a need for families and small communities to be self-sufficient in energy through use of locally-available natural resources. This need still exists and has become intense and difficult to satisfy in the developing world where fuelwood is often the only available but rapidly dwindling energy source. Here tremendous efforts are needed if sustainability is to be achieved. The cultural tradition in the industrialized world remains, for example in the Nordic countries where the relatively important role of fuelwood as a source of energy can be attributed at least in part to continued interest in careful use of the environment for basic needs. In this tradition, sustainability is inherent and almost instinctive.

Sometimes cultural traditions need to be revived. In the boreal forest region of Canada, many native communities have no year-round road or electricity grid connection to the rest of the country. They are dependent on diesel fuel flown or barged in at very high cost for power, including space heating. These remote northern communities are often surrounded by forest which could provide fuelwood in a system that would simultaneously increase self-sufficiency, reduce costs, provide employment, and relate to the forest-based culture of the native people. Examples already exist where a shift to locally-produced bioenergy has been very successful.

Urban attitudes to fuelwood production are related to urban attitudes to conventional forestry systems and to broader concerns for nature and the environment. As a renewable resource, bioenergy has a much lower public profile than wind or solar energy, although associated technology is considered to be more mature, with a much higher level of use in many countries. Better communication among all stakeholders will help to improve this situation. The general public in the Western world, led by the environmental movement, is now seeking certification for forest management and forest products. Such certification is, in effect, an independent attestation that products are generated from forests managed in accordance with criteria associated with sustainability. It is to be expected that the production of bioenergy from conventional forestry systems will sooner or later be included. Prospective purchasers of 'green power' will require the assurance that can be provided by certification.

8.5   CONCLUDING THOUGHTS

Wood for energy is a by-product of the growth of natural forest stands and plantations and as such is recognized and used world-wide. In some developing countries, fuelwood utilization may not be sustainable due to pressures which cannot be addressed without political will and the application of science and technology. Scientifically-based guiding principles have been developed which can help to ensure the economic, environmental and social sustainability of fuelwood production systems. All aspects must be considered, since implementation, socio-economics and the environment are completely interdependent.

This book is based on the premise that the unifying concept of sustainability must underlie all components of conventional forestry systems designed for energy production. Further extended scientific study will be required before we fully understand the principles involved. The technical details of individual system components are complex and vary with geographical location, ecosystem characteristics and socio-economic conditions. Guiding principles, supported repeatedly by scientific evidence reviewed in this book, are simple.

The concept of sustainability is summed up in the North American aboriginal idea of 'a forest for seven generations'. The greatest span of time that any individual is likely to know about personally, is that represented by the seven generations from great-grandparents to great-grandchildren. Very few, if any of us, will have direct experience of human contact beyond a period of 250 years. The challenge of a 'forest for seven generations' is that it must be the same for our great-grandchildren as it was for our great-grandparents. The forest ecosystem must be the same. The intrinsic values and benefits that can be derived from the forest must remain unchanged. That is the concept of sustainability. If we each individually strive to uphold that principle from the centre of our own seven-generation span, learning holistically from sound experimental studies and past experience, and projecting scientifically into the future on the basis of rigorous analysis and careful modelling, we may achieve sustainability. The guiding principles can assure us of bioenergy production from sustainable forestry.

Appendix 1. Units, Equivalents and Conversion Factors

Units and equivalents

a	year (Latin *annum*)
Btu	British thermal unit
°C	degree on Celsius scale
cal	calorie
d	day
°F	degree on Fahrenheit scale
ft	foot
g	gram
gal	gallon
$G_{15}h$	productive time (hours) including disturbances <15 minutes/occasion
h	hour
ha	hectare
hp	horsepower
in	inch
J	joule (basic SI unit of energy) = 1 Nm
kWh	kilowatt-hour
l	liter = 1 dm^3
lb	pound
m	meter
MC	moisture content (percent of water in total mass)
mi	mile
min	minute
N	newton (basic SI unit of force) = 1 kg ms^{-2}
Nm	newtonmeter = 1 J
Nms^{-1}	newtonmeter/second = 1 Js^{-1} = 1 W
oz	ounce
PMH	productive machine hour
ppm	parts per million
s	second
t	tonne (metric ton) = 1000 kg = Mg
toe	metric ton of oil equivalent
W	watt (basic SI unit of power) = 1 Nms^{-1} = 1 Js^{-1}
W_c	calorimetric heating value of dry biomass
W_{ea}	effective heating value of dry biomass
W_{em}	effective heating value of biomass with moisture
yr	year
θ	volumetric water content

Metric conversion factors

Length

1 mm	=	0.0393701 in	1 in	=	25.400 mm
1 cm	=	0.393701 in	1 in	=	2.5400 cm
1 m	=	3.28084 ft	1 ft	=	0.3048 m
1 m	=	1.09361 yd	1 yd	=	0.9144 m
1 km	=	0.621371 mi	1 mi	=	1.60934 km

Area

1 mm^2	=	0.0015500 in^2	1 in^2	=	645.16 mm^2
1 cm^2	=	0.15500 in^2	1 in^2	=	6.4516 cm^2
1 m^2	=	10.763915 ft^2	1 ft^2	=	0.0929030 m^2
1 m^2	=	1.19599 yd^2	1 yd^2	=	0.836127 m^2
1 ha	=	2.47105 acres	1 acre	=	0.404686 ha
1 km^2	=	0.386102 mi^2	1 mi^2	=	2.58999 km^2

Volume

1 cm^3	=	0.061024 in^3	1 in^3	=	16.38706 cm^3
1 m^3	=	35.3147 ft^3	1 ft^3	=	0.0283168 m^3
1 m^3	=	1.30795 yd^3	1 yd^3	=	0.764555 m^3
1 m^3	=	0.353147 cunit	1 cunit	=	2.83168 m^3
1 m^3 stacked	=	0.275896 cord stacked	1 cord stacked	=	3.62456 m^3 stacked $\approx$ 2.5 m^3 solid
1 l dm^3	=	0.264172 gal (US)	1 gal (US)	=	3.7854121

Mass

1 g	=	0.0352740 oz	1 oz	=	28.3495 g
1 kg	=	2.20462 lb	1 lb	=	0.453592 kg

Density

1 kg m^{-3}	=	0.062428 lb ft^{-3}	1 lb ft^{-3}	=	16.0185 kg m^{-3}

Energy

1 J	=	0.94782 $\times$ 10^{-3} Btu	1 Btu	=	1.05506 $\times$ 10^3 J
1 J	=	0.27778 $\times$ 10^{-6} kWh	1 kWh	=	3.6 $\times$ 10^6 J
1 J	=	0.37767 $\times$ 10^{-6} metric hph	1 hph	=	2.6478 $\times$ 10^6 J

Miscellaneous forestry units

1 m^2 ha^{-1}	=	4.35600 ft^2 acre^{-1}	1 ft^2 acre^{-1}	=	0.229568 m^2 ha^{-1}
1 m^3 ha^{-1}	=	14.2913 ft^3 acre^{-1}	1 ft^3 acre^{-1}	=	0.069972 m^3 ha^{-1}
1 m^3 ha^{-1}	=	0.1429 cunits acre^{-1}	1 cunit acre^{-1}	=	6.9973 m^3 ha^{-1}
1 m^3 stacked/ha	=	0.111651 cord stacked acre^{-1}	1 cord stacked acre^{-1}	=	8.95647 m^3 stacked ha^{-1} $\approx$ 6.2 m^3 solid ha^{-1}

Metric prefixes with exponent values

Prefix	Symbol	Exponent
peta	P	10^{15}
tera	T	10^{12}
giga	G	10^9
mega	M	10^6
kilo	k	10^3
hecto	h	10^2
deca	da	10^1
		10^0
deci	d	10^{-1}
centi	c	10^{-2}
milli	m	10^{-3}
micro	μ	10^{-6}
nano	n	10^{-9}

Conversion factors between energy units

	toe	MWh	GJ	Gcal
toe	1	11.63	41.868	10
MWh	0.086	1	3.6	0.86
GJ	0.02388	0.2778	1	0.2388
Gcal	0.1	1.163	4.1868	1

Net heat contents and carbon dioxide conversion factors (CV)

Fuels	Unit	Net heat content		Density	CV
		GJ	MWh	$t\,m^{-3}$	$gCO_2\,MJ^{-1}$
Crude oil	t	41.8	11.6	0.86	
Heavy fuel oil	t	40.6	11.3	0.99	77.4
Light fuel oil	t	42.3	11.7	0.85	74.1
Diesel oil	t	42.5	11.8	0.85	73.5
Jet fuel	t	43.1	12.0	0.80	71.5
Motor gasolines	t	43.1	12.0	0.75	72.5
LPG	t	45.6	12.7	0.51	63.1
Hard coal	t	25.5	7.1		94.6
Coke	t	29.3	8.1		108.1
Natural gas (0°C)	1000 m^3	36.0	10.0		56.1
Black liquor (dry matter)	t	11.7	3.3		110.0
Chips (MC 40 %)	loose m^3	3.3	0.9	0.26	109.6
Chips (MC 40 %)	solid m^3	8.2	2.2	0.65	109.6
Milled peat	t	10.1	2.8	0.32	106.0
Sod peat	t	12.3	3.4	0.38	106.0

Appendix 2. Acronyms

4WD	four wheel drive
AEAT	AEA Technology (Company)
AF&PA	American Forest and Paper Association
AFB-Nett	Agriculture and Forestry Biomass Network (European Union)
BFB	bubbling fluidized bed (combustion)
BIOSEM	Biomass Socio-Economic Multiplier (socio-economic model)
BMPs	best management practices
C&I	criteria and indicators
CFB	circulating fluidized bed (combustion)
CHP	combined heat and power
CIFOR	Center for International Forest Research
CRL	composite residue log
dbh	stem diameter over bark at breast height (1.3 m above ground)
DKK	Danish crown
EC	European Commission
ECE	Economic Commission for Europe (United Nations)
EFI	European Forestry Institute
EIA	Energy Information Administration (USA)
EMS	Environmental Management System
EPA	Environmental Protection Agency (USA)
ET	evapotranspiration
EU	European Union
EUREC	European Renewable Energy Centres
EW	energywood
FAO	Food and Agriculture Organization (of the United Nations)
FBC	fluidised bed combustion
FCCC	Framework Convention on Climate Change (United Nations)
FIM	Finnish mark
FSC	Forest Stewardship Council
GHG(s)	greenhouse gas(es)
ICRAF	International Centre for Research in Agroforestry
IEA	International Energy Agency
IFF	Intergovernmental Forum on Forests
IGCC	integrated gasification combined cycle
IPCC	Intergovernmental Panel on Climate Change
IPF	Intergovernmental Panel on Forests
ISO	International Standards Organization
ITTO	International Tropical Timber Organization
LPG	liquefied petroleum gas

Acronyms continued ...

NFFO	non-fossil fuel obligation
NL	The Netherlands
NGOs	non-governmental organizations
NIMBY	"not in my backyard" syndrome
ODTS	organic dust toxic syndrome
OECD	Organization for Economic Cooperation and Development
OPEC	Organization of Petroleum Exporting Countries
PCF	Prototype Carbon Fund
PEFC	Pan-European Forest Certification
PL	Public Law (USA)
PURPA	Public Utility Regulatory Policies Act (USA)
PW	pulpwood
RPS	Renewable Portfolio Standard (USA)
RWEDP	Rural Wood Energy Development Program (United Nations)
SEK	Swedish crown
SFM	Sustainable Forest Management
SFS	svensk Forfattningssamling (Swedish law publication)
SME	small and medium enterprises
SOM	soil organic matter
SRC	short-rotation culture/crop
SW	solidwood
UBECA	United BioEnergy Commercialization Association
UK	United Kingdom
UNCED	United Nations Conference on Environment and Development
UNECE	United Nations Economic Commission for Europe
UNESCO	United Nations Educational, Scientific and Cultural Organization
UNFCCC	United Nations Framework Convention on Climate Change
USA	United States of America
USD	United States dollar
USDA	United States Department of Agriculture
VAT	value-added tax
VOC	volatile organic compounds
WCED	World Commission on Economic Development

Subject Index[1]

---

[1] Section references shown for topics in this list are intended to direct the reader to the primary treatment of the subject rather than to every mention.